Lecture Notes in Computer Science   16046

Founding Editors

Gerhard Goos
Juris Hartmanis

Editorial Board Members

Elisa Bertino, *Purdue University, West Lafayette, IN, USA*
Wen Gao, *Peking University, Beijing, China*
Bernhard Steffen, *TU Dortmund University, Dortmund, Germany*
Moti Yung, *Columbia University, New York, NY, USA*

The series Lecture Notes in Computer Science (LNCS), including its subseries Lecture Notes in Artificial Intelligence (LNAI) and Lecture Notes in Bioinformatics (LNBI), has established itself as a medium for the publication of new developments in computer science and information technology research, teaching, and education.

LNCS enjoys close cooperation with the computer science R & D community, the series counts many renowned academics among its volume editors and paper authors, and collaborates with prestigious societies. Its mission is to serve this international community by providing an invaluable service, mainly focused on the publication of conference and workshop proceedings and postproceedings. LNCS commenced publication in 1973.

Robert Wrembel · Gabriele Kotsis · A Min Tjoa ·
Ismail Khalil
Editors

Database and Expert Systems Applications

36th International Conference, DEXA 2025
Bangkok, Thailand, August 25–27, 2025
Proceedings, Part I

Editors
Robert Wrembel
Poznan University of Technology
Poznań, Poland

Gabriele Kotsis
Johannes Kepler University Linz
Linz, Austria

A Min Tjoa
Vienna University of Technology
Vienna, Wien, Austria

Ismail Khalil
Johannes Kepler University Linz
Linz, Austria

ISSN 0302-9743 ISSN 1611-3349 (electronic)
Lecture Notes in Computer Science
ISBN 978-3-032-02048-2 ISBN 978-3-032-02049-9 (eBook)
https://doi.org/10.1007/978-3-032-02049-9

© The Editor(s) (if applicable) and The Author(s), under exclusive license
to Springer Nature Switzerland AG 2026

This work is subject to copyright. All rights are solely and exclusively licensed by the Publisher, whether the whole or part of the material is concerned, specifically the rights of translation, reprinting, reuse of illustrations, recitation, broadcasting, reproduction on microfilms or in any other physical way, and transmission or information storage and retrieval, electronic adaptation, computer software, or by similar or dissimilar methodology now known or hereafter developed.
The use of general descriptive names, registered names, trademarks, service marks, etc. in this publication does not imply, even in the absence of a specific statement, that such names are exempt from the relevant protective laws and regulations and therefore free for general use.
The publisher, the authors and the editors are safe to assume that the advice and information in this book are believed to be true and accurate at the date of publication. Neither the publisher nor the authors or the editors give a warranty, expressed or implied, with respect to the material contained herein or for any errors or omissions that may have been made. The publisher remains neutral with regard to jurisdictional claims in published maps and institutional affiliations.

This Springer imprint is published by the registered company Springer Nature Switzerland AG
The registered company address is: Gewerbestrasse 11, 6330 Cham, Switzerland

If disposing of this product, please recycle the paper.

Preface

We present the proceedings of the 36th International Conference on Database and Expert Systems Applications (DEXA 2025). DEXA was established in 1990 under the name International Conference on Database and Expert Systems Applications, which has persisted until today. The conference has been running annually without a break for over three decades, serving as a premier international forum for researchers, practitioners, and industry experts in the fundamental fields of data modeling, databases and data storage systems, data engineering, data analytics, data science, and recently - machine learning and artificial intelligence, for standard and big data.

This year, DEXA was held on 25–27 August, 2025 in Bangkok, Thailand. The conference received 123 submissions. From this set, 35 were accepted as regular papers (giving an acceptance rate of 28%).

DEXA 2025 is proud to have accepted also 22 short papers. They offer a vital platform for presenting innovative projects and preliminary results, novel ideas, ongoing research, or concise technical contributions that may not yet be mature enough for a full paper but have significant potential for future development and promise vibrant discussions. The inclusion of short papers encourages broader participation and facilitates the timely dissemination of emerging research.

The selection of all papers was based on evaluations by the Program Committee members. Each paper was single-blindly evaluated by three members. Here we express our gratitude to the PC members of DEXA 2025, for their timely and thorough evaluations.

As in the past, the DEXA 2025 proceedings consist of two volumes with regular and short papers. The accepted papers cover a variety of research topics on both theoretical and practical aspects. The papers cover among others the following topics: (1) large language models, (2) data quality, (3) applications of machine learning and artificial intelligence, (4) classification techniques, (5) image processing, analytics, and vision systems, (6) recommender techniques, (7) data integration techniques, (8) optimization methods, (9) graph applications, (10) data analytics methods, (11) security and privacy, and (12) benchmarks and surveys.

This year, DEXA introduced so-called short invited talks, with the goal to present trends in data and knowledge engineering in a less formal setting. Four such talks were accepted for the conference.

Also, this year, for the first time, the best papers will be published in a special issue of the Data & Knowledge Engineering (DKE, Elsevier) journal, entitled *Integrating Machine Learning and Data Engineering for Advanced Data Science*. This special issue will also include the best papers from the DAWAK 2025 conference. Taking the opportunity, the PC-chair would like to thank the DKE Editor-in-Chief, Carson Woo, for his approval of the special issue. Special gratitude goes to Ismail Khalil - a Steering Committee member of DEXA/DAWAK and the main organiser of these events. His invaluable help in all tasks related to organizing DEXA 2025 materialized in these two volumes of the proceedings and the event itself.

August 2025,

Robert Wrembel
Gabriele Kotsis
A Min Tjoa
Ismail Khalil

Organisation

Program Committee Chair

Robert Wrembel   Poznań University of Technology, Poland

Publicity Chairs

Aziz Nanthaamornphong   Prince of Songkla University, Thailand
Putu Wuri Handayani   University of Indonesia, Indonesia

Steering Committee

Gabriele Kotsis   Johannes Kepler University Linz, Austria
A Min Tjoa   Vienna University of Technology, Austria
Lukas Fischer   Software Competence Center Hagenberg, Austria
Bernhard Moser   Software Competence Center Hagenberg, Austria
Christine Strauss   University of Vienna, Austria
Ismail Khalil   Johannes Kepler University Linz, Austria

Program Committee Members

A Min Tjoa   Vienna University of Technology, Austria
Abdelkader Hameurlain   IRIT, Toulouse University, France
Abdessamad Imine   Loria, France
Adam Przybylek   Gdańsk University of Technology, Poland
Adriana Marotta   Universidad de la República, Uruguay
Aida Omerovic   SINTEF and NTNU, Norway
Allel Hadjali   LIAS/ENSMA, France
Andreas Ekelhart   Secure Business Austria, Austria
Anne Kayem   University of Exeter, UK
Bala Srinivasan   Monash University, Australia
Bartosz Bebel   Poznań University of Technology, Poland
Bettina Fazzinga   University of Calabria, Italy
Brahim Ouhbi   ENSAM, Morocco
Cedric Du Mouza   CNAM, France

Christian Thomsen	Aalborg University, Denmark
Dawid Wiśniewski	Poznań University of Technology, Poland
Deborah Dahl	Conversational Technologies, USA
Ela Pustulka	FHNW University of Applied Sciences and Arts, Switzerland
Elio Masciari	University of Naples Federico II, Italy
Erich Neuhold	University of Vienna, Austria
Eunika Mercier-Laurent	University of Reims Champagne Ardenne and IFIP, France
Flavio Ferrarotti	Software Competence Centre Hagenberg, Austria
Flavius Frasincar	Erasmus University Rotterdam, The Netherlands
Florence Sedes	IRIT - University of Toulouse, France
Franck Morvan	RIT - University of Toulouse, France
Giovanna Guerrini	University of Genova, Italy
Gheorghe Cosmin	Babeş-Bolyai University, Romania
Silaghi Hamidah Ibrahim	Universiti Putra Malaysia, Malaysia
Hendrik Decker	Ludwig Maximilian University of Munich, Germany
Hiroyuki Toda	Yokohama City University, Japan
Idir Amine Amarouche	USTHB, Algeria
Ionut Iacob	Georgia Southern University, USA
Isao Echizen	University of Tokyo, Japan
Ismael Navas-Delgado	University of Málaga, Spain
Ivan Izonin	Lviv Polytechnic National University, Ukraine
Ivanna Dronyuk	Jan Dlugosz University, Poland
Javier Nieves	Azterlan, Spain
Jean-Paul Kasprzyk	University of Liège, Belgium
Jérôme Darmont	Université Lyon 2, France
Jianwei Zhang	Iwate University, Japan
Johann Gamper	Free University of Bozen-Bolzano, Italy
Jorge Lloret	University of Zaragoza, Spain
Josef Küng	Johannes Kepler University Linz, Austria
Jun Miyazaki	Tokyo Institute of Technology, Japan
Kamonluk Suksen	Chulalongkorn University, Thailand
Karim Benouaret	Université Claude Bernard Lyon 1, France
Lars Moench	University of Hagen, Germany
Laura Erhan	Free University of Bozen-Bolzano, Italy
Laurent d'Orazio	IRISA, France
Lenka Lhotska	CVUT, Czech Republic
Luca Caviglione	IMATI-CNR, Italy
Manfred Hauswirth	TU Berlin, Germany
Manolis Gergatsoulis	Ionian University, Greece

Marcin Paprzycki	IBS PAN and WSM, Poland
Marinette Savonnet	University of Burgundy, France
Markus Endres	Munich University of Applied Sciences, Germany
Massimo Guarascio	ICAR-CNR, Italy
Maude Manouvrier	Université Paris Dauphine - PSL, France
Michal Kratky	VSB-Technical University of Ostrava, Czech Republic
Michael Sheng	Macquarie University, Australia
Mizuho Iwaihara	Waseda University, Japan
Mustafa Atay	Winston-Salem State University, USA
Nazha Selmaoui	University of New Caledonia, New Caledonia
Noura Faci	Université Lyon 1, France
Olivier Teste	IRIT, France
Pavlo Radiuk	Khmelnytskyi National University, Ukraine
Paweł Misiorek	Poznań University of Technology, Poland
Peiquan Jin	University of Science and Technology of China, China
Petra Asprion	FHNW University of Applied Sciences and Arts, Switzerland
Rachid Anane	Coventry University, UK
Riad Mokadem	University of Toulouse, France
Riccardo Albertoni	CNR-IMATI, Italy
Samira Maghool	Università degli Studi di Milano, Italy
Sergio Ilarri	University of Zaragoza, Spain
Soon Chun	City University of New York, USA
Srinath Srinivasa	International Institute of Information Technology Bangalore, India
Stéphane Jean	University of Poitiers, ISAE-ENSMA, LIAS, France
Sven Groppe	University of Lübeck, Germany
Talel Abdessalem	Télécom Paris, France
Toshiyuki Amagasa	University of Tsukuba, Japan
Traian Marius Truta	Northern Kentucky University, USA
Vincenzo Deufemia	University of Salerno, Italy
Vitaliy Yakovyna	University of Warmia and Mazury in Olsztyn, Poland
Wojciech Macyna	Wrocław University of Technology, Poland
Yan Zhu	Southwest Jiaotong University, China
Yang-Sae Moon	Kangwon National University, South Korea

External Reviewers

A. K. M. Tauhidul Islam	Informatica, USA
Amna Rizvi	University of Sydney, Australia
Anand Kumar	Amazon, USA
Andre Kashliev	Eastern Michigan University, USA
Davide Costa	Altilia.ai, Italy
Eleftherios Kalogeros	Ionian University, Greece
Fayçal Saidani	Université Mouloud Mammeri de Tizi Ouzou, Algeria
Feng Yu	Youngstown State University, USA
Francesco Granata	Altilia.ai, Italy
Gautier Filardo	Centre de Recherche de la Gendarmerie Nationale, France
Luca De Grandis	Altilia, Italy
Maryam Mozaffari	Free University of Bozen-Bolzano, Italy
Matthew Damigos	Ionian University, Greece
Muhammad Umair	University of Sydney, Australia
Thilina Lokuruge	University of Sydney, Australia
Vinu Venugopal	International Institute of Information Technology, Bangalore, India

Organisers

From Data Silos to Data Mesh: A Case Study in Financial Data Architecture (Industrial Talk)

Mariusz Sienkiewicz

Director of Supervisory Data Analysis Center, Polish Financial Supervision Authority, Poland

Abstract. Successful data analytics implementation requires seamless access to both data and related metadata. In many organizations, analytics challenges arise from Data Silos, which impede cross-functional access to data and knowledge sharing across the organization. This talk presents practical insights from a data architecture transformation project conducted at a large institution with over 1,400 employees and overseeing over 2,000 market entities. The organization faced significant analytical and operational challenges due to the presence of Data Silos—isolated repositories associated with specific business areas. To address these limitations, the institution initiated a transition to a Data Mesh architecture to improve data availability and enhance analytical capabilities. This talk explains the rationale behind the persistence of silos, evaluates alternative architectural models, and justifies the choice of Data Mesh based on organizational context. Key elements of the transformation include developing a data management framework, implementing a data catalog, creating a data lake to provide data input flexibility, and establishing a common analytics platform based on Data Domains. While the project is still ongoing, the talk describes the methods being implemented and shares early results, key learnings, and practical recommendations for institutions undertaking similar architectural transitions.

Invited Talks

Injured Talk

Blending Contextual Data with Heterogeneous Time Dimensions for Improved Time Series Analysis

Anton Dignös

Free University of Bozen-Bolzano, Italy

Abstract. In modern industrial settings, sensors continuously generate vast amounts of time series data critical for automation and process optimization. However, analyzing this data in isolation limits its effectiveness, as it often lacks integration with contextual factors that influence outcomes but are not directly observable. While traditional data fusion techniques aim at combining multi-modal data such as images or videos, contextual factors in industrial environments frequently differ not in modality but in temporal structure. We identify four distinct time dimensions - constant, time series, events, and intervals - that commonly characterize contextual data in these settings. By transforming diverse time structures into a unified format, we enable the application of conventional machine learning techniques, enhancing the depth and accuracy of industrial data analysis. This talk presents a case study and initial work on a foundational approach for systematically integrating such temporally heterogeneous contextual factors into time series analysis.

A Hybrid Data Model to Support Transportation Analytics of Emergency Service Vehicles

Carson K. Leung

University of Manitoba, Winnipeg, Canada

Abstract. Using a single type of database solution to support real-world applications is becoming more and more challenging because of the volume and variety of data. For instance, the data collected for the transportation industry comprise both structured and unstructured data. Using solely a single type of database solution—relational database system-only or graph database-only—to store and manage data can be challenging. As real-world applications ask even more complex questions related to data, the database solution should be able to facilitate answering these questions in a reasonable time. Hence, in this talk, I present a hybrid model, which integrates data to support transportation analytics. The model consists of relational databases and non-relational databases (namely, graph databases), pooling their strengths to support the demands of the modern application. I also demonstrate this hybrid data model as a practical solution with a case study on improving emergency services—such as emergency medical services (EMS)—response times by having the support of the presented platform.

Contents – Part I

Industrial Keynote

From Data Silos to Data Mesh: A Case Study in Financial Data Architecture ... 3
  Mariusz Sienkiewicz

Invited Talk

Blending Contextual Data with Heterogeneous Time Dimensions
for Improved Time Series Analysis 23
  Saifullah Burero, Anton Dignös, Jerry W. Sangma, and Johann Gamper

A Hybrid Data Model to Support Transportation Analytics of Emergency
Service Vehicles 35
  Carson K. Leung

Large Language Models

Automated Archival Descriptions with Federated Intelligence of LLMs 53
  Jinghua Groppe, Andreas Marquet, Annabel Walz, and Sven Groppe

Entropy-Guided Probing for Predicting LLM Hallucinations
with Knowledge Graph Features 68
  Ushtar Ali, Steven Lynden, Akiyoshi Matono, and Toshiyuki Amagasa

Towards Automating RDF Extraction for Archaeological Knowledge
Graphs with LLMs 83
  Ali Hariri, Stéphane Jean, and Mickaël Baron

Ontology-Based Forest Fire Management Using Complex Event
Processing and Large Language Models 98
  Ritesh Chandra, Sonali Agarwal, and Sadhana Tiwari

Table Annotation Utilizing Large Language Model and Knowledge Graph 113
  Ying Zhang and Mizuho Iwaihara

Improving Software Security Through a LLM-Based Vulnerability
Detection Model 122
  Syeda Sadia Alam, Mst Shapna Akter, and Alfredo Cuzzocrea

SysResolve: Study on In-Context LLM Generation of Resolution Scripts 130
   Harsh Borse, Utkalika Satpathy, Mainack Mondal, and Bivas Mitra

Data Quality

A Novel Unsupervised Anomaly Detection Method Based
on TCN-LSTM-CMA Autoencoder 139
   Jiaji Feng, Yongpan Zhang, Cheng Ding, and Su Pan

Behaviour Modelling and Wayfinding Error Detection in Low Mountain
Hiking 154
   Masaharu Inoue, Hidekazu Kasahara, and Qiang Ma

Explainable Time Series Anomaly Detection by Dynamic Mode
Decomposition 169
   Shun Kawakami, Toshiyuki Amagasa, and Savong Bou

Exploring Quantum Bootstrap Sampling for AQP Error Assessment:
A Pilot Study 184
   Feng Yu and Raya Jahan

AI-Driven Semantic Data Quality Assessment and Scoring for Relational
Databases 199
   *Antony Seabra, Claudio Cavalcante, Nicolaas Ruberg,
   and Sergio Lifschitz*

Network Anomaly Detection Using Gramian Angular Field Transformation
and Vision Transformer 207
   Jaroslaw Kobiela

Machine Learning/Artificial Intelligence Applications

Identifying Multimodal Sarcasm Based on Incongruous Knowledge
Capturing and Contrastive Learning 215
   Yan Zhu, Chang Liu, and Yiqiang Peng

Ensemble ToT and Its Application to Automatic Grading 230
   Yuki Ito and Qiang Ma

Improving Prompt-Based Learning Framework for Mental Health Aspect
Detection from Social Media 245
   Jia-Ling Koh, Hsiao-Ting Huang, and Yin-Ju Lien

DInos: A Deep Reinforcement Learning Approach to Generalizable
Autoscaling in Stateless Cloud Applications 260
  *Constantinos Bitsakos, Dimitrios Tsoumakos, Ioannis Konstantinou,
  and Nectarios Koziris*

Influential Slot and Tag Selection in Billboard Advertisement 276
  Dildar Ali, Suman Banerjee, and Yamuna Prasad

Speech-Scenario Generation Based on the Philosophy of a Prominent
Leader Within a Small Community 291
  Tetsuya Kitahata, Kazuhiro Seki, and Akiyo Nadamoto

VarCGAN: Variational Cyclic Generative Adversarial Network For Music
Genre Style Transfer 307
  Pooja Singh, Dhruv Mishra, and Ankita Khandelwal

Innovative Framework for Early Estimation of Mental Disorder Scores
to Enable Timely Interventions 322
  *Himanshi Singh, Sadhana Tiwari, Ritesh Chandra, Sonali Agarwal,
  Sanjay Kumar Sonbhadra, and Vrijendra Singh*

A Hybrid Approach to Estimating AI Carbon Emissions 329
  Salvatore Borraccia, Elio Masciari, and Enea Vincenzo Napolitano

A Data Product Classification by Technical and Machine Learning Aspects 338
  Laura Schuiki, Ulf Schreier, Holger Schwarz, and Bernhard Mitschang

Classification Techniques

Discovering Voting Power for Ensemble Methods 347
  Pratik Karmakar, Angelo Saadeh, Pierre Senellart, and Stéphane Bressan

Classifying Public and Private Documents Using Context-Based
Predictions 363
  Abrar Hasin Kamal and Anne V. D. M. Kayem

Author Index 381

Contents – Part II

Image Processing, Analytics, and Vision Systems

Relationship Analysis of Image-Text Pair in SNS Posts 3
   Takuto Nabeoka, Yijun Duan, and Qiang Ma

Enhancing Segmentation of Irregular Microstructural Elements Using
Extended Channel Information and Transfer Learning 19
   *Łukasz Marcjan, Sandra Gajoch, Dorota Wilk-Kołodziejczyk,
Marcin Małysza, Krzysztof Jaśkowiec, and Grzegorz Gumienny*

Deep-RVT: A Residual Vision Transformers for Human Action
Recognition 34
   Sayda Elmi, Morris Bell, and Sai Karthik Navuluru

Recommender Techniques

Food Recommendation With Balancing Comfort and Curiosity 51
   Yuto Sakai and Qiang Ma

ONFOODS: A Substitute Recommendation System in Food Recipes 66
   *Maryam Mozaffari, Anton Dignös, Oswald Lanz, Dominik Matt,
Gabriele Pasetti Monizza, Matthias Gauly, and Johann Gamper*

Inspire Me with Your Questions: Repurposing Historical Questions
for New Documents 80
   Yifan Liu, Yixuan Cao, and Ping Luo

Data Integration

MRF-JOIN: Differentially Private Vertical Data Synthesis via Federated
Marginal Join on Shared Attributes 99
   Marin Matsumoto, Tsubasa Takahashi, Shun Takagi, and Masato Oguchi

Efficient Source Selection for Federated SPARQL Queries Using Adjacent
Predicate Information 115
   Yudai Ogura, Tadashi Masuda, and Toshiyuki Amagasa

Empathetic Response Generation in Emotional Support Conversation
via Multi-stage Cascading Information Fusion 130
  Jianwei Zhang, Shota Sato, Yuta Sasaki, and Yuhki Shiraishi

Unified Schema-Driven Graph Polystore: Achieving Transparency
in Multi-model Integration and Migration 136
  Fumihiro Yamashita, Qiong Chang, and Jun Miyazaki

Optimisation Methods

Group Trip Planning Query Problem with Multimodal Journey 147
  Dildar Ali, Suman Banerjee, and Yamuna Prasad

A Model-Based Approach for Simple Construction and Efficient
Evaluation of Dataframes 163
  *Konstantina Zouni, Ioanna Moraiti, Sotirios Angelopoulos,
  Damianos Chatziantoniou, and Verena Kantere*

Energy and Performance Evaluation of Serverless and Serverful Models
on Spark for Database Join Operations 178
  *Phan-An-Truong Tran, Laurent D'orazio, Thuong-Cang Phan,
  and Le Gruenwald*

Graph Applications

The Missing Link: Joint Legal Citation Prediction Using Heterogeneous
Graph Enrichment 197
  *Lorenz Wendlinger, Simon Alexander Nonn, Abdullah Al Zubaer,
  and Michael Granitzer*

Graph Patterns in Fine-Grained Access Control for Graph-Structured Data 212
  *Daniel Schmid, Aya Mohamed, Dagmar Auer, Bahara Muradi,
  and Josef Küng*

An Efficient Point-of-Interest Placement Method Based on Betweenness
Centrality 228
  Ryuta Shiraishi, Ryusei Ohtani, Yuko Sakurai, and Satoshi Oyama

Analytics

Analytics Modelling over Multiple Datasets Using Vector Embeddings 237
  Andreas Loizou and Dimitrios Tsoumakos

Towards IoT-Based Smart Mobility Framework for Proactive Road Stress
Detection in Individuals with ASD 254
   *Barry Amadou Djoulde, Nawal Guermouche, Viviane Kostrubiec,
   and Pierre Vincent Paubel*

A Divisive Unsupervised Feature Selection Approach for Explainable
Remaining Useful Life Prediction 270
   *Mouhamadou Lamine Ndao, Genane Youness, Ndèye Niang,
   and Gilbert Saporta*

Data Storytelling to Unlock the Communicative Power of Digital Twins 287
   Faten El Outa, Hugo Breuillard, and Guillaume Dechambenoit

Queueing Theory for Verifying the Utilization Rate of an Image Processing
System 293
   *Jaqueline Donin Noleto Noleto, Thiago Germano do Nascimento,
   and Pedro Henrique Malheiros Costa Martins*

Effect of Frequency Features of ELA Maps on the Detection Performance
of Image Manipulation Based on DCT and FFT Basis Features 299
   Jarosław Kobiela and Piotr M. Dzierwa

ALPHA: A Multi-Attention Enhanced YOLO Framework for Robust
Photovoltaic Defect Detection 305
   Bechir Ben Tekfa, Amira Mouakher, and Naeem Ayoub

Security/Privacy

Secure Approach for Blockchain-Based Anonymous Attribute-Based
Searchable Encryption Scheme for Data Sharing 313
   Dhruv Kalambe, Nish Shah, Payal Chaudhari, and Priyanshi Manglani

Incremental k-Anonymization for Continuously Growing Big Databases 329
   Akifumi Kurumatani, Hiromasa Yoshimoto, and Kazuo Goda

Post Quantum Cryptographic Schemes and Libraries Selection 338
   *Shubhro Roy, Mangesh Gharote, Pankaj Sahu, Sutapa Mondal,
   M. A. Rajan, and Sachin Lodha*

Benchmarks and Surveys

Workload-Based Clustering of Large Number of Database-as-a-Service
Instances 347
   Maciej Zakrzewicz

Accelerating Python Code with Parallel I/O 360
  Robin Varghese, Hashirul Quadir, Ladjel Bellatreche,
  and Carlos Ordonez

Benchmarking Embedding Techniques for Modeling User Navigation
Behavior on Task-Oriented Software 367
  Ikram Boukharouba, Florence Sèdes, Benoit Verhaeghe,
  and Christophe Bortolaso

The Wrecking SQL Incremental Validation Methodology 376
  Ruanitto Docini, Eduardo C. de Almeida, and Luiz S. Oliveira

A Survey of Control Technologies for Autonomous Underwater Vehicles 383
  Janette Christin Kaspar

Author Index 389

Industrial Keynote

From Data Silos to Data Mesh: A Case Study in Financial Data Architecture

Mariusz Sienkiewicz

Polish Financial Supervision Authority, Piękna 20, 00-549 Warsaw, Poland
mariusz.sienkiewicz@knf.gov.pl

Abstract. Successful data analytics implementation requires seamless access to both data and related metadata. In many organizations, analytics challenges arise from Data Silos, which impede cross-functional access to data and knowledge sharing across the organization. This article presents practical insights from a data architecture transformation project conducted at a large institution with over 1,400 employees and overseeing over 2,000 market entities. The organization faced significant analytical and operational challenges due to the presence of Data Silos–isolated repositories associated with specific business areas. To address these limitations, the institution initiated a transition to a Data Mesh architecture to improve data availability and enhance analytical capabilities. This article explains the rationale behind the persistence of silos, evaluates alternative architectural models, and justifies the choice of Data Mesh based on organizational context. Key elements of the transformation include developing a data management framework, implementing a data catalog, creating a data lake to provide data input flexibility, and establishing a common analytics platform based on Data Domains. While the project is still ongoing, the paper describes the methods being implemented and shares early results, key learnings, and practical recommendations for institutions undertaking similar architectural transitions.

Keywords: Data Mesh · Data Silos · data integration · data warehouse · data architecture

1 Introduction: the Data Silos Architecture

Modern organizations increasingly rely on data as a basis for decision-making, process optimization and building competitive advantage - *data-driven organization*. However, despite the growing availability of analytical tools and integration technologies, many companies struggle with a fundamental problem - the existence of Data Silos. This phenomenon consists in the fact that data is stored in isolated systems or departments, without the possibility of easy sharing it with other parts of the organization [12].

Data Silos not only limit the flow of information, but also lead to duplication of resources, reduced data quality, inconsistencies, and difficulties in conducting coordinated analyses [7]. This type of fragmentation of the data infrastructure is becoming a significant barrier to the implementation of data-driven strategies and also slows down the digital transformation of enterprises [9].

In response to these challenges, organizations are increasingly pursuing the elimination of Data Silos by adopting architectural models that promote decentralized data ownership while upholding interoperability standards [4]. A thorough understanding of the origins, implications, and remediation strategies for Data Silos is essential for organizations aiming to effectively leverage data assets across the enterprise.

1.1 Origin of the Need for Business Transformation

At the outset, it is essential to underscore that the organization under study is a complex entity, established several years ago through the merger of three distinct institutions–each operating within separate market segments and engaging with different stakeholder groups. As a result, each organizational unit developed its own data repositories and independent data acquisition processes, tailored to its respective operational context.

The demand for advanced, cross-domain analytical capabilities has primarily emerged at the executive level of institutional governance. However, the preparation of such analyses by specialized departments requires the retrieval of data sets from discrete data custodians–commonly referred to as Data Silos. This process is frequently time-consuming and largely dependent on informal interpersonal networks, rather than on standardized, institutionalized data-sharing mechanisms.

Historically, there has been no strong organizational impetus to integrate data across silos. At present, however, there is a growing imperative to streamline the execution of cross-sectional analyses and to construct a unified and comprehensive knowledge base concerning the entities monitored by various departments.

1.2 The Source of Data Silos

In contemporary organizations, diverse data organization paradigms may emerge as a result of multiple structural, technological, and cultural factors. Typically, data architectures are designed to support the operational and business processes intrinsic to the organization. However, under the influence of these factors, such architectures may evolve organically and in an ad hoc manner, ultimately resulting in fragmentation, operational inefficiencies, and governance challenges. A prominent manifestation of this architectural fragmentation is the phenomenon of Data Silos.

The challenges associated with Data Silos are pervasive, transcending both industry sectors and organizational models. This observation is substantiated by the discussions held during the CIONET Data Excellence conference [2], where strategies for transitioning from Data Silos to Data Mesh architectures were presented. These discussions were further enriched by the authors' empirical insights derived from several large-scale financial institutions, each employing over one thousand staff members. In all observed cases, despite the deployment of centralized information management systems–such as Data Warehouses–Data Silos continued to persist.

Data Silos refer to the condition in which data is isolated within discrete, autonomous repositories. These silos typically arise when individual organizational units independently design, implement, and maintain their own data infrastructures. As a result, a Data Silo constitutes a repository whose accessibility and utility are restricted to the owning department, rendering it largely inaccessible or irrelevant to other organizational domains.

A Data Silo refers to a data repository that is functionally and logically isolated, operating independently from other repositories or information systems within the organization. The root causes contributing to such isolation–and consequently, to the emergence of Data Silos–can be broadly categorized as follows:

- technological factors - arising from deficiencies in the design and implementation of data systems, leading to isolated and incompatible data structures,
- organizational factors - stemming from organizational structures and governance models that inhibit data sharing and interoperability across departmental boundaries.,
- cultural factors - reflecting an organizational culture that fosters departmental autonomy or competition, thereby discouraging collaborative data sharing initiatives.

Naturally, the root causes of Data Silos are often multifactorial, with several contributing factors frequently coexisting. A particularly significant structural driver of Data Silos is organizational mergers and acquisitions. In such contexts, silos tend to arise when the constituent entities of a newly consolidated organization retain their legacy systems, and no coordinated effort is undertaken to unify or modernize the technological infrastructure. Alternatively, the merged entities may continue to operate within distinct business domains, each governed by its own operational processes and regulatory frameworks [10,20]. This scenario accurately reflects the institutional context examined in the present transformation project.

Data Silos is an architecture in which:

- data is divided into separate data groups,
- data is stored in separate, dedicated repositories,
- processing is carried out by separate processes, limited to a given group and repository,
- access to individual data groups is carried out using a tool dedicated to a given repository,
- access to data may be granted separately at the level of individual repositories. As a result, access to general enterprise data is limited.

Any architectural approach that facilitates seamless data sharing and integration across organizational units is considered fundamentally antithetical to the concept of a Data Silo. Such architectures typically involve centralized data repositories, including data lakes–designed for storing unstructured data–and data warehouses–optimized for structured data storage and analytical processing [16].

Nevertheless, Data Silos may still arise within organizations that formally adopt system architectures intended to mitigate them, including those centered around centralized repositories. In practice, entire data warehouses may become encapsulated within departmental silos, where control over the data and its associated processes is retained by the originating departments. This protective stance extends to both the ownership of the data itself and the knowledge of its associated processing workflows. As a result, the lack of transparent access to metadata and data lineage perpetuates the existence of isolated, closed repositories, accessible only to a narrowly defined group of internal personnel. This scenario accurately reflects the organizational setting explored in the present transformation project.

Within the context of the institution in which this project is being implemented, a particularly promising alternative to traditional Data Silos is the Data Mesh paradigm. As demonstrated by examples shared by the Data Mesh Learning community, this architectural model has already been successfully adopted across a wide range of industries–including finance and payments, professional services, pharmaceuticals, and retail.

This article delineates the experience of implementing a data architecture transformation project, where:

- the data architecture was organized in the form of data warehouses closed in silos,
- the premises determining the finding of an alternative data architecture in the form of Data Mesh,
- a set of solutions that serve to transform to the new architecture.

The transformation project has been in progress for several months within a financial sector institution employing over 1,400 personnel and acquiring extensive datasets from more than 2,000 entities operating across diverse segments of the financial market. Given the inherent complexity of the organization's existing processes and systems–as well as the challenges encountered throughout implementation–it is evident that the transition to a fully integrated architectural model will constitute a long-term, iterative undertaking.

1.3 The Detrimental Impact of Data Silos

Data Silos exert a deleterious impact on organizational effectiveness, primarily by impeding interdepartmental collaboration through restricted access to critical information, limiting the ability of management to obtain comprehensive, organization-wide insights, and complicating the work of analysts engaged in cross-domain data analysis. These barriers–stemming from fragmented and isolated data repositories–significantly prolong the time required to generate accurate and actionable management information. This inefficiency often necessitates the involvement of a larger number of personnel compared to scenarios that utilize integrated information architectures.

In today's data-intensive business environment, organizations are increasingly dependent on the information they generate. As the volume of data produced across diverse operational domains continues to grow, and as analytical capabilities–particularly those incorporating Artificial Intelligence and Machine Learning (AI/ML) techniques–become more advanced, organizations seek to capitalize on these technologies to optimize business processes. However, the persistence of Data Silos severely constrains–and in some cases entirely prevents–the development of robust, enterprise-wide AI/ML solutions and data-driven decision-support systems.

In addition to structural impediments such as restricted data access and limited metadata transparency, the elevated costs associated with developing new analytical solutions to address centralized organizational requirements must also be acknowledged. Moreover, the duplication of data across silos contributes to delays in the preparation of comprehensive analytical outputs.

The presence of Data Silos has a pervasive and detrimental effect on both operational efficiency and analytical capabilities across the organization [20]. As a result, the institution faces numerous constraints and challenges, including:

- constructing a cohesive management information framework by integrating disparate data sources,
- identifying the most accurate and current data,
- ensuring reliable access to complete data sets,
- enabling comprehensive (360°) analytical perspectives,
- managing operational and analytical processes within complex technological and organizational environments,
- addressing the absence of unified, enterprise-wide knowledge to accurately interpret data,
- overcoming limitations in leveraging advanced analytical approaches, including ML and AI methodologies,
- mitigating increased IT infrastructure costs arising from redundant repositories and suboptimal resource utilization,
- reducing the extended timelines required to develop deep data expertise within the organization,
- and addressing regulatory and compliance processes, such as those mandated by GDPR.

The most critical challenges arising from the fragmentation of data across multiple repositories pertain to the adverse effects on the development and deployment of Artificial Intelligence (AI) and Machine Learning (ML) models within the organization, as well as their subsequent integration into business processes [17].

The pervasive presence of Data Silos poses a substantial impediment to the execution of advanced analytical initiatives aimed at synthesizing insights across the diverse segments of the financial market in which the institution operates.

The operational constraints and systemic barriers identified to date underscore the critical need for transformative architectural changes and the establishment of a unified, enterprise-wide information platform capable of effectively supporting the institution's strategic, analytical, and decision-making objectives.

2 Design Method

The primary objective of the project is to develop a common, uniform information platform to support the operations of financial institutions. The project commenced with a comprehensive analysis of the current data architecture and the associated operational processes. This analytical phase encompassed the following key activities:

- conducting a thorough review of the data management solutions and systems deployed within individual departments,
- interviewing departmental representatives to understand domain-specific data practices and challenges,
- examining the production and operational processes managed by the institution's IT department,
- and evaluating all available technical documentation, including architectural diagrams and system maps.

These activities led to the identification of several key observations:

- significant data duplication, with identical data sets redundantly stored across multiple repositories,
- absence of standardized tools and processes for data-driven workflows,
- lack of a shared business glossary, resulting in inconsistent data interpretations across silos,
- limited availability and accessibility of metadata and knowledge regarding the collected data,
- insufficient and fragmented documentation describing the data models and associated processes,
- and consolidation of critical data and processing knowledge into domain-specific expert knowledge, impeding organization-wide understanding and collaboration.

A schematic representation of the institution's data architecture at the inception of the project is provided in Fig. 1.

Data Silos are aligned with the operational domains of the institution's business areas. Each silo is managed by a dedicated team of analysts who work exclusively with data housed within a single, isolated repository. The lack of standardized analytical tools and differential access to data have contributed to significant disparities in analytical capabilities across these teams. The most frequently cited challenges include: (1) difficulties in interdepartmental data exchange, (2) restricted access to critical data knowledge, and (3) pervasive data redundancy, particularly concerning reference and dictionary data.

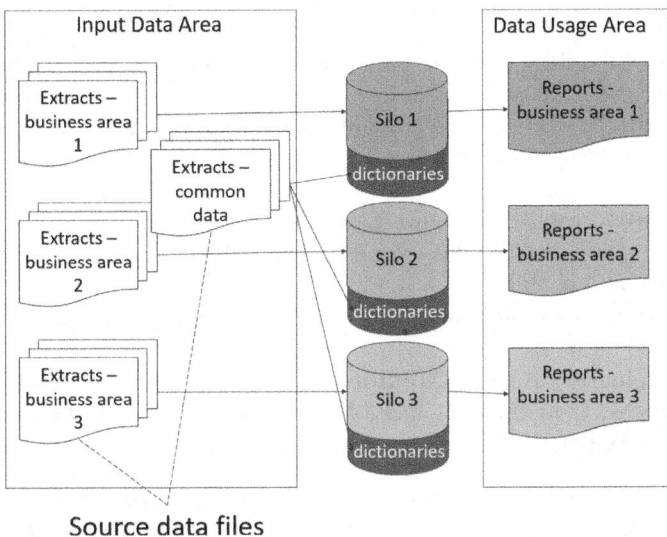

Fig. 1. Data architecture diagram at the beginning of the project - Data Silos

3 Alternative Architectures to Data Silos A Comparative Review

As presented earlier, the data architecture in the form of Data Silos brings many negative effects. At the same time, the goal of the implemented project was to build a uniform information platform. Another aspect that had to be taken into account was the large separation of business areas of individual departments of the institution. An architectural structure was sought that would fit the organization. Below is an overview of data architectures.

Data Warehouse - a centralized repository designed to consolidate, integrate, and transform data from disparate sources to support analytical and reporting requirements.

- Advantages:
  - structured data models that streamline analytical workflows,
  - enhanced data quality achieved through systematic ETL (Extract, Transform, Load) processes,
- Disadvantages:
  - centralized architecture can inadvertently foster the emergence of new data silos,
  - limited adaptability and flexibility in managing unstructured data sets,
- Silo Elimination Rating: **Low** - due to the risk of centralization-induced silos [6,11].

Data Lake - a data repository designed to store raw data in its native format, accommodating both structured and unstructured data sets.

- Advantages:
    - high scalability and flexibility in accommodating heterogeneous data types and structures,
    - capacity to store vast volumes of data without immediate transformation or structuring requirements,
- Disadvantages:
    - absence of robust data governance frameworks can result in the proliferation of a so-called "data swamp," characterized by poor data discoverability and usability,
    - challenges in maintaining data quality, consistency, and lineage across diverse data assets,
- Silos Elimination Rating: **Medium** - while data lakes integrate disparate data sources, their inherent lack of structure can impede data accessibility and usability [13].

Data Fabric - an integrated data architecture that interconnects heterogeneous data sources through a unified metadata and services layer, facilitating seamless and consistent data access across the enterprise.

- Advantages:
    - automated data integration and management capabilities, reducing manual intervention and enhancing operational efficiency
    - consistent and transparent data accessibility across diverse technological environments,
- Disadvantages:
    - elevated complexity in implementation and ongoing management, requiring sophisticated data governance and orchestration capabilities,
    - potential challenges related to data security and privacy due to the broad accessibility of integrated data sources,
- Silos Elimination Rating: **High** - by providing a unified data access framework, Data Fabric architectures significantly reduce data silos across the organization [13].

Data Mesh - a data architecture paradigm founded on the principles of decentralized data governance, wherein individual domain teams assume ownership of their data, treating it as a product to be shared and maintained.

- Advantages:
    - enhanced accountability for data stewardship within domains, leading to improved data quality and contextual relevance,
    - greater scalability and flexibility in data management through decentralized responsibility and localized data expertise,
- Disadvantages:
    - necessitates a fundamental cultural transformation within the organization, along with the development of new domain-specific competencies,
    - requires robust standardization and interoperability frameworks to ensure effective cross-domain data sharing and integration,

- Silos elimination rating: **High** - by promoting domain-level accountability and a product-centric view of data, Data Mesh architectures actively mitigate data isolation [1].

Lakehouse - combination of data warehouse and data lake features, offering a unified environment capable of storing both raw and processed data while supporting advanced analytical workflows.

- Advantages:
  - accommodates heterogeneous data states, including both raw and processed data assets,
  - enables transactional integrity (ACID properties) and facilitates advanced analytics on large-scale data sets,
- Disadvantages:
  - as an emerging architectural approach, it may be constrained by the limited availability of mature tools and standardized best practices,
  - potential complexities associated with integrating Lakehouse solutions into existing data infrastructure landscapes,
- Silos Elimination Rating: **Medium** - while Lakehouse architectures foster data integration, effective data governance remains essential to avoid fragmentation and ensure data usability [6].

Table 1 provides a comparative analysis of various data architectures in the context of their potential to replace Data Silos.

Table 1. Comparing Data Architectures to Eliminate Silos

Architecture	Silos Elimination	Centralization	Flexibility
Data Warehouse	Low	High	Low
Data Lake	Medium	High	High
Data Fabric	High	Medium	High
Data Mesh	High	Low	High
Lakehouse	Medium	Medium	High

Based on the comparative analysis presented above, it is difficult to clearly indicate the data architecture that should be selected to replace Data Silos. In order to determine what the target architecture should be, the specifics of the organization in which the given architecture is to be created should be taken into account. Major architectural changes, and thus systemic changes, entail significant changes in the organizational culture and in the organization itself, including the functioning business processes. Such a major change in the organization may disrupt its functioning and meet with very strong opposition.

4 Data Mesh as an Alternative to Data Silos

As discussed in Sect. 1.1, the newly proposed architecture must not substantially disrupt existing operational processes within the institution. These business imperatives are further reinforced by organizational observations: there exists a pronounced attachment of Data Owners (i.e. business process owners) to the data resources amassed within their respective decentralized Data Silos. No internal drivers were identified within the institution that would incentivize business areas to centralize their data repositories. The only potential motivator for centralization is the need for unified data sets describing external entities with which the institution interacts. However, such shared entities constitute a relatively minor portion–approximately 5%–of the overall data landscape at the business area level. These contextual considerations suggest adopting an architectural solution that allows for the replacement of Data Silos while preserving a high degree of decentralization. Consequently, the Data Mesh paradigm was selected as the target architecture.

In the project, a decision was made that the existing repositories currently represented as Data Silos will ultimately be transformed into Data Domains, managed by individual business areas, which will not disturb the current organizational order. The implementation of mechanisms that increase knowledge about the collected data combined with tools that enable the uniform use of all Data Domains will allow for achieving the expected result in the form of a uniform information platform.

5 Benefits of Transitioning to a Data Mesh Architecture

The transition from Data Silos to a Data Mesh architecture confers several substantial organizational and technical benefits. The most prominent advantages include:

1. **Increased organizational scalability** - Data Mesh supports scaling data across large organizations with decentralized management and domain-oriented ownership.
2. **Increased agility and speed of data delivery for analytics** - domain teams can manage and publish data themselves, reducing the time from query to response.
3. **Increased data quality** - Data as a Product is a key concept that forces teams to take care of data quality, documentation, and availability.
4. **Reduced bottlenecks for central teams** - by shifting responsibility to domain teams, central teams are no longer a bottleneck, which accelerates development.
5. **Better alignment of data with business needs** - the proximity of domain teams to the business context leads to better data modeling and more relevant metrics.

The academic and industry literature consistently underscores that the Data Mesh architecture enhances organizational scalability through the decentralization of data ownership and stewardship. The adoption of the "Data as a Product" paradigm further promotes improved data quality and ensures that data assets are more effectively aligned with evolving business requirements. Moreover, by redistributing data responsibilities to domain-level teams, Data Mesh alleviates the workload of centralized teams and fosters greater operational flexibility and adaptability [3,14,15,19].

6 Challenges and Risks Associated with Data Mesh Adoption

The adoption of a Data Mesh architecture introduces a series of organizational and technical challenges. The most salient issues include:

1. **Insufficient organizational maturity** - many organizations lack sufficient data culture, DevOps practices, and domain competencies, making it difficult to implement a decentralized model.
2. **Increased complexity of Data Management** - Data Mesh assumes domain responsibility for data quality, availability, and security, which can lead to inconsistencies, data duplication, and integration difficulties.
3. **Issues with standardization and interoperability** - in the absence of central standards and oversight, it can be difficult to connect data across domains.
4. **Implementation and maintenance costs** - transforming a data architecture toward Data Mesh requires significant investments in technology, training, and team reorganization.
5. **Responsibility and compliance challenges** - shifting responsibility to domain teams can make it difficult to manage compliance with regulations (e.g. GDPR) and security policies.
6. **Potential inter-domain conflicts** - in a distributed model, there may be disputes over resource availability, scope of responsibility, or integration priorities.

While the Data Mesh paradigm presents substantial potential, it also introduces a distinct set of challenges. Organizations frequently grapple with the necessary levels of technological and cultural maturity to support such an architectural transition. Decentralized data management practices inherent to Data Mesh can give rise to data inconsistencies and integration complexities. Furthermore, the absence of standardized tooling, the high costs associated with comprehensive transformation efforts, and the possibility of conflicts among domain teams constitute significant impediments to successful adoption [4,14,18].

The difficulties mentioned in the Data Mesh project were related to the specifics of the organization. The large separation of business areas gathered in Data Silos induces domain-based action and thinking about data in the organization. The greatest difficulty in a situation of high attachment to data may be the separation of the shared data domain. The challenges encountered during the implementation of the Data Mesh project are inherently linked to the organizational context. The pronounced segmentation of business areas, each encapsulated within Data Silos, reinforces a domain-centric approach to data stewardship and analytical practices within the institution. The primary challenge in this environment, where strong data ownership and attachment prevail, is the delineation and management of shared data domains. Challenges related to data standardization and semantic consistency have been mitigated by the adoption of a framework grounded in the Data Point Model (DPM). Furthermore, a central Data Governance team has been designated to ensure data quality standards are upheld. This central team is tasked with developing solutions to support the Data Domains and their associated analytical tools, thereby maintaining tool standardization and containing associated costs.

7 Transitioning to Data Mesh: A Case Study of a Financial Institution

The transformation of Data Silos, which cause a multitude of operational and analytical challenges, into a modern data architecture paradigm such as Data Mesh is a highly compelling proposition–particularly when considered in the context of the institution's structural characteristics and the current organization of Data Silos aligned with specific market sectors. However, this transformation process is inherently complex, necessitating a series of deliberate actions aimed at mitigating existing obstacles and progressively evolving the Data Silos towards fully functional data domains. The conceptual framework for implementing the Data Mesh architecture within this financial institution is illustrated in Fig. 2.

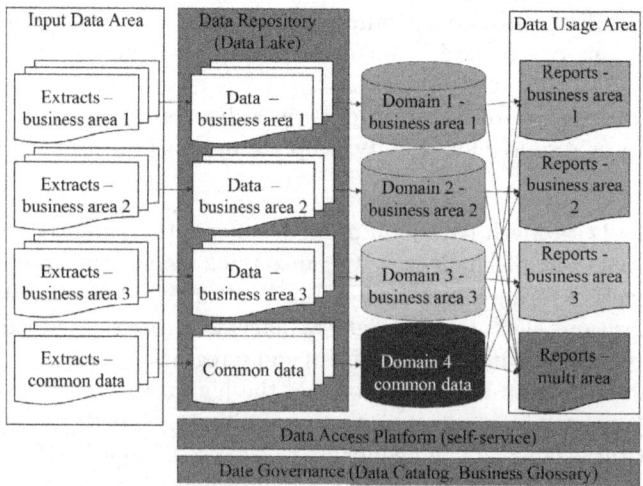

Fig. 2. The concept of a new data architecture inspired by the Data Mesh architecture

7.1 Data Knowledge

A critical step in transitioning to a Data Mesh architecture involves addressing the challenges associated with data knowledge accessibility. The existence of Data Silos inherently limits data availability and significantly restricts the pool of individuals possessing comprehensive knowledge of the data assets. The lack of consistent, clear, and up-to-date documentation describing data models–often dispersed across numerous, disparate documents within the institution–further exacerbates these limitations, creating barriers for data practitioners.

In efforts to dismantle Data Silos, it is imperative to establish a comprehensive data knowledge framework, encompassing detailed metadata that describes the data stored within repositories, as well as the processes applied to that data. The development of this data knowledge foundation can be anchored in robust data management practices. For instance, methodologies proposed by organizations such as DAMA (Data Management Association) provide structured approaches for defining corporate data models, identifying data owners, and establishing data quality standards. Furthermore, data governance tools that leverage metadata derived from database systems and ETL processes enable the discovery of: (1) the nature and scope of data assets within the organization, (2) the processing activities performed on this data, and (3) the data's inherent characteristics and quality [5,8].

The knowledge captured through data governance tools–built upon comprehensive metadata encompassing all existing silos within the organization–enables:

- systematic documentation of the scope of data collected and processed across the enterprise,
- identification of redundant data assets,
- mapping of data flows and facilitation of data impact analysis.

This structured and consolidated knowledge repository is referred to as the Data Catalog [8]. While the establishment of a data catalog and accompanying data governance processes may not immediately eliminate Data Silos from the institution's IT architecture, it does provide critical transparency and accessibility regarding: (1) the breadth and scope of data assets, (2) data quality metrics, (3) data ownership and stewardship responsibilities, (4) mechanisms of data access and utilization, and (5) data lineage and flow dynamics. Within the context of the ongoing project, the development of the data catalog was deemed the foundational step in the broader data architecture transformation effort. The state achieved by building knowledge about data significantly improves the operation of the institution and eliminates a number of the previously mentioned difficulties. The current status of activities pertaining to the development of data knowledge includes:

- completion of the corporate data model framework,
- identification and formalization of data ownership structures,
- definition of key data governance processes,

- initiation of work on mechanisms to evaluate and ensure data quality, and
- commencement of efforts to identify and procure a data governance toolset.

At present, the suite of tools to support data governance processes has not yet been finalized; the selection process for an appropriate solution remains ongoing.

7.2 Input Data Area

As previously noted, the existing Data Silos within the organization are constructed upon heterogeneous technological platforms, including Oracle, MS SQL, and DB2. Data is ingested by the institution in a wide range of formats, with varying frequencies, sizes, and levels of structure–often encompassing a significant proportion of unstructured data. To enhance the institution's ability to accommodate this data heterogeneity and to effectively organize the ingestion of data into newly defined data domains, the decision was made to integrate an input data repository in the form of a data lake, leveraging big data technologies.

The Data Lake architecture was selected for its inherent versatility and scalability, accommodating diverse data formats and volumes. As part of the ongoing project, the installation of Big Data tools within a Cloudera environment has been successfully completed to support the implementation of the Data Lake. The subsequent phases will focus on constructing data processing pipelines within the Data Lake and redesigning the data ingestion processes to align with the emerging data domain structures.

7.3 Data Usage Area

The reorganization of data structures–specifically, transforming Data Silos into data domains–does not, in isolation, complete the comprehensive overhaul of the institution's data architecture from Data Silos to a Data Mesh paradigm. Given that the current Data Silos are built upon disparate technological platforms (including DB2, Oracle, and MS SQL), leveraging native analytical tools presents significant challenges. Therefore, it is imperative to enable analysts to access and utilize data across all repositories–i.e., the emerging data domains. This requires the augmentation of the institution's data architecture with an integrated analytical platform that unifies access to data from these diverse repositories.

The envisioned analytical platform will also encompass the critical capability of managing data access rights, an aspect of paramount importance due to the sensitive and confidential nature of a substantial portion of the institution's data assets. Although the detailed governance of data access rights falls beyond the scope of this document, it remains a central consideration. As part of the ongoing project, a systematic review of commercially available data integration platforms is currently underway. In an effort to streamline and expedite analytical activities, particular interest is being directed towards "low-code" platforms that facilitate the rapid development of analytical models.

In parallel with the analytical platform that will provide standardized data access and reporting capabilities (building upon existing business intelligence frameworks), the data analytics teams within the institution will also be supported by dedicated analytical sandboxes. These sandboxes will function as isolated environments equipped with programming tools–such as Python, R, or Scala–enabling data analysts to conduct exploratory analyses and develop advanced models independently.

7.4 Change in Data Structures

Rebuilding the data architecture is a challenging process, due to the complexity and cost of such an operation. Transforming a data silo architecture, existing in an institution into a Data Mesh involves defining data domains [4]. Existing Data Silos are focused on collecting data on specific sectors of the financial market. They are functionally independent data systems, which means that some data, in particular dictionaries and lists of entities and people in the form of registers, are present in all identified silos. The step in transforming the data architecture is to separate part of the shared data (dictionaries, registers of entities and people) as a separate shared data domain. The remaining part will correspond to the business part of the siloed financial sector data, which brings the changed architecture closer to the concept described as Data Mesh.

The project has initiated efforts to develop a shared data domain encompassing dictionaries and a register of entities and individuals. This initiative leverages dedicated Master Data Management (MDM) tools to ensure consistent governance and data stewardship. At present, the MDM repository integrates data spanning two sectors of the financial market.

Subsequent phases of the transformation will focus on refining the data domains associated with individual financial market sectors–specifically, by extracting shared data elements from the legacy Data Silos and completing the cataloging of all data assets using data governance frameworks. The conceptual model underpinning the transition from Data Silos to a Data Mesh architecture is depicted in Fig. 3.

7.5 Working Environment for Analysts

As previously indicated, the institution's analytical teams are currently embedded within Data Silos, possessing extensive expertise specific to the siloed data they manage. Presently, interactions among these teams are minimal, and data exchange or sharing is highly formalized and encumbered by bureaucratic processes. Consequently, obtaining data from another silo for cross-domain analyses involves lengthy and often cumbersome formal procedures. Given the project sponsors' emphasis on minimizing operational disruptions, the concept of establishing a data science network team within the institution has been adopted.

The data science network team model is designed to harness the analytical capabilities of domain-specific teams for intra-domain analyses. For cross-domain analyses that necessitate data integration from multiple domains, these tasks will

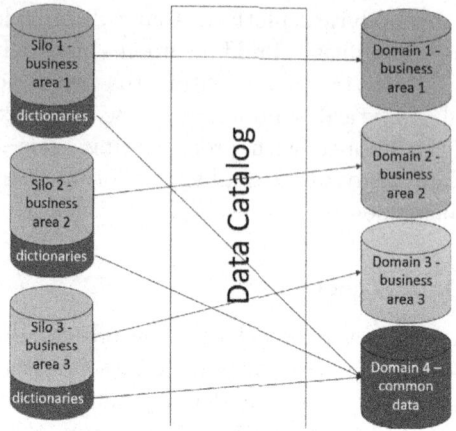

Fig. 3. Transformation of Data Silos into business domains

be executed by the central team, with domain-specific analysts providing critical support and contextual expertise.

The envisioned working environment for all analytical teams will be composed of the following key elements:

- a unified analytical platform that enables seamless data integration across domains, or alternatively, a dedicated analytical sandbox for exploratory work,
- a comprehensive data catalog serving as the repository of metadata and information on data locations, ownership, and quality,
- and a business glossary to standardize and translate organizational terminology related to data concepts.

The project has thus fostered an environment conducive to enhanced collaboration and knowledge sharing among analytical teams throughout the institution.

8 Conclusions

The example of the implemented project illustrates the inherent complexity and multi-dimensional nature of transforming a data architecture from siloed structures to a Data Mesh paradigm. This transformation encompasses several interrelated dimensions: (1) restructuring the organization and management of data repositories, (2) systematically capturing and governing institutional data knowledge, and (3) implementing organizational and procedural reforms to enhance the effectiveness of analytical operations.

This case study further underscores the pivotal role of comprehensive data knowledge–comprising the full spectrum of outputs generated by robust data governance frameworks–as a foundational enabler of a successful transition from

Data Silos to a Data Mesh architecture within an operational context. Ensuring the long-term sustainability of the new architecture will require the ongoing maintenance and continuous refinement of the corporate data model, data catalog, and business glossary, thereby fostering a shared and consistent understanding of data concepts across the enterprise.

Although the transformation project remains ongoing, the results achieved to date are promising. The initiative has garnered strong support from the institution's executive leadership, and data-driven decision-making has been effectively embedded into the broader strategic framework of the organization.

References

1. Amazon Web Services: What is a data mesh? (2024). https://aws.amazon.com/what-is/data-mesh/. Accessed 28 Apr 2025
2. CIONET: Data excellence: Czwarta sesja (2025). https://www.cionet.com/dataexcellence/4sesja. dost?p: 2025-05-01
3. Cruickshank, E., Patel, R.: A practical guide to implementing data mesh at scale. In: International Conference on Data Architecture (2022)
4. Dehghani, Z.: Data Mesh: Delivering Data-Driven Value at Scale. O'Reilly Media, Sebastopol (2020)
5. Earley, S., Henderson, D., Association, D.M.: Dama-dmbok: data management body of knowledge. second ed. (2017)
6. IBM: Data lakehouse vs. data fabric vs. data mesh (2023). https://www.ibm.com/think/topics/data-lakehouse-vs-data-fabric-vs-data-mesh. Accessed 28 Apr 2025
7. Jarke, M., Otto, B.: Data governance and silos: a review of academic and industry perspectives. Inf. Syst. J. **29**(6) (2019)
8. Khatri, V., Brown, C.V.: Designing data governance **53**(1) (2010)
9. Oswald, G., Wenzel, M.: Data mesh in practice: how decentralization combats data silos. J. Data Archit. **9**(4) (2022)
10. Patel, J.: Bridging data silos using big data integration. Int. J. Database Manag. Syst. **11**(3) (2019)
11. Priebe, T., Neumaier, S., Markus, S.: Finding your way through the jungle of big data architectures. arXiv preprint arXiv:2201.04233 (2022)
12. Robertson, S.: Data silos and how to break them. J. Digit. Soc. Media Mark. **6**(2) (2018)
13. SAP: What is data architecture? (2024). https://www.sap.com/resources/what-is-data-architecture. Accessed 28 Apr 2025
14. Smith, J., Kumar, P.: Challenges and opportunities of data mesh in enterprise data architectures. J. Data Eng. **9**(3) (2022)
15. Streitenberger, D., Erek, K.: Evaluating the adoption of data mesh in practice. J. Data Manag. **15**(3) (2022)
16. Tsidulko, J.: What are data silos? Why are they problematic? (2024). https://www.oracle.com/database/data-silos/
17. Vial, G., Jiang, J., Giannelia, T.: The data problem stalling AI (2020). https://sloanreview.mit.edu/article/the-data-problem-stalling-ai/
18. Wang, L., Becker, J.: Data mesh implementation: organizational and technical challenges. In: Proceedings of the 18th International Conference on Data Management, pp. 211–220 (2022)

19. Wang, Y., Schmidt, A.: Adoption challenges of data mesh in enterprise environments. In: Proceedings of the ACM Symposium on Cloud Computing (2023)
20. Wilder-James, E.: Breaking down data silos (2016). https://hbr.org/2016/12/breaking-down-data-silos

Invited Talk

Invited Talk

Blending Contextual Data with Heterogeneous Time Dimensions for Improved Time Series Analysis

Saifullah Burero, Anton Dignös(✉), Jerry W. Sangma, and Johann Gamper

Free University of Bozen-Bolzano, 39100 Bozen-Bolzano, Italy
saifullah.burero@student.unibz.it,
{anton.dignoes,jerry.wattresangma,johann.gamper}@unibz.it

Abstract. In modern industrial environments, sensors play a crucial role for automation by continuously analyzing large volumes of time series data vital for process optimization. However, analyzing this data in isolation poses significant challenges, particularly in time series analysis, due to the influence of external contextual factors that are not always directly observable. Integrating these is essential for time series analysis. While, data fusion is a technique that aims at integrating or blending data with *different modalities* for time series analysis, such as images or videos, contextual factors may not always be heterogeneous in modality, but rather *heterogeneous in time dimension*, which makes its integration challenging. Therefore, we identified four different types of time dimensions that often appear in industrial environments, namely *constant*, *time series*, *events*, and *intervals*, and we aim at introducing the foundation towards a systematic approach for integrating contextual factors with heterogeneous time dimensions. This enables the transformation of data with heterogeneous time dimensions into a format that can be effectively processed by traditional machine learning models for time series analysis.

Keywords: Time series data · Contextual information · Heterogeneous time dimensions

1 Introduction

In the industrial domain, sensors play a pivotal role in automation applications, such as for instance IoT [17], IIoT [22], and Industry 4.0 [9]. These applications heavily rely on time series data captured by sensors, which is often stored in relational databases, such as PostgreSQL with the Timescale extension or Azure SQL Edge, for integration with other data and to facilitate efficient retrieval and analysis [23]. This data is essential for optimizing processes, predicting demand, and detecting equipment failures [14,16,21]. In such a setting, time series data ingested from sensors is organized in tables, where each sensor is identified by a key, e.g., sensor ID. In addition to these sensor measurements, other relevant information is stored in its associated tables and can be *directly or indirectly* linked with sensor measurements as illustrated in Fig. 1. This, for instance includes sensor metadata (e.g., sensor type, location, and properties) and operational logs (e.g., maintenance, replacements, and calibrations) which are mostly directly linked to a sensor as well as for instance external sensory data (e.g., environmental conditions) which are indirectly linked to a sensor through for instance its location. These contextual factors may be of great value when analyzing sensor data as they leads to more robust, accurate, and interpretable analysis.

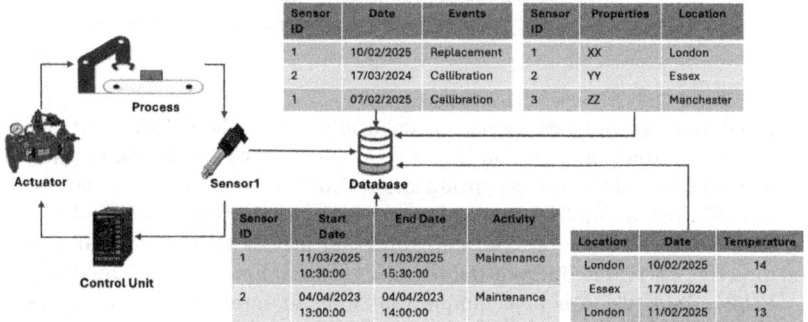

Fig. 1. Database Systems

While many approaches for processing and analyzing time series analysis exist such as missing value imputation [11], time series forecasting [12], similarity search [7] and time series classification [10], they typically rely on input data that is exclusively either a time series or a data sequence. This poses a challenge, since contextual information may not necessarily be in the form of a time series. The most straightforward example is the sensitivity of a sensor, which is not changing over time, but when blended with the sensor measurements, can provide context when compared to other sensor. In such cases, the sensitivity value can be transformed into a time series by simply repeating its value for each time point, and in case of non-metric properties, one-hot encoding may be applied first. We denote such type of data as *static data*, as its temporal validity differs from the validity of the time series and thus is heterogeneous in time dimension.

In a more general setting, we identified four different types of data that often appear in industrial applications and are heterogeneous in time dimension [4]. *Static data* that is not changing with time, e.g., sensor location, sensitivity, and type. *Event data* that is recoded irregularly at some time points, e.g., system errors, sensor replacements, calibrations, and cleaning. *Interval data* that is typically over a period of time and is represented with a start and end time, e.g., system maintenance periods, sensor calibration periods, and machine tasks. *Secondary time series* that comes from sensor with mostly different frequency or sampling rate, e.g., temperature, pressure, humidity. All these types of data have in common that their validity of values differs from the one of the main time series that we intent to analyze and thus cannot be directly fed into an algorithm that expects uniform data in the form of a time series, data sequence or sequence vector.

To address this issue, this talk provides the main intuitions towards a systematic integration approach that transform data with heterogeneous time dimensions into a homogeneous format over regular time steps that can be used by time series algorithms, machine learning and deep learning models.

2 Related Work

In the literature across different application domains and tasks such as event detection [1,5,6,15], classification [18,19], and forecasting [2,3,13,20], there is an emphasis on information fusion to enhance model performance. Data fusion provides rich information that contributes to improved model accuracy and effectiveness by integrating external or contextual data.

Recently, Manuel et al. [15] categorize fusion techniques for multi-modal event detection into three types: data characterization-based, representation-based, and decision-based. These methods fuse modalities such as text, images, time series, and audio, typically assuming perfect temporal alignment. Most data fusion studies adopt one of these strategies.

Bryan et al. [13] introduces Temporal Fusion Transformers (TFT). TFT integrates diverse types of time series data, static (unchanging), historical (observed), and known future inputs to produce accurate, interpretable forecasts. It dynamically selects important variables affecting forecast accuracy through gating, uses attention to capture temporal relationships, and leverages historical patterns along with future-known events. A Gated Residual Network (GRU) ensures optimal information flow, refining predictions.

In [24], the authors proposed a probabilistic fusion approach based on Dempster-Shafer evidence theory. The approach explicitly quantifies uncertainty individually for each variable including both the primary time series and exogenous variables and explicitly merges these uncertainties across variables (channel fusion) and across multiple historical time steps (temporal fusion). Thus, effectively leveraging complementary information from exogenous sources and capture temporal dependencies.

Sameep et al. [8] proposed a fusion methodology that employs a cross-attention mechanism to explicitly merge contextual features (categorical, continuous, and textual data) with the primary time series, selectively emphasizing the most relevant context, thereby significantly improving forecasting accuracy and robustness.

Several studies [2,3,20] demonstrate the benefit of incorporating contextual features (e.g., weather, holidays) into forecasting models. However, they often do not explicitly address how alignment across different time resolutions is achieved. This is possibly due to the common practice of resampling or interpolating time series to a uniform frequency prior to modeling.

Most studies assume either naturally aligned time series contexts or fuse time-derived internal features (e.g., time of day, month, season) that are easy to extract and consistently aligned. In contrast, multi-modal fusion studies focus on integrating semantically complementary information across modalities, where alignment is often presumed rather than explicitly managed.

Our work differs in that it emphasizes contextual data integration, enriching a primary time series with temporally heterogeneous external factors such as static metadata, event based data, interval based data and secondary time series. We address the practical challenge of aligning data with heterogeneity in time dimensions, transforming them into a homogeneous tabular format aligned with a user defined sampling rate. This process facilitates and leverages downstream machine learning and deep learning workflows.

3 Blending Contextual Data with Heterogeneous Time Dimensions

Figure 2 shows a typical pipeline for time series analysis. In our case the data is composed of time series data and contextual data with heterogeneous time dimensions. This data needs first, to be transformed into a homogeneous format during the representation stage, followed by feature extraction, where time based and contextual features are derived. A subsequent feature selection then retains the most relevant features for the task. Finally, the task is executed. The goal of our approach is to integrate contextual features heterogeneous in time dimension with time series in the stages of data representation, homogeneous representation and feature engineering and selection, such that it is transparent to the final time series analysis task.

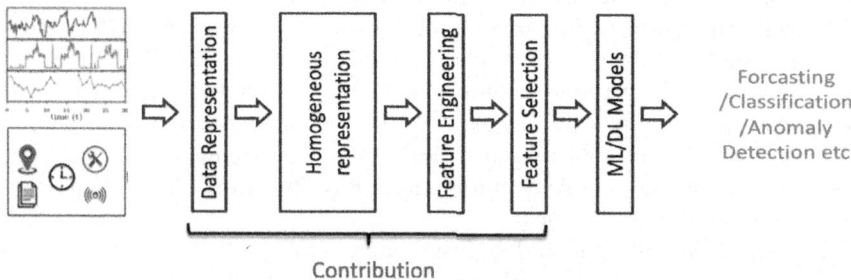

Fig. 2. Data integration with down stream task

3.1 Data with Heterogeneity in Time Dimensions

Machine learning models, such as Random Forest, SVM, and LSTM require input data that is uniform in structure, i.e., numerical vectors, and work best if recorded over regular time steps. However, contextual data often comes with heterogeneous time dimensions and effectively integrating such data requires tailored strategies for each type to ensure effectiveness. We now review the different types of data and their differences in time dimension.

Time Series Data. A time series consist of numerical observations recorded over time, either at regular or irregular time steps. Formally, given a time series P with n observations we have

$$P = \{(t_i, v_i) \mid i = 1, \ldots, n\}, \tag{1}$$

where t_i represent a timestamp, and v_i the corresponding values of the observation. In our integration process, we typically have a main signal, e.g., water or energy consumption, or GPU temperature, that constitutes a primary time series, while contextual signals, e.g., weather, temperature, or GPU utilization are secondary time series and while they are of the same structure, they often use different time steps due to data collection or different sampling rates.

Static Data. Static data is constant over time. Formally, given static data ST and a time domain we have

$$ST = \{(t_i, v) \mid t_i \in \text{time domain}\}, \tag{2}$$

where v denotes the value. Static data, such as sensor id, location, sensor properties, or user type, is constant over time and may be associated with every time point in the data integration process.

Event Data. Event data typically consists of non-numerical states recorded very irregularly over time. Formally, given event data E with q observations we have

$$E = \{(t_i, e_i \mid i = 1, \ldots, q\}, \tag{3}$$

where t_i is the timestamps and e_i is the corresponding event. Event data, such as sensor calibration, sensor replacement, bank holiday, or GPU interrupts, captures happenings at specific time points.

Interval Data. Interval data is characterized by a time period rather than a time point. Formally, given interval data I with r entries we have

$$I = \{(s_i, e_i, v_i) \mid i = 1, \ldots, r\}, \tag{4}$$

where s_i and e_i are the start and end timestamps of a data entry and v_i its corresponding value. Interval data, such as drought periods, tourism seasons, school vacations, sensor maintenance periods, inflation and interest rates periods, are characterized by start and end time points for which a particular value or label is valid.

3.2 Data Representation with Heterogeneous Time Dimensions

In the data representation stage, time series and contextual data with heterogeneous time dimensions is transformed into a uniform or homogeneous format suitable for machine learning models. These models typically require numerical vectors and work best if recorded over regular time steps. The transformation is performed based on a user-defined frequency f_{user}, which may also correspond to the one of the primary time series.

To achieve a unified homogeneous representation we convert the data into numerical representations over fixed-length windows, aka tumbling windows, with window length f_{user}, start time t_{start} and end time t_{end} of the primary series, where each window is defined as

$$w_k = [t_{start} + (k-1) \cdot f_{user},\ t_{start} + k \cdot f_{user}]$$

where $k = 1, 2, \ldots, K$, and $K = \left\lceil \frac{t_{end} - t_{start}}{f_{user}} \right\rceil$ is the total number of windows.

Figure 3 illustrates the different types of data using different colors: blue for primary time series, green and red for secondary series, yellow for interval data, purple for static data, and orange for event data.

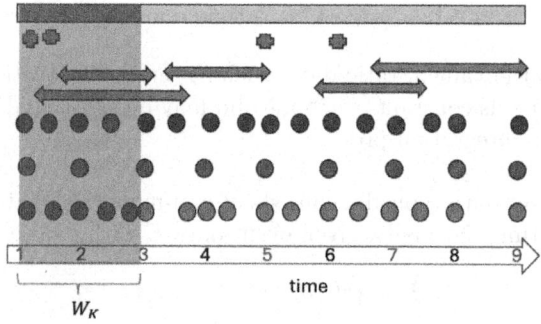

Fig. 3. Illustration of the data integration process

The figure also illustrates an example when f_{user} is 2 time units, resulting in 4 windows. While representing a time series using this window-based approach is common using either up- or downsampling with different aggregation functions (min, max, average, standard deviation, etc. or linear center interpolation etc.) other data types provide opportunities for additional representations. Static data can be encoded with its value in case of metric numerical value or using techniques like one-hot encoding. Event data can be encoded by counting occurrences of events within each window. Interval data can be represented by features such as the number of starting, ending and overlapping intervals, as well as the average, maximum, minimum, and total duration so far in each window.

3.3 Feature Extraction and Selection

After data representation, feature extraction aims at derives meaningful features from the unified data. This process is sensitive to the nature of the time dimensions. For time series data, lag features capture historical patterns essential for analysis. For event data, features such as number of previous occurrences or elapsed time since the last occurrence can be derived. For instance, many calibration events in the past may indicate an unreliable sensor, while a very recent maintenance event may indicate an uncalibrated signal. For interval data, the number of past data entries as well as the duration so far may be of value. For instance, the duration of a drought so far may be very relevant in analyzing water consumption pattern.

Feature extraction may result in a large number of features, making feature selection essential to reduce computational cost and prevent overfitting. Features are selected based on their contribution to model performance, using an iterative process that refines the selection based on the model's output. This approach not only improves efficiency but also enhances interpretability by identifying the most relevant features for decision-making.

3.4 Time Series Models

After feature extraction and selection, machine learning models are used for time series analysis, such as for instance forecasting. After transforming the data into a homogeneous representation standard algorithms can be used at this step.

4 Use Case and Dataset

As a use case we consider the processed version of the Smart Meters in London dataset, publicly available on Kaggle[1], which builds upon the dataset provided by UK Power Networks as part of the Low Carbon London project[2]. It includes additional preprocessed information such as household metadata, weather data, and UK bank holidays. We further incorporated school vacations as an additional source of contextual data, from the London school calendar[3], which is particularly relevant for understanding changes in energy consumption. The dataset provides the electrical power consumption from smart meters from 5,567 London households, recorded at a half-hour frequency from November 2011 to February 2014. It includes 112 CSV files, each representing a group of consumers categorized by block. Within each file, the LCLid field identifies the unique consumer ID, tstp indicates the timestamp, and energy(kWh/hh) records the half-hourly energy consumption for each consumer. We merged the 112 files vertically into a single CSV file, and as this study focuses specifically on London, relevant contextual data including weather, holidays, and school vacations were linked using the Location and LCLid fields. The overall schema is illustrated in Fig. 4.

---

[1] https://www.kaggle.com/datasets/jeanmidev/smart-meters-in-london.
[2] https://data.london.gov.uk/dataset/smartmeter-energy-use-data-in-london-households.
[3] https://portesbery.surrey.sch.uk/docs/2011-2012.pdf.

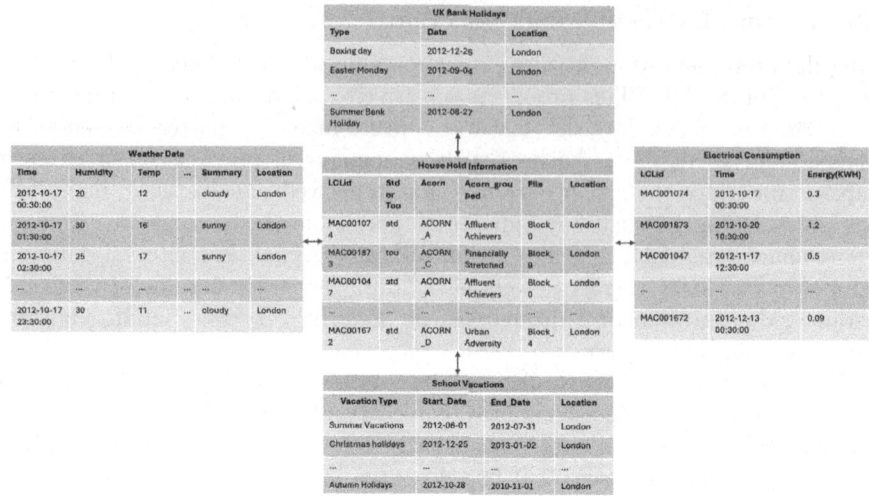

Fig. 4. Data set and contextual data with heterogeneous time dimensions

We are interested in the total energy consumption, serving as the primary time series, and thus aggregated individual consumers as shown in Fig. 5. When aggregating, we retained the number of consumers as an additional time series categorized into `StdorTou` (standard vs. time-of-use tariff). Temperature and humidity from the weather data serve as secondary time series. UK bank holidays were incorporated as event data, and school vacations were included as interval data.

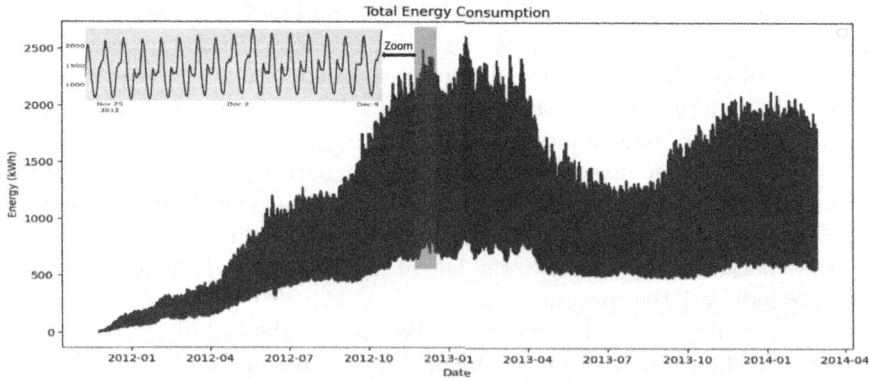

Fig. 5. Total Energy Consumption

When transforming the data, we use the 30 min frequency from the primary time series. Weather data, including temperature and humidity, is available at an hourly frequency, thus values are forward filled (interpolated) to align with the primary series, without additional statistical transformations. From event-based data (bank holidays), we derived feature whether a holiday occurs within each resampling window. From school vacations (interval data), we computed features including interval starts and ends, as well as the duration of active intervals within each window.

Once this unified format is obtained, we derived additional temporal features. Specifically, for bank holidays (event data), we included the recent holiday in the past and elapsed time since last holiday. For school vacations (interval data), we derived features, including how many days have passed since the vacation started, how many days have passed since the last vacation ended, and the duration of the most recent interval. These enriched features enables the model to capture deeper temporal dependencies and better understanding of the evolving impact of contextual factors on energy consumption.

5 Experimental Setup and Results

In the experiments we show the impact of contextual features with heterogeneous time dimensions on time series analysis tasks. We train three Random Forest regressors (from `scikit-learn`[4]) with default parameters and perform the simple yet challenging task of next value (30 min ahead) prediction, 24th value (12 h ahead) prediction, and 48th (24 h ahead) value prediction by only using features from a single (current) time point.

As a dataset we use the dataset from Sect. 4 enhanced with contextual features with heterogeneous time dimensions according to our approach. We split the data by time, i.e., data from 2011-11-23 09:00:00 to 2013-06-30 00:00:00 is used for training and 2013-06-30 00:30:00 to 2014-02-28 00:00:00 is used for testing. For the three regressors, we compare eight different settings based on the used features. "E" uses only the electrical consumption time series data (baseline). The remaining seven settings use this baseline feature and add additional features from the contextual data, where by additional features we mean all derived features based on the contextual data according to our data transformation and feature extraction process described in Sect. 4. "ES" adds the number of consumers per group (Std and ToU), "EW" adds the weather data (temperature and humidity), "EWS" adds weather and consumer group data, "EE" adds bank holidays, "EEWS" adds weather, consumer group, and bank holidays, "EH" adds school vacations, and "All" contains all features.

As evaluation metrics we use Mean Squared Error (MSE) and Mean Absolute Error (MAE). The results for our experiments are shown in Table 1. We can see that the additional features derived from contextual data with heterogeneous time dimension have the potential to substantially improve the performance of a time series analysis task compared to the baseline. In our case of regression, the improvements are up to 28% and 16% for MSE and MAE, respectively for

---

[4] https://scikit-learn.org/stable/modules/generated/sklearn.ensemble.RandomForestRegressor.html.

next value prediction, which corresponds to an improved mean square percentage error from 0.71% to 0.53% and improved mean absolute percentage error from 6.59% to 5.61%. The respective improvements on MSE and MAE for 24th value prediction are 44% and 27%, and for 48th value prediction 27% and 15%. In all cases adding features improves the performance, even if only by a small amount in some cases, but adding all features (All) provides the best results.

Table 1. Prediction accuracy (MSE and MAE) and improvement in % when using contextual features for next value prediction (30 mins ahead), 24th value prediction (12 h ahead), and 48th value prediction (24 h ahead).

Features	Next Value		24th Value		48th Value	
	MSE	MAE	MSE	MAE	MSE	MAE
E	0.001175	0.026301	0.036294	0.151177	0.002966	0.040323
ES	0.000933	0.023134	0.024732	0.125947	0.002354	0.036054
EW	0.001026	0.024406	0.026082	0.125752	0.002583	0.037274
EWS	0.000864	0.022235	0.020371	0.110471	0.002169	0.034270
EE	0.001173	0.026287	0.036277	0.151146	0.002964	0.040316
EEWS	0.000865	0.022234	0.020359	0.110495	0.002174	0.034339
EH	0.000973	0.023732	0.028809	0.133083	0.002449	0.036480
ALL	0.000846	0.022054	0.019989	0.109103	0.002159	0.034036

	Improvement (%)		Improvement (%)		Improvement (%)	
Features	MSE	MAE	MSE	MAE	MSE	MAE
E	-	-	-	-	-	-
ES	20.59	12.04	31.85	16.68	20.63	10.58
EW	12.68	7.20	28.13	16.81	12.91	7.56
EWS	26.46	15.45	43.87	26.92	26.87	15.01
EE	0.17	0.05	0.04	0.02	0.06	0.01
EEWS	26.38	15.46	43.90	26.91	26.70	14.84
EH	17.19	9.76	20.62	11.96	17.43	9.53
ALL	28.00	16.14	44.92	27.83	27.20	15.59

6 Conclusion

We provided the intuition towards blending contextual data with heterogeneous time dimensions with time series for analysis tasks. To this end, we identified data with four different types of time dimensions that often appear in industrial environments, namely constant, time series, events, and intervals and show how these can be transformed into a homogenous format such that traditional machine learning algorithms can be employed. Based on an evaluation on a real-world use case, we show that such an approach can provide promising results.

Acknowledgments. This work was supported in part by the DIADEM project, funded by the Free University of Bozen-Bolzano.

References

1. Banerjee, T., Whipps, G., Gurram, P., Tarokh, V.: Sequential event detection using multimodal data in nonstationary environments. In: FUSION, pp. 1940–1947. IEEE (2018)
2. Behmiri, N.B., Fezzi, C., Ravazzolo, F.: Incorporating air temperature into midterm electricity load forecasting models using time-series regressions and neural networks. Energy **278**, 127831 (2023)
3. Brahim, G.B.: Weather conditions impact on electricity consumption in smart homes: machine learning based prediction model. In: ICEEE, pp. 93–98. IEEE (2021)
4. Burero, S.: Integrating heterogeneous contextual data for enhanced time series analysis. In: EDBT/ICDT Workshops. CEUR Workshop Proceedings, vol. 3946. CEUR-WS.org (2025)
5. Cai, H., Yang, Y., Li, X., Huang, Z.: What are popular: exploring twitter features for event detection, tracking and visualization. In: MM, pp. 89–98 (2015)
6. Cecaj, A., Mamei, M.: Data fusion for city life event detection. J. Ambient Intell. Humaniz. Comput. **8**, 117–131 (2017)
7. Charane, A., Ceccarello, M., Gamper, J.: Shapelets evaluation using silhouettes for time series classification. In: DOLAP. CEUR Workshop Proceedings, vol. 3653, pp. 36–44. CEUR-WS.org (2024)
8. Chattopadhyay, S., Paliwal, P., Narasimhan, S.S., Agarwal, S., Chinchali, S.P.: Context matters: leveraging contextual features for time series forecasting. CoRR **abs/2410.12672** (2024)
9. Duan, L., Da Xu, L.: Data analytics in industry 4.0: a survey. Inf. Syst. Front. 1–17 (2021)
10. Farahani, M.A., McCormick, M.R., Harik, R.F., Wuest, T.: Time-series classification in smart manufacturing systems: an experimental evaluation of state-of-the-art machine learning algorithms. Robot. Comput. Integr. Manuf. **91**, 102839 (2025)
11. Kazijevs, M., Samad, M.D.: Deep imputation of missing values in time series health data: a review with benchmarking. CoRR **abs/2302.10902** (2023)
12. Kim, J., Kim, H., Kim, H., Lee, D., Yoon, S.: A comprehensive survey of time series forecasting: architectural diversity and open challenges. CoRR **abs/2411.05793** (2024)
13. Lim, B., Arik, S.Ö., Loeff, N., Pfister, T.: Temporal fusion transformers for interpretable multi-horizon time series forecasting. CoRR **abs/1912.09363** (2019)
14. Menapace, A., Zanfei, A., Felicetti, M., Avesani, D., Righetti, M., Gargano, R.: Burst detection in water distribution systems: the issue of dataset collection. Appl. Sci. **10**(22), 8219 (2020)
15. Mondal, M., Khayati, M., Sandlin, H., Cudré-Mauroux, P.: A survey of multimodal event detection based on data fusion. VLDB J. **34**(1), 9 (2025)

16. Moretti, M., Fiorillo, D., Guercio, R., Giugni, M., De Paola, F., Sorgenti degli Uberti, G.: A preliminary analysis for water demand time series. Environ. Sci. Proc. **21**(1), 7 (2022)
17. do Nascimento, N.M., Alencar, P., Cowan, D.D.: Context-aware data analytics variability in iot neural network-based systems. In: IEEE Big Data, pp. 3595–3600. IEEE (2021)
18. Oh, S., et al.: Multimedia event detection with multimodal feature fusion and temporal concept localization. Mach. Vis. Appl. **25**, 49–69 (2014)
19. Peng, H., et al.: Fine-grained event categorization with heterogeneous graph convolutional networks. arXiv preprint arXiv:1906.04580 (2019)
20. Prabakar, A., Wu, L., Zwanepol, L., Van Velzen, N., Djairam, D.: Applying machine learning to study the relationship between electricity consumption and weather variables using open data. In: ISGT-Europe, pp. 1–6. IEEE (2018)
21. Qian, K., Jiang, J., Ding, Y., Yang, S.: Deep learning based anomaly detection in water distribution systems. In: ICNSC, pp. 1–6. IEEE (2020)
22. Rodríguez, M., Tobón, D.P., Múnera, D.: Anomaly classification in industrial internet of things: a review. Intell. Syst. Appl. **18**, 200232 (2023)
23. Shah, B., Jat, P.M., Sashidhar, K.: Performance study of time series databases. arXiv preprint arXiv:2208.13982 (2022)
24. Zhan, T., He, Y., Deng, Y., Li, Z., Du, W., Wen, Q.: Time evidence fusion network: multi-source view in long-term time series forecasting. arXiv preprint arXiv:2405.06419 (2024)

A Hybrid Data Model to Support Transportation Analytics of Emergency Service Vehicles

Carson K. Leung[✉]

University of Manitoba, Winnipeg, MB, Canada
`Carson.Leung@UManitoba.ca`

Abstract. Using a single type of database solution to support real-world applications is becoming more and more challenging because of the volume and variety of data. For instance, the data collected for the transportation industry comprise both structured and unstructured data. Using solely a single type of database solution—relational database system-only or graph database-only—to store and manage data can be challenging. As real-world applications ask even more complex questions related to data, the database solution should be able to facilitate answering these questions in a reasonable time. Hence, in this paper, we present a hybrid model, which integrates data to support transportation analytics. The model consists of relational databases and non-relational databases (namely, graph databases), pooling their strengths to support the demands of the modern application. We also demonstrate this hybrid data model as a practical solution with a case study on improving emergency services—such as emergency medical services (EMS)—response times by having the support of the presented platform.

Keywords: Database applications · Big data analytics · Knowledge discovery · Big data · Data model · First responders · Emergency services · Emergency medical services · Relational database · Graph database

1 Introduction and Related Works

With advances in technology, big data are everywhere. Embedded in these big data is valuable information and knowledge. Data science—which makes good use of data management [1–5], data mining (e.g., frequent or rare pattern mining [6–11]), and machine learning [12] (e.g., unsupervised learning [13], supervised learning [14]) techniques—analyzes big data and discovers knowledge and useful information for various real-world application areas such as finance analysis [15], healthcare informatics [16, 17], and social network analysis [18]. In this paper, we focus on transportation and urban analytics [19–23].

Data science aims to discover useful information from large amounts of data [24]. In recent years, the advances in data storage and communication technologies have led to increased availability and diversity of data [25]. In the transportation industry alone, multiple data sources come from—but not limited to—global positioning systems (GPS), Wi-Fi/Bluetooth sensors, road condition assessment reports, and data associated with land use. These many forms of data underscore the need to extract previously unknown and potentially useful information. Hence, we need new capabilities for data extraction, storage/management, and dissemination systems [26].

Knowledge discovered from these rich sources of information can significantly affect the results of any analysis. For example, when dealing with *emergency medical services* (*EMS*) transportation, knowing of a new factor that might affect emergency responders' travel time can be crucial when making decisions. Hence, it would be very beneficial to researchers and other stakeholders for managing traffic systems and conducting traffic analysis [25].

Unsurprisingly, the process of data mining strives towards deriving useful information from raw data. The information gained helps to make better predictions and identify patterns to facilitate the decision-making process [27]. Such facilitation in decision-making could yield substantial benefits in EMS. In general, EMS administrators seek methods to minimize response time (i.e., the time between reporting an incident and when first responders arrive at the scene). According to some studies [28, 29], the risk of fatality remains very low if affected individuals receive medical attention within 4–8 min of an emergency. Hence, following the National Fire Protection Association (NFPA) 1710 standard[1], many first responders—such as fire and paramedics services (FPS)—set their aim that the first EMS unit responding to an emergency must arrive at the scene within 4 min.

There are several challenges [28–31] that could hinder an EMS vehicle from arriving at the scene within this established time frame. For instance, a challenging factor is delay or blockages at active grade-level crossings (subsequently referred to as crossings) when responding to calls. An *active crossing* is a crossing blocked by a moving or stopped train. At any rate, emergency responders are delayed daily by trains at active crossings, costing them precious minutes in a time-critical situation. Delays at crossings are usually caused by freight trains stopping or moving slowly at those crossings [32, 33]. That being the case, one would think a trivial solution is to avoid active crossings if there were predictions for the blockages. However, predicting the arrival times of freight trains at crossings is also difficult [34, 35] to avoid. As a result, these blockages are a significant cause for concern in many cities. For instance, the Canadian city of Winnipeg in central Canada has the highest number of crossings among major Canadian cities[2]. While being delayed at a crossing could be nothing more than an inconvenience for many, emergency cases (as it usually is for EMS) could mean life or death. As another instance, Fire Department in another Canadian city of Niagara Falls are aware of the risk of being delayed at crossing. Because of this risk, their default response navigation strategy is to use an alternative route that does not interact with crossings. Although the alternative route may add an extra two or three minutes to their travel time, they consider it acceptable. However, in cardiac arrest emergencies, two minutes could be the difference between life and death [29]. Hence, the FPS increasingly faces the need for a personalized mobility solution that ensures lower risks of delays at crossings when responding to emergencies. So, having a model to assess their risks of being exposed to such delays on their route would immensely reduce travel time, and thus saving lives.

---

[1] https://www.nfpa.org/codes-and-standards/nfpa-170-standard-development/170.

[2] https://tc.canada.ca/en/rail-transportation/grade-crossings/grade-crossings-inventory.

Knowing the risk of exposure would benefit their decision-making when navigating to call locations. As a result, a data management system that can support this type of analysis would be crucial in helping the FPS mitigate their risks of exposure. This data management system would allow for quick and efficient generation and storage of these models.

We contend that using one kind of database is not a satisfactory solution. While it is possible to use a non-relational database to perform transportation analysis, this thesis shows that it is inefficient. However, by leveraging the strengths of relational and non-relational databases, we demonstrate this as a model solution to support modern-day use for EMS transportation analytics. Our *key contributions* of this paper are the design and development of a hybrid model. To elaborate, we:

- design a conceptual model of a hybrid database management system (DBMS) that combines the strengths of relational and non-relational databases to support transportation analytics;
- implement the hybrid database for supporting transportation analytics;
- develop an algorithms to extract and convert connected linear shape files into a graph network; and
- conduct case studies on improving emergency medical service response times.

2 Background

2.1 Emergency Response Process

The emergency response-time process involves call processing time (the time from receiving a call on an emergency phone number—such as 911 in Canada/USA, 112 in many European countries, or 191 in Thailand—to alerting emergency response stations or units), turnout time (the time from receiving the emergency alert to boarding the response vehicle), and travel time (the time that the vehicle begins travelling to the call location to the arrival time). Emergency response processes in North America commonly involve three parties—namely, call takers, dispatchers and responders [36]. Call takers are in charge of inbound requests and communicate with an emergency caller. Dispatchers are in charge of outbound calls to the ambulance teams, fire services and hospitals. It is important to note that dispatchers support an emergency event from when they are assigned to them until the case is closed. Current challenges in many cities are:

- *EMS specialization*: Different first responder units may have different licenses, allowing them to administer certain medications in emergencies. Certain specialized units must remain available for deployment.
- *Equipment access*: Some fire trucks are more equipped for certain emergency scenarios than others. For example, only five ladder-equipped fire trucks servicing the entire city, making their availability equally important.

Dispatchers aim to maintain reasonable response times and fleet capacity while responding to emergencies. The presence of an active crossing presents a challenging dilemma for dispatchers. For example, if a responder is stuck at a train crossing, the dispatchers can instruct the unit to either:

1. turn around and go another way, or
2. stay while the dispatcher sends another unit from the opposite side of the train to respond to the event.

In the first scenario, the responder would have increased their response time. Even worse, in the second scenario, we have two units responding to a call that only needed one unit, reducing fleet capacity by two. For crossings notorious for causing delays, Fire Paramedic Service sends multiple units from different directions to respond just in case the crossing is active upon the unit's arrival.

2.2 Composition of an EMS Trip

To further the understanding of the analysis that our proposed database will support, let us give a high-level description of FPS' composition of responder trips. We can further break an EMS trip down into multiple legs. A leg is a physical trip of a responder from one point to another. Depending on the call, a trip can compose two or more legs. In order to understand the concept of legs, let us first review the spatial components of completing a call. An emergency responder completing a call can comprise the following spatial components:

- Emergency response station: The station they assign the responder unit.
- Call location: The location of the emergency response event.
- Hospital: A facility a responder can deliver a patient to for transferring care.

There are three primary trip compositions a responder would follow when responding to emergencies:

1. Trip originates from station: When a trip originates from a station, there are three possibilities:
   a. the unit arrives at the location and then returns to the station (if the call does not require advanced medical care and they do not dispatch the unit to another call)
   b. the call does not require advanced medical care, but they dispatch the unit to another call when the call is closed at the call location
   c. the call requires advanced medical care, so they transport patient(s) to the hospital. When responders transfer care over to the hospital, they either (i) return to the station or (ii) are assigned to a new call. Then, they begin a new trip.
2. Trip originates from a previous call location: Here, the trip composition will have the responder dispatched after the completion of the previous call. The trip composition also has one of three aforementioned possibilities (i.e., possibilities 1a, 1b, and 1c).
3. Trip originates from hospital: Here, the trip composition will have the responder dispatched after completing a previous call at the hospital. Again, the rest of the trip will follow one of the three aforementioned possibilities (i.e., possibilities 1a, 1b, and 1c).

As outlined, three leg types exist for emergency responder trips:

1. Origin leg: the first leg of a trip associated with a call;
2. Waypoint leg: the leg of the trip connecting the call location to a hospital;
3. Return leg: the leg of the trip returning the responder to a station.

2.3 Shapefiles

Classically, spatial data can be considered as raster or vector data. There are three types of vector data: points, lines, and polygons. A shapefile contains structured data, which are in rows and columns. Thus, they can be stored in a relational database. A shapefile has a required column called "geometry"—where we store the geometric data as a Binary Large Object commonly referred to as a "BLOB". A shapefile can only store one type of geometric feature per file. Hence, if one were to represent a road network, they may need one shapefile to store all the roads as a linear feature. Then, another shapefile to store all intersections, bus stops, and other points of interest as point features.

3 Our Hybrid Data Model for Transportation Analytics

3.1 Essential Features for Performing Transportation Analysis

We intend to build a model to support analyzing railway and road transportation. On the most basic level, if we intend to analyze any form of transportation, we need to build that transportation network model to proceed. Hence, we build railway and road transportation networks, which are the skeleton of our analysis process.

3.1.1 Transportation Network

Road and rail have geospatial features. To build a network for these components, we need to ingest geospatial data—and thus, shapefiles. It is important to note that, although shapefiles are available as open data in many cities, the specific features and attributes vary between cities. To make our model as easily adaptable as possible, we use shapefiles from OpenStreetMap (OSM). A reason is that they are the most common providers worldwide. Unfortunately, many jurisdictions still maintain and use their own spatial data sets, with their own unique fields and structures. Shapefiles combine both structured and unstructured data. Other than the geometry column, they strictly conform to a structure. Hence, we consider shapefiles semi-structured data. In addition, all layers of the OSM data set have the following common attributes:

- id: a unique identifier for the feature
- osm id: an OSM identifier for the feature
- name: the name of the feature
- code: a four-digit code defining the feature class
- fclass: the class name of the feature
- geometry: the geospatial record of this feature (e.g., POINT, POLYGON, LINESTRING)

Let us describe the shapefiles for road and rail. First, roads shapefile holds linear features of all kinds of roads, from streets and avenues to gravel tracks and lanes. For example, depending on the size of the location, the data set can contain between 200,000 to 500,000 rows of data. In addition to the attributes mentioned earlier, this shapefile also contains the following attributes:

- ref: A reference number for the road segment, expressed as a character string.

- oneway: A single-character string to answer the question "Is this a one-way road?": 'T' (true/yes) represents a one-way road, 'B' to represent a bi-directional road, and 'F' (false/no) to represent that only driving in the direction of the line-string is allowed.
- maxspeed: The maximum allowed speed in km/h, expressed as an integer.
- layer: The relative layering of roads, expressed as an integer.
- bridge: A single-character string containing 'T' or 'F' to answer the question "Is this road on a bridge?".
- tunnel: A single-character string containing 'T' or 'F' to answer the question "Is this road in a tunnel?".

Railways shapefile holds linear features of the railway tracks. For example, depending on the size of the location, the data set can contain between 3,000 to 5,000 rows of data. In addition to the attributes mentioned earlier, this shapefile also contains the following attributes:

- layer: The relative layering of railways, expressed as an integer.
- bridge: A single-character string containing 'T' or 'F' to answer the question "Is this rail track on a bridge?".
- tunnel: A single-character string containing 'T' or 'F' to answer the question "Is this rail track in a tunnel?".

Traffic shapefile holds traffic-related points information from stops and crossings to traffic signals and speed cameras. The feature class of interest in this data set is "crossings". Although OSM does not differentiate between railway or pedestrian crossings, we filter out pedestrian crossings as it is not a feature of interest for our analysis described in this paper. Again, for example, depending on the size of the location, the data set can contain between 2,000 to 50,000 rows of data.

3.1.2 Supplemental Data Sets

Shapefiles on their own may not contain sufficient information upon which to build our analysis. More often than none, supplemental data sets rich in information may be required to complement the data in the shapefiles. For example, say one intends to group their analysis by postal code or other regions of interest, one would need to integrate those data sources. We can include other grade crossing information as a structured comma separated values (CSV) file for this case. It is standard practice worldwide for transportation authorities to generate and maintain statistics on transportation infrastructure. We use this database approach to perform analysis in the USA and Canada. Our findings reveal that the US Federal Railway Authority (FRA) and Transport Canada provide important statistical information about grade crossings in the United States and Canada, respectively.

In addition to the name, latitude and longitude, the inventory also contains statistical information. Core information like the number of accidents, fatalities, and trains per day are all important for analysis. For example, there are 26 attributes contained in the Transport Canada data set and 255 in FRA data set. Let us consider four important attributes:

- Protection: Type of warning at the crossing, expressed as a character string.

- Vehicles daily: An estimate of the number of road vehicles per day over the crossing, expressed as an integer.
- Total trains daily: An estimate of the number of freight (and passenger) train movements per day over the crossing, expressed as an integer.
- Train max speed (mph): An estimate of the maximum train speed over the grade crossing in miles per hour, expressed as an integer.

Here, bidirectional road network, train tracks inventory, and traffic related points are owned by OSM, and can be store as *semi-structured* data in Shapefile format. In contrast, grade crossing inventory are owned by Transport Canada or FRA and can be stored as structured data in the CSV file.

3.2 Problems with Using Only a Relational Database Model

A potential model to handle the aforementioned data sets is using only a relational database with entities and relationships. As such, road segment, railway, and crossing can be modeled as entities. As train crossings are an intersection between rail and road segments, they can be modeled as a relationship:

In this relational model, we extract the contents of the shapefile and put them into a table. The caveat of having this data stored in a table form is that we cannot infer certain spatial information with the way we stored the data. For example, without performing several queries, one cannot look at a row in the road segment table and infer the road segments that are spatially next to it. Similarly, if one intends to find a route from one road segment to another on the road network, it would prove cumbersome. We first must build a relationship between each road segment and all other road segments. This process is computationally expensive, hence inefficient, not to mention impractical.

With the relational database model, we must join several tables to pull in all those properties to see the characteristics of a feature, say, a road segment. For example, if we needed to determine the number of casualties/accidents on a road segment, we would need to join with the crossings-roads-rail table and then use the joined table to examine the grade crossing inventory table where that record lies.

Imagine the effort it would take to aggregate the number of trips interacting with each crossing. We need to generate another table. Essentially, we need to generate another table or view to analyze each new metric of interest. At the very generic level, the number of tables is already growing. Conveying these multiple tables might be a steep learning curve.

3.3 Problems with Using Only a Graph Database Model

As discussed in Sect. 3.2, it may be impractical to model the emergency response data set in only relational database. Since we are dealing with highly connected data, storing it in a *graph database* may be more effective.

On the basis of which we are conducting our analysis, the first step would be to extract the content and features of the shapefiles. The elements in the shapefile already store connected information. We then add other data sources that complement the attributes of the shapefiles. These become additional properties of the edges and nodes in the graph.

The purpose of this graph database is to support studies that analyze the interaction between road and railway transportation. Our graph is composed of a network of road and railway edges and nodes; that is, nodes and edges can be road or rail components. To build such a graph, we first extract the road and rail features and build them as individual layers. Then, in another step, we join both layers into a fused network of nodes and edges. Recall that the traffic-related points shapefile, among other things, holds records of grade crossings, i.e., locations where the rail lines interact with the road network. Those points are the fusing locations to merge both layers into a single graph layer, as shown in Fig. 1. In other words, we create two separate graphs: one for roads and the other for the railway. We then fuse both graphs at their geographical meeting points. After forming individual graphs, we then join the layers of the graph. Finally, we build the property graph by adding supplemental data containing crucial statistical information from the grade crossing inventory file. We use the coordinates from both data sets to try and find their match. We make a simple circle around the coordinates of both data sets with a diameter of 8 m to 10 m. Then, we consider intersecting circles from both data sets as matches. Note that data sets for roads and for railway were from different sources. As such, they may not align as a 1:1 match in several spots. We scale them before the integration.

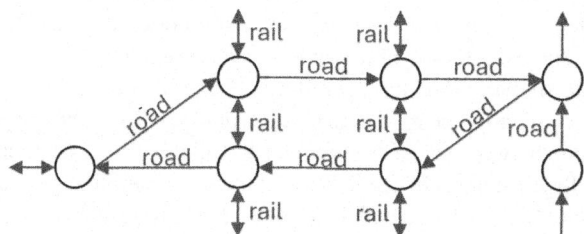

Fig. 1. Graph database representation.

We need to store the composite graph's meta sources somewhere. We store the meta sources to reproduce the graph in the future. We also keep the meta sources if we need to isolate a graph layer and perform an analysis only on one layer. Building these individual layers, then combining them, may be cumbersome. When we run analysis on the graph, the question that comes to mind is how we store the results from that analysis. With the graph-only approach, we would have to store the results from graph operations directly on the graph. A typical operation includes pathfinding routes between two nodes. Over time, storing those results on the graph would increase the size of the nodes and edges and may cause searches to become slower.

3.4 Building Our Hybrid Model

By looking at the database information without considering the data involved in the analysis, the relational database seems to have some advantages. For instance, it has an index which makes it more efficient to search through some related data, something we cannot achieve with the graph database. The relational database also has mass support for visualization tools, making it an easy choice for relaying results from analysis to users. Since relational databases are best at handling transactional records, they will be the preferred storage choice to hold and process transactional transportation data. These data include travel logs, commuter schedules, and terminal information. Storing abstract data types such as lines, polygons, and bit maps allow relational databases to store and retrieve spatial information efficiently.

The structure of the transportation network is inherent to a graph structure. Geographical information in shapefiles makes a strong case for storing it on the graph as the data are highly connected. In just the tabular relational model, the network's complexities have been represented using multiple relationship tables and joins. The graph structure would eliminate this overhead if we were to store the data in the structure of the road and rail networks. A challenge of solely using a graph database is slower query times when analysis data are stored directly on the graph. We explore the possibility of performing analyses on the geospatial data on the graph database, then storing the results and statistical summarizations on the relational database.

It becomes inevitable that we must retrieve the data from shapefiles and store them in tables or a graph database. With tables, we can hold transactional records that run on the graph or hold results retrieved from the graph. With a graph, we maintain the same class structure, the extraction becomes cleaner, and the data features are all in one place. Ultimately, we need to use a relational database for some data types, while we need the graph database for others. As a result, there is a trade-off between the relational model and the graph database model—we do not have a clear winner. To take advantage of both worlds, a *hybrid model* would be logical.

Specifically, we present a hybrid model design comprising a relational database and a property graph database. A two-component hybrid model helps us take full advantage of the strengths of each component database. We discuss the roles of the individual components as follows:

- Graph database: We use a property graph for this component. Then, we have all the key/essential data attributes in one place, at the node or edge. Since the structure works best for connected data, we use this database to perform most of the analysis. The graph aids us in discovering connections and patterns between nodes and edges. It will also help us answer questions we did not know to ask at the beginning of the database design. Hence, the graph database is essential to the hybrid database design. Then, the relational and document databases exist to support the capabilities of the graph. Since the graph database is a property graph, the contents of the shapefiles and crossing inventory are stored—once extracted—as key-value attributes in the graph.
- Relational database: This is the second of the two components of the data warehouse. It is crucial to take advantage of the relational database's quick look-up functionality, especially when building the graphs. With the contents of the shapefiles loaded into a relational database, we can save time when building the graph. Finding these corresponding geometries would be cumbersome. Those built-in functions take constant time to yield results, whereas the alternative would be to loop through the data. Additionally, the relational database can hold summarization statistics about our analyses, that is, the output of the analyses that run on the graph. A strong reason for using relational databases over graphs is their strong support for visualization. Holding

summary data from analyses in the relational database makes using readily available visualization tools seamless. In the future, when graph databases develop strong support for visualization, this may not be a required feature. To improve the efficiency of the graph database, we can store graph traversals between two nodes in the relational database. That way, instead of repeating a traversal, we easily retrieve the path from a table.

4 Evaluation

4.1 Features

First, we evaluated our hybrid model when compared with relational database-only model and graph database-only model. Results are shown in Table 1.

Table 1. Comparison on features provided by our hybrid model when compared with relational database-only model and graph database-only model.

Criteria	Relational DB	Graph DB	Hybrid DB
Support visualization	Yes	No	Yes
Store and manage structured data	Yes	No	Yes
Store and manage semi-structured data	No	Yes	Yes
Store and manage unstructured data	No	Yes	Yes
Easily modify schema	No	Yes	Yes
Easily model geo-spatial data	No	Yes	Yes

4.2 Runtime

In addition to features, we also compared their runtime. Results are shown in Tables 2 and 3. To elaborate, we ran all the tests using a computer running Ubuntu 20.04 LTS as the primary operating system. The CPU was an Intel Core i7 with eight cores clocked at 2.8 GHz. The machine also had 32 GB of RAM and a Solid State Drive. The total data obtained were 168,419 WFPS trip records from the Winnipeg Fire and Paramedics in 2018. Hence, the data comprise trips with different emergencies and road conditions from all four seasons. After the data pruning, we broke down the remaining trips into legs, resulting in 196,265 leg records.

Here, Table 2 shows the runtime for preliminary steps, which include the following:

1. Runtime to find the fastest path (i.e., best travel time for each leg);
2. Runtime to examine paths and flag interacted legs (i.e., legs going over rail crossings as interacted with road crossings);
3. Runtime to cross-reference flagged legs with historical blockage event records at crossings to see if there were trains at the time of travel. If so, these legs are flagged as potential impacted legs;

4. Runtime to flag the legs as confirmed impacted by a train if the actual travel time of the leg is beyond a threshold higher than the expected travel time.

Note that, for both the graph database and our hybrid database, there is no runtime in flagging interacted legs in Step 2 because this step is included in Step 1 when finding the fastest travel-time path for these databases. In Step 3, both relational database and our hybrid database excel due to the existence of indices. Moreover, in Step 4, both relational database and our hybrid database excel again due to the transactional nature of the task of flagging confirmed impacted legs.

Table 2. Comparison on runtime for preliminary steps required by our hybrid model when compared with relational database-only model and graph database-only model.

Step	Relational DB	Graph DB	Hybrid DB
Fastest path, travel time	738 min	100 min	74 min
Flag interacted legs	459 min	-	-
Flag potential impacted legs	414 min	750 min	415 min
Flag confirmed impacted legs	54 min	139 min	54 min
Total	**1665 min**	**989 min**	**543 min**

Table 3. Comparison on runtime for the result generation required by our hybrid model when compared with relational database-only model and graph database-only model.

Operation	Relational DB	Graph DB	Hybrid DB
Crossing risk score	117 min	51 min	36 min
Area risk score	157 min	49 min	40 min
#responders delayed per week	60 min	492 min	62 min
Average delay per trip	45 min	482 min	44 min
Total	**379 min**	**876 min**	**182 min**

Then, Table 3 shows the runtime for the result generation, which include the following:

1. Runtime for computing the crossing risk scores of top-3 crossings, which involves (a) evaluating every trip undergone by first responders and (b) counting the frequency of the crossings that interacted with the trips. Crossings with the top-3 highest frequencies are selected.
2. Runtime for computing the area risk scores of top-3 city areas, which involves the product of two distributions: the crossing occupany distribution and the responder-crossing interaction distribution. The latter is a subset of the first responder destinations. Areas with the top-3 highest product are selected.

3. Runtime for aggregating (e.g., summing) the weekly numbers of responders who were delayed by a train on the confirmed impacted legs.
4. Runtime for aggregating (e.g., averaging) the delay time per trip.

4.3 Comparisons with Related Works

Moreover, we also compared with related works. Comparisons are shown in Table 4.

Table 4. Comparison on features provided by our hybrid model when compared with related works.

Criteria	Sokolova et al. [37]	Bessa [38]	Villalobos et al. [39]	Our hybrid model
Disjoint data between DBs	No	Yes	No	Yes
Contribute to transportation industry	No	Yes	Yes	Yes
Easily obtainable data	No	No	Yes	Yes
Easy learning curve to use	Yes	No	Yes	Yes
Use real-world data set	Yes	Yes	No	Yes

4.4 A Case Study: Transportation Analytics in Winnipeg, Canada

We conducted a user study in the Canadian city of Winnipeg. With our hybrid model, we identified trips that went over crossings: We have a start and end position for each trip. We then use the graph to find the fastest path between the start and end position nodes on the graph. The path is a collection of nodes and edges on the graph. Suppose any of the nodes is a grade crossing. In that case, we flag the trip row on the relational database table as "possibly impacted". In turn, we increase the count attribute for the number of trips the crossing node encountered.

We also identified delayed trips. The expected travel time is 240 s, as the standard requires for any EMS trip. Identifying delayed trips is a multi-stage process:

1. Retrieve travel time information of "possibly impacted" trips from the trips table.
2. Pick those trips with travel time above their expected travel time.
3. Retrieve their route from the origin-destination table.
4. Correlate the route to the spatial blockage impacts table to verify if the route was within any geofence within an active crossing blockage time window. If the route is within the geofence, we flag that trip in the trips table as "confirmed impacted".

We also rerouted trips. The last phase of the analysis is rerouting trips confirmed as impacted by an active crossing, then calculating the travel time difference. We selected an active crossing node to avoid, then ran the traversal on the graph. We then calculated the expected travel time for this new route. For example, at the Shaftesbury crossing, a senior complex near the rail crossing is a regular first responder destination with 11 calls per week. First responders that respond to emergencies at this complex and other locations north of the rail track typically travel north on Kenaston Blvd. They turn west on Sterling Lyon Pkwy and then north on Shaftesbury Blvd over the tracks. As shown in Fig. 2, when the rail crossing is clear, this is usually the fastest route. However, when the crossing is blocked, an alternate route would be continuing north on Kenaston Blvd (driving under the rail overpass/overbridge), turning west on Grant Ave, and then turning south onto Shaftesbury Blvd. This alternate route is nearly 2 miles further and takes about four minutes longer to reach the senior complex. However, if there were a 5-min train at the Shaftesbury crossing, it would be at least one minute faster.

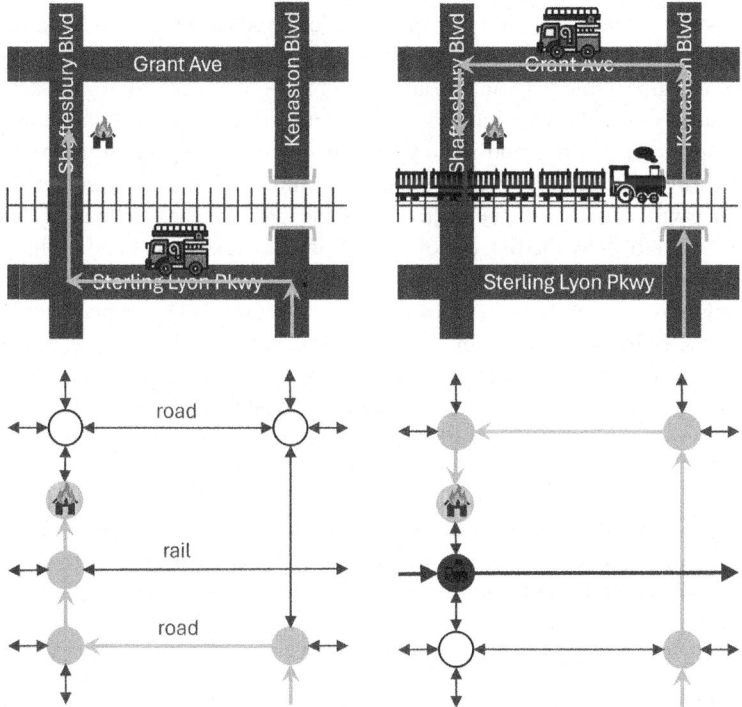

Fig. 2. Top row shows simplified road maps and bottom row shows the corresponding graph database representations. Green paths showing the faster route when the rail crossing is (a) clear (on the left column) and (b) blocked (on the right column).

5 Conclusions

In this paper, we presented a hybrid data model to support transportation analytics of emergency service vehicles. It demonstrates that a single-style database architecture (e.g., only relational database, only graph database) was no longer an acceptable solution for modern-day transportation analytics. We observed that relational databases alone could not handle unstructured data. Similarly, non-relational databases (e.g., graph databases) alone would underperform when handling structured data. To the best of our knowledge, existing related works in this area of research focused heavily on demonstrating the power of NoSQL database solutions. Specifically, they focused on NoSQL's ability of handling unstructured data while ignoring the obvious benefits of relational databases. Some bodies of related work identified the need to use both styles of databases for handling both data structures. However, they showed little or no real-world application on a large enough data set. Our hybrid model addressed both concerns by leveraging the strengths of relational and non-relational (e.g., graph) databases in our hybrid architecture. We also went further to show its applicability in the real-world urban and transportation analytics. It helps improve patient outcomes (e.g., increase survival rates, reduce severity of injuries, prompt faster recovery, reduce long-term health impacts) and enhance efficiency and resource optimization (e.g., improve resource allocation, enhance community safety, reduce strain on first responders/fire paramedic services).

Additionally, we showed that our hybrid data model is practical as a case study were conducted on data in a mid-sized Canadian city of Winnipeg. Furthermore, such a generic hybrid database architecture is a sound solution for transportation analytics. Its performance was shown to be better than a single-style database. We used this hybrid database design to study the risk of delay to emergency responders and evaluated alternative routing solutions. The results are expected to go a long way to helping cities understand their risks of delay. It would positively transform their emergency response process and save lives. From a business perspective, we uncovered several essential factors about notorious crossings and areas in Winnipeg. Such revelations can help resource planners focus resources on areas that need more immediate attention. As *ongoing and future work*, we explore ways to further enhance our model by incorporating additional information.

Acknowledgments. This work is partially supported by Mitacs, NSERC (Canada) and University of Manitoba. Thanks M. Ghaffari Dolama, J.D. Regehr, N. Ternowetsky, and B. Wodi, for their domain expertise in the transportation domain.

References

1. Amagata, D., et al.: Efficient algorithms for top-k stabbing queries on weighted interval data. In: DEXA 2024, Part I. LNCS, vol. 14910, pp. 146–152 (2024)
2. Coelho, G.M.C., et al.: Improving the accuracy of text-to-SQL tools based on large language models for real-world relational databases. In: DEXA 2024, Part I. LNCS, vol. 14910, pp. 93–107 (2024)
3. Eom, C.S., et al.: Effective privacy preserving data publishing by vectorization. Inf. Sci. **527**, 311–328 (2020)
4. Nakano, M., et al.: Extension of parallel primitives and their applications to large-scale data processing. In: DEXA 2024, Part II. LNCS, vol. 14911, pp. 248–253 (2024)
5. Roy, K.K., et al.: Mining sequential patterns in uncertain databases using hierarchical index structure. In: PAKDD 2021, Part II. LNCS (LNAI), vol. 12713, pp. 29–41 (2021)

6. Capillar, E., et al.: Bitwise vertical mining of minimal rare patterns. In: DaWaK 2023. LNCS, vol. 14148, pp. 135–141 (2023)
7. Chowdhury, M.E.S., et al.: A new approach for mining correlated frequent subgraphs. ACM Trans. Manag. Info. Syst. (TMIS) **13**(1), 9 (2022)
8. Czubryt, T.J., et al.: Q-VIPER: quantitative vertical bitwise algorithm to mine frequent patterns. In: DaWaK 2022. LNCS, vol. 13428, pp. 219–233 (2022)
9. Leung, C.K., et al.: Scalable vertical mining for big data analytics of frequent itemsets. In: DEXA 2018, Part I. LNCS, vol. 11029, pp. 3–17 (2018)
10. Leung, C.K., Hayduk, Y.: Mining frequent patterns from uncertain data with MapReduce for big data analytics. In: DASFAA 2013, Part I. LNCS, vol. 7825, pp. 440–455 (2013)
11. Madill, E.W., et al.: Enhanced sliding window-based periodic pattern mining from dynamic streams. In: DaWaK 2022. LNCS, vol. 13428, pp. 234–240 (2022)
12. Leung, C.K., et al.: Health analytics on COVID-19 data with few-shot learning. In: DaWaK 2021. LNCS, vol. 12925, pp. 67–80 (2021)
13. Brown, P.O., et al.: Mahalanobis distance based k-means clustering. In: DaWaK 2022. LNCS, vol. 13428, pp. 256–262 (2022)
14. Leung, C.K., et al.: Effective classification of ground transportation modes for urban data mining in smart cities. In: DaWaK 2018. LNCS, vol. 11031, pp. 83–97 (2018)
15. Camara, R.C., et al. Fuzzy logic-based data analytics on predicting the effect of hurricanes on the stock market. In: FUZZ-IEEE 2018, pp. 576–583 (2018)
16. Leung, C.K., et al.: Big data science on COVID-19 data. In: IEEE BigDataSE 2020, pp. 14–21 (2020)
17. Olawoyin, A.M., et al.: Privacy-preserving spatio-temporal patient data publishing. In: DEXA 2020, Part II. LNCS, vol. 12392, pp. 407–416 (2020)
18. Braun, P., et al.: MapReduce-based complex big data analytics over uncertain and imprecise social networks. In: DaWaK 2017. LNCS, vol. 10440, pp. 130–145 (2017)
19. GhaffariDolama, M., et al.: Quantifying emergency response system risk caused by grade crossing blockages. Transp. Plan. Technol. (2025). https://doi.org/10.1080/03081060.2025.2480692
20. Kawabata, T., Toda, H.: Exploring of STGNN for traffic forecasting at expanding traffic network. In: DEXA 2024, Part II. LNCS, vol. 14911, pp. 116–121 (2024)
21. Leung, C.K., et al.: Urban analytics of big transportation data for supporting smart cities. In: DaWaK 2019. LNCS, vol. 11708, pp. 24–33 (2019)
22. Wang, Z., Li, B.: Traffic flow prediction based on deep spatio-temporal domain adaptation. In: DEXA 2024, Part II. LNCS, vol. 14911, pp. 110–115 (2024)
23. Wodi, B.H.: A hybrid database solution to support transportation analytics with case studies on improving emergency medical service response times. M.Sc. thesis, University of Manitoba (2022)
24. Fayyad, U., et al.: From data mining to knowledge discovery in databases. AI Mag. **17**(3), 37 (1996)
25. Cui, Z., et al.: Establishing multisource data-integration framework for transportation data analytics. J. Transp. Eng. Part A Syst. **146**(5), 04020024 (2020)
26. Gettman, D., et al.: Integrating emerging data sources into operational practice: opportunities for integration of emerging data for traffic management and TMCs. In: Technical report FHWA-JPO-18-625, US Department of Transportation (2017)
27. Khan, S., et al.: Educational intelligence: applying cloud-based big data analytics to the Indian education sector. In: IC3I 2016, pp. 29–34 (2016)
28. Blackwell, T.H., Kaufman, J.S.: Response time effectiveness: comparison of response time and survival in an urban emergency medical services system. Acad. Emerg. Med. **9**(4), 288–295 (2002)

29. Picado-Aguilar, G., Aguero-Valverde, J.: Emergency response times and crash risk: an analysis framework for Costa Rica. J. Transp. Health **16**, 100818 (2020)
30. Park, P.Y., et al.: First responders' response area and response time analysis with/without grade crossing monitoring system. Fire Saf. J. **79**, 100–110 (2016)
31. Mukhopadhyay, A., et al.: An online decision-theoretic pipeline for responder dispatch. In: ACM/IEEE ICCPS 2019, pp. 185–196 (2019)
32. Chen, Y., Rilett, L.R.: Train data collection and arrival time prediction system for highway–rail grade crossings. Transp. Res. Rec. **2608**(1), 36–45 (2017)
33. Ogden, B.D., Copper, C.: Highway-rail crossing handbook, 3rd Edn. US Department of Transportation - Federal Highway Administration (FHWA) (2019)
34. Ashok, V., et al. A secure freight tracking system in rails using GPS technology. In: ICONSTEM 2016, pp. 47–50 (2016)
35. Prokhorchenko, A., et al.: Forecasting the estimated time of arrival for a cargo dispatch delivered by a freight train along a railway section. Eastern-Europ. J. Enterprise Technol. **3**, 30–38 (2019)
36. van Buuren, M., et al.: EMS call center models with and without function differentiation: a comparison. Oper. Res. Health Care **12**, 16–28 (2017)
37. Sokolova, M.V., et al.: Migration from an SQL to a hybrid SQL/NoSQL data model. Journal of Management Analytics **7**(1), 1–11 (2020)
38. Bessa, P.F.M.: Transportation management in an era of big data: from data to knowledge. Master's thesis, Universidade do Porto, Portugal (2022)
39. Villalobos, M.T., et al.: Evaluation of the response time of a geoservice using a hybrid and distributed database. Revista Colombiana de Comput. **23**(1), 34–42 (2022)

Large Language Models

Automated Archival Descriptions with Federated Intelligence of LLMs

Jinghua Groppe[1], Andreas Marquet[2], Annabel Walz[2], and Sven Groppe[1]

[1] IFIS, University of Lübeck, Ratzeburger Allee 160, 23562 Lübeck, Germany
{jinghua.groppe,sven.groppe}@uni-luebeck.de
[2] Friedrich-Ebert-Stiftung e.V., Godesberger Allee 149, 53175 Bonn, Germany
{Andreas.Marquet,Annabel.Walz}@fes.de

Abstract. Enforcing archival standards demands specialized expertise, and manually creating metadata descriptions for archival materials requires considerable manual effort and is error-prone. This work aims to explore the potential of agentic AI and large language models (LLMs) in addressing the challenges of implementing a standardized archival description process. To this end, we introduce an agentic AI-driven system for the automated generation of high-quality metadata descriptions of archival materials. We develop a federated optimization approach that unites the intelligence of multiple LLMs to construct optimal archival metadata. We also suggest methods to overcome the challenges associated with using LLMs for consistent metadata generation. To evaluate the feasibility and effectiveness of our techniques, we conducted extensive experiments using a real-world dataset of archival materials, which covers a variety of document types and formats. The evaluation results demonstrate the feasibility of our techniques and highlight the superior performance of the federated optimization approach compared to single-model solutions in metadata quality and reliability.

Keywords: information extraction · file repositories · archive · metadata generation · federated · LLM · agentic AI

1 Introduction

Despite the availability of numerous specialized applications and document management systems, file repositories remain widely used technical environments that often merit preservation. From an archival perspective, the main challenge stems from their inherently flexible use by both individuals and organizations. This flexibility frequently leads to unstructured storage practices, inconsistent file naming conventions, redundancies, and incomplete or missing metadata, and these factors complicate efforts to maintain order and accessibility of file repositories. Additionally, the absence of a standardized procedure for file creation often results in the loss of contextual information, making it difficult to determine the origins and significance of stored files [8,16,21]. In addition, file repositories generally lack quantitative and qualitative constraints, meaning that relevant information can be stored in unrestricted and highly variable forms [1,5].

Given the vast and practically unlimited volume of data [7], coupled with the fundamentally different nature of digital media, traditional archival methods - such as appraisal, arrangement, and description - are only partially applicable in their current form. As a result, methodological and technical adaptations are necessary to ensure effective archival practices. Initial pilot projects relied predominantly on manual and intellectual efforts to process file repositories, further confirming the assumption that the associated workload is overwhelming, given the sheer size of the records. Considering these challenges and the immense number of files to be managed, research topics such as the automatic extraction of metadata have become increasingly significant.

For efficient subject-specific processing, (semi-) automatic approaches - computational methods initiated, supervised, and refined by archivists - appear to be the most promising (e.g., [15]). These (mostly) automated methods help manage the large-scale processing of files. However, intellectual input remains essential, particularly for contextualizing individual documents, understanding the workflows of the organizations or individuals creating them, and integrating files within existing archival holdings. These challenges are further exacerbated by the prevalence of hybrid record-keeping practices, in which analogue and digital documents coexist and their proportions shift over time. As a result, file systems present multiple complexities, introducing potential ambiguities regarding their origin, acquisition, and subsequent processing.

The use of artificial intelligence (AI) in archival practice has recently been tested for a variety of purposes, demonstrating significant potential - particularly in recognizing objects, people, or buildings in photographs and in content indexing of audiovisual material. Research conducted within the British LUSTRE network highlights both the promising applications of AI and the organizational prerequisites tied to adopting this emerging technology [9]. In a related initiative, the British National Archives evaluated several AI-based tools to support the selection and acquisition of digital records [19]. Building on these developments, this paper, centered on the file storage scenario, seeks to offer a more flexible and adaptive AI framework that advances the current state of the art.

Large language models (LLMs), owning to their remarkable capacity to produce human-like, context-aware responses across a wide range of tasks such as translation, summarization, question-answering, creative writing, and code generation, are the basis of chatbots (for general purposes like ChatGPT, Grok, DeepSeek and Gemini, and for special purposes like [12]) and are becoming an integral part of many aspects of daily life [13]. Their ability to perform tasks with little to no prior examples - known as zero-shot or few-shot learning [2] - makes them highly adaptable for use in data pipelines [10,11]. This flexibility coupled with other strengths is particularly valuable in generating archival metadata across heterogeneous documents, where LLMs can automate much of the process while allowing archivists to intervene with minimal effort to refine results and streamline subsequent workflows.

In this work, we investigate the abilities of agentic AI and LLMs in addressing challenges in implementing archival standards and propose a universal LLM-agent-driven framework for automated generation of archival metadata. Our main contributions include:

- an agentic AI-based metadata generation system, which utilizes the federated intelligence of multiple LLMs to automatically generate high-quality metadata for archival materials,
- a methodology for determining best LLMs for archival metadata extraction,
- techniques to address the challenges of using LLMs in generating consistent metadata descriptions,
- a federated optimization approach, which synthesizes an optimal metadata description from the results of an ensemble of LLMs, and
- an extensive experimental evaluation on real-world archival materials containing documents of various types and in various formats. The evaluation results confirm the practicality of the LLM-agent-driven framework and demonstrate the superior performance of the federated approach compared to the use of individual LLMs alone.

2 Related Work

Large language models have already been used very successfully for information extraction tasks in other application areas: For example, Wang et al. [20] and Parekh et al. [17] used OpenAI's large language models to extract structured data about events from unstructured text. Goel et al. [6] combined large language models with human expertise to annotate patient-related information in medical texts. By using state-of-the-art large language models such as GPT-4, Schimmenti et al. [18] achieved very high accuracies for extracting metadata from historical texts (title: 98%, type: 94%, date: 89%, location: 95%, author: 79%). Although the results of these studies come from partly different application areas, they indicate that large language models can also be successfully used for information extraction in archives, with their advantages in terms of language understanding and few-shot learning.

The transfer of document content into knowledge graphs can also be largely automated using large language models [4,14], whereby archivists, for example, gain better control (compared to e.g. retrieval augmented generation) over what is freely accessible to archive users in a subsequent step, or which documents are subject to a protection period, by post-processing the knowledge graph.

The authors in [3] discuss the related work about LLM ensembles. In this work, we propose a variant of LLM ensembles that uses an ensemble after inference approach to re-generate in case of unsuccessful validation and to generate a synthesized output from all outputs of the LLMs in the ensemble, tailored to the needs of archives by providing specialized contexts and a specialized validator.

None of the existing approaches investigates the use of large language models in the context of supporting archivists for archiving files.

3 LLM-Agent-Driven Automatic Generation of Archival Metadata

We suggest an agentic AI-based metadata generation system, which employs the federated intelligence of LLMs to automatically create a complete and precise metadata description of archival materials. We first introduce the agentic AI architecture of the LLM-based metadata generation system in Sect. 3.1, then present the methodology to determine the best LLMs for the metadata generation task in Sect. 3.2. Sect. 3.3 describes the techniques to address the challenges of LLMs in constructing consistent metadata descriptions and Sect. 3.4 presents the federated optimization approach, which leverages the intelligence of multiple LLMs to create high-quality archival metadata.

3.1 System Architecture

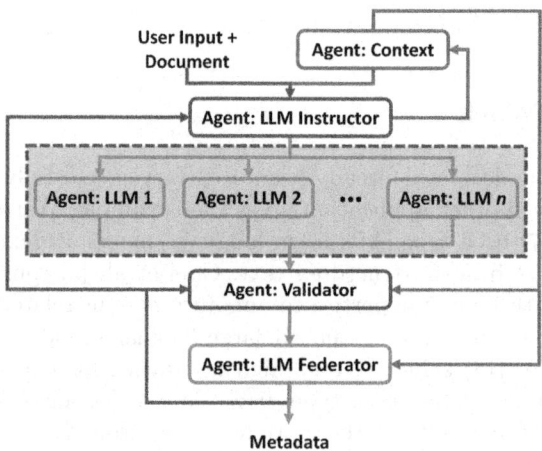

Fig. 1. Agentic AI-driven system architecture for automated generation of optimal archival metadata

Figure 1 depicts the architecture of the automatic metadata generation system composed of agents powered by LLMs. After getting the user input, the LLM Instructor agent first analyses the input content and retrieves the corresponding context (like metadata specification of ISAD(G), Records in Contexts, and alike) from the Context agent (See Sect. 3.3 for details about contexts). Based on the user input and the context, the LLM Instructor then constructs instructions to generate metadata descriptions. These instructions and the archival material provided by the user are then fed into an ensemble of agents powered by various LLMs, and each LLM will generate a metadata description for the archival material according to the instructions. All the metadata descriptions generated are then checked by a Validator agent in terms of, e.g. the structure, completeness of extraction, and format adherence according to the context information.

If validation fails, the validator agent will notify the LLM instructor agent of which LLMs output incorrect metadata descriptions. The LLM instructor then instructs the LLM agents to re-extract the metadata description. When all metadata descriptions created by the LLM agents in the ensemble are successfully validated, the validator agent sends all data and information to the LLM Federator agent. The Federator agent will construct synthesis instructions based on the requirements given in the context information and then synthesize an optimal metadata description from all individual ones according to the instructions. The new metadata description will be validated according to the context as well by the Validator agent. If validation fails, it will inform the LLM Federator agent to perform the synthesis again until validation is successful. Finally, the agentic AI system will output a high-quality metadata description for the given archival material, which maximizes the completeness and accuracy of the content and is optimally aligned with the archiving standard.

3.2 Determining Candidate LLMs

A large number of LLMs are available for data extraction, and this presents an initial challenge: selecting the most suitable models to generate high-quality, reliable metadata descriptions. This section proposes a systematic methodology to identify the optimal candidate LLMs.

Given a collection of LLMs:

1. For each LLM, check whether it knows the standard to be used for the archival description. If this is not the case, it is removed from the collection.
2. For each LLM, ask it to create a metadata description according to the archiving standard for some samples of the archival materials. If it is unable to do so or the quality of the extracted metadata description is low, it will be removed.
3. Design a prompt that exactly describes the format and structure of the metadata. For each LLM, we test with examples to see if it can extract metadata according to the requirements. If it cannot, it is removed.
4. Evaluate the LLMs in the collection with samples of archival materials. If the performance of an LLM is lower than a threshold, it is removed.
5. Finally, if there are multiple LLMs from the same series in the collection, we will only keep the latest or more powerful one.

The website https://lmarena.ai/[1] collects 94 LLMs for researchers and practitioners to compare and test their performance, and so we use the collection as the initial one. After applying the selection methodology, we determine four LLMs: Grok 3 built by xAI, GPT-4-turbo by OpenAI, DeepSeek-V3 from DeepSeek, and Gemini 2.0 Flash from Google.

---

[1] visited on 31.3.2025.

3.3 Computing Consistent Metadata Descriptions

Many archival description standards provide descriptive elements, and most of them are made up of several words. This causes several problems: i) Some data store formats (like XML) require one-word tags. ii) Using long descriptive elements (or their concatenation) as tags is inefficient. For these reasons, simplified one-word tags are applied instead of descriptive names, and LLMs also follow this practice. However, different LLMs could adopt different ways to tag these elements, and even the same LLM might use various tags at different time. Furthermore, although we use the LLMs, which are familiar with the archival description standards, they do not always strictly follow them, and the metadata descriptions created by LLMs could have different structures and elements at different times. Metadata descriptions with varying structures and different tags are useless and need significantly post-processing. However, an automatic post-processing, which converts them into a consistent format, is almost impossible because the structure and terminology that LLMs use could be unlimited. Therefore, we need a solution to ensure that metadata descriptions generated by any LLMs at any time are strictly consistent in terms of structure and tags.

In this work, we suggest a context-based solution: We use a context that contains the necessary information to create consistent metadata. We provide the context to the LLMs, which use it to create metadata descriptions. We also provide the context to the validator agent, which will validate if the generated metadata descriptions comply with the specification given in the context. If validation fails, we ask the LLMs to re-regenerate the metadata description. This process is repeated until validation succeeds as shown in Fig. 1. In extreme cases, it could occur that LLMs cannot create a metadata description according to the requirements. But this can be easily handled by setting up a limited number of repetitions. If this number is exceeded, the system notifies the user.

To create a clear reference for LLMs to follow, such a context should include the following pieces of information:

- the structure of the metadata description, which LLMs should follow.
- the purpose of each element, which provides LLMs with information on how to extract its value.
- examples for each element, which illustrate the use of the element.
- one-word tag for each element, which enables the use of uniform vocabulary.

3.4 Creating Optimal Metadata Descriptions

According to the methodology suggested in Sect. 3.2, we can determine the best LLMs for automated generation of archival metadata. We want to utilize all of their intelligence to create an optimal archival metadata description. To realize this, we first ask each of the LLMs to analyze the archival material and generate metadata descriptions for it. Next, we aim to integrate the strengths of these metadata descriptions to synthesize a new, optimized version–one that maximizes completeness and precision of content while aligning closely with the metadata standard and the overarching intent of the archive.

To create the optimal metadata set from the source metadata ones, we suggest a systematic optimization strategy as follows:

1. Analyze source metadata sets and identify their strengths and weaknesses.
   - assess against the archival description standard: check if the metadata sets reflect the purposes of each element and follow the rules and conventions of the standard.
   - compare against the archival material: cross-reference the metadata sets with the archival material to check if they reflect the archival intent and verify factual details like dates, creators, and content.
   - identify variations, differences, and contradictions across the source metadata sets.
2. Element-by-element optimization For each element:
   - select the best base: Choose the most accurate or detailed entry from the source metadata sets as a starting point.
   - enhance with details: Add specifics from other sets or the archival material.
   - resolve discrepancies: cross-reference the metadata with the archival material and the archival standard.
   - correct errors: Adjust based on evidence (e.g., OCR errors given in notes).
   - standardize format: Ensure compliance with the rules of the archival standard.
3. Validation and refinement
   - Cross-check with the archival standard and format requirements: Ensure that all elements follow the hierarchical structure and use the correct tags.
   - Align with archival material: Verify if the metadata reflects the intent and focus of the material.
   - Eliminate redundancy: Streamline overlapping details.
   - Enhance utility: Add details to aid users and systems if necessary.

For automation, we need an agent to carry out these tasks. This agent should have knowledge of the archiving process, be familiar with the archival description standard, and have the ability to analyze and reason. No doubt, LLMs are currently an optimal candidate for such an agent. In order to find an LLM that can best perform the optimization task, we tested and evaluated a number of LLMs, and the results of the evaluation show that Grok3 from xAI is superior to other LLMs.

4 Experimental Evaluation

4.1 Metrics and approach

The quality of archival metadata is determined by many aspects: factual accuracy, content completeness, equivalence of intent, avoidance of contradiction, and alignment with the archival description standard. To evaluate these aspects, we need to compare the metadata sets against both the ground truth and the standard. The existing evaluation techniques, from the purely statistical scorers (Blue, ROUGE, Meteor and Levenshtein distance) to embedding models (like BertScore and MoverScore) to Natural Language Inference models (like NLI scorer) and BLEURT (which uses pre-trained models like BERT to score LLM outputs on ground truth), are not able to do this.

Recently, LLM-as-a-Judge is emerging to perform complex evaluation due to their superior reasoning capabilities and knowledge about the world, and so we also use LLMs as evaluators. We instruct LLMs to evaluate the quality of the automatically generated metadata sets based on the ground truth and the standards, and compute a score for each element on a scale from 0 to 1 based on the factual accuracy, content completeness, contextual consistency, equivalence of intent, avoidance of contradiction and alignment with the archival description standard, where 0 indicates no similarity and 1 indicates identical or near-identical meaning, with increments reflecting nuanced differences. More concretely:

- 1: Exact or near-exact match in meaning and detail.
- 0.75–0.99: High similarity with minor omissions or slight rephrasing.
- 0.5–0.74: Moderate similarity with noticeable differences or partial matches.
- 0.25–0.49: Low similarity with significant deviations or missing key details.
- 0–0.24: Complete mismatch or entirely missing relevant information.

In order to find the most appropriate LLM as evaluator, we test and evaluate different LLMs, and the test results show that Grok3 is currently the best LLM for this task.

4.2 Data and Ground Truth

In our evaluation, we use real-world archive materials from Deutscher Gewerkschaftsbund (DGB)[2] held at the Archive of Social Democracy of the Friedrich-Ebert-Stiftung (AdsD) (https://www.fes.de/archiv-der-sozialen-demokratie/). To reflect the diversity and heterogeneity of archival materials and evaluate the capabilities of LLMs in generating metadata descriptions, we choose 22 representative archival units from the archival corpus as evaluation data, which contain different document types (minutes, speeches, letters, memos, articles, e-mails, newsletters, presentations, notes, calendar, wage and income tax statistics) and cover various topics in different data formats (PDF, MS Word, PowerPoint, Excel). Figure 2 presents the word statistics of the documents in the evaluation data set. The number of words is on average 2037.25, the median 1003.5, the minimum 189, and the maximum 18141. Most documents have between 500 and 1500 words.

---

[2] The DGB is an umbrella organization for eight member unions in Germany. Please see https://en.dgb.de/ (visited on 31.3.2025) for further information.

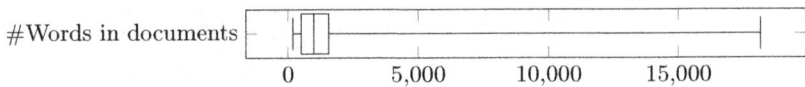

Fig. 2. Box plots of number of words of each document in the evaluation dataset

Table 1. ISAD(G) areas and elements considered in this evaluation with abbreviations used in this paper

Area/Element	Description	Abbreviation
1. Identity Statement Area	This area provides essential information to identify the archival unit being described	Id
1.2 Title	The name given to the archival unit, either a formal title (if one exists) or a descriptive title created by the archivist	Id:Title
1.3 Date(s)	The date range or specific dates of the archival unit, including creation or accumulation dates	Id:Date
1.4 Level of Description	Indicates the hierarchical level of the archival unit (e.g., fonds, sub-fonds, series, file, item)	Id:DescLev
1.5 Extent and medium of the unit of description (quantity, bulk, or size)	Describes the physical or digital size and format of the unit (e.g., "3 boxes," "0.5 linear meters," "10 digital files (PDF)")	Id:Extent
3. Content and Structure Area	This area describes the intellectual content and organization of the archival materials	Cont
3.1 Scope and Content	A summary of the subject matter, themes, and types of records included in the unit, helping users assess its relevance	Cont:Scope
4. Conditions of Access and Use Area	This area outlines the terms and conditions for accessing and using the materials	AccessUse
4.3 Language(s) and Script(s) of Material	Identifies the language(s) and script(s) used in the materials (e.g., "English, with some documents in French; Latin script")	AccessUse:Lang
4.4 Physical Characteristics and Technical Requirements	Describes any physical or technical conditions affecting use, such as fragility or digital file formats	AccessUse:PhysTech

Table 2. Improvements of the LLM scores of the federated approach in comparison to direct use of LLMs

Element/Area	LLM 1	LLM 2	LLM 3	LLM 4
Id:Title	0%	11%	0%	9%
Id:Date	15%	10%	6%	9%
Id:DescLev	0%	0%	12%	10%
Id:Extent	-2%	4%	-9%	5%
Cont:Scope	1%	20%	13%	16%
AccessUse:Lang	0%	0%	-4%	-5%
AccessUse:PhysTech	3%	8%	0%	42%
Id	3%	7%	1%	8%
Cont	1%	20%	13%	16%
AccessUse	2%	4%	-2%	13%
All	2%	10%	4%	12%

The ground truth was manually created by professional archivists of the AdsD and includes a full set of ISAD(G) metadata for archival materials in the dataset. ISAD(G) stands for General International Standard Archival Description, a framework developed by the International Council on Archives (ICA) to standardize the description of archival materials. It is widely used by archivists globally to make archival records more accessible and understandable. ISAD(G)

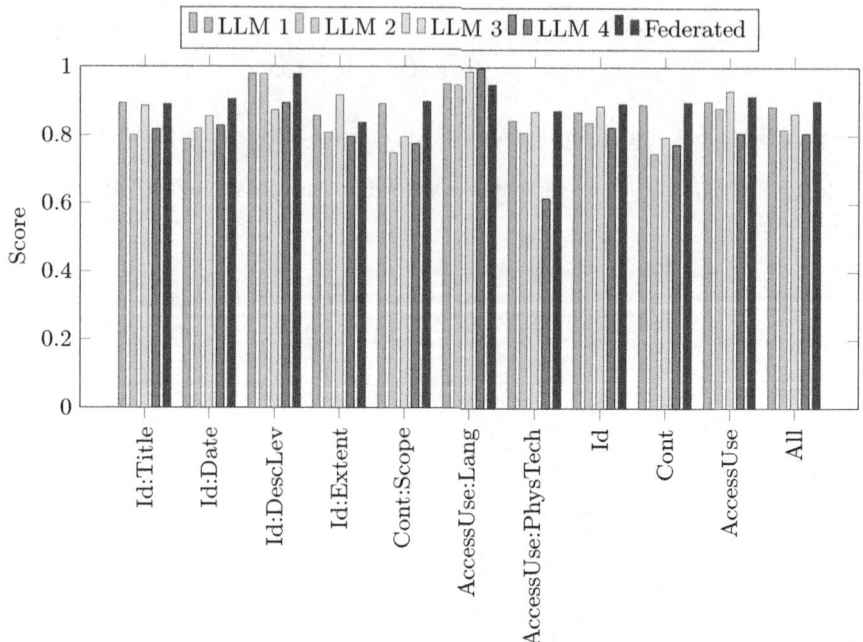

Fig. 3. LLM scores for information extraction (average over all scores for all documents)

organizes metadata elements into several areas, and each element serves a specific purpose in creating a comprehensive and standardized archival description. These elements can be divided into two groups: one group of elements whose values can be determined directly or by inferring from the archive material itself; another group of elements whose values need information from external sources or are assigned by an archivist. As a result, the elements in the second group are not suitable for evaluation. We list the areas and elements of ISAD(G) considered in our evaluation in Table 1, and also introduce their one-word tags, which we use in the succeeding figures (Fig. 3 and Fig. 4) and tables (Table 2).

4.3 Analysis and Discussion

We have determined the best LLMs for metadata extraction in Sect. 3.2 and for metadata optimization in Sect. 3.4. We will refer to them as LLM 1 (Grok 3 from xAI), LLM 2 (GPT-4-turbo from OpenAI), LLM 3 (DeepSeek-V3 built by DeepSeek), and LLM 4 (Gemini 2.0 Flash from Google) in the analysis.

We present in Figure 3 the average LLM scores over all the documents in the dataset for each element, each area, and the whole metadata according to the ISAD(G) standard. Looking at the performance of the single LLMs, LLM 1 performed the best, followed by LLM 3, LLM 2, and finally LLM 4. It is remarkable that the open source LLM 3 is the second-best LLM in our evaluation of the single LLMs. Even the last LLM (LLM 4) also achieves a score of 0.81 for

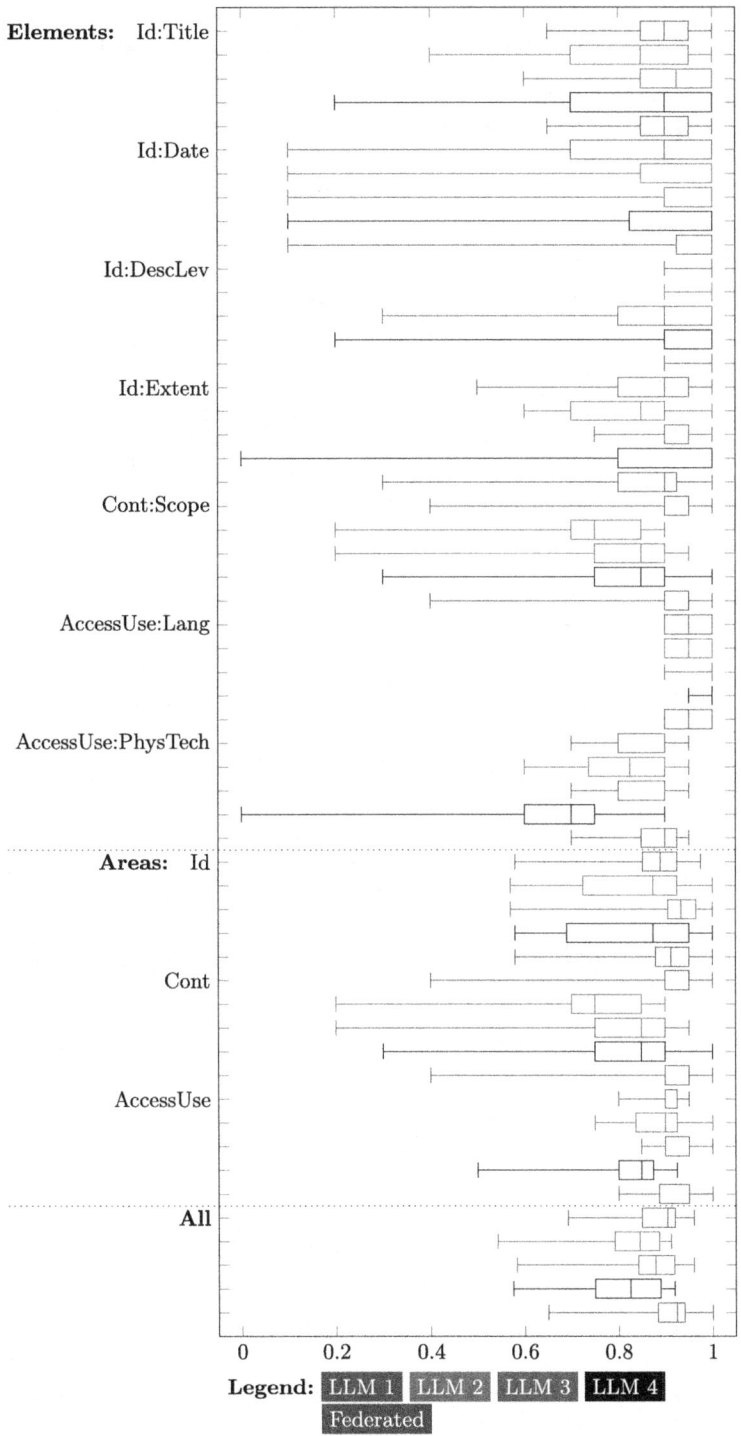

Fig. 4. Box plots of LLM scores for information extraction

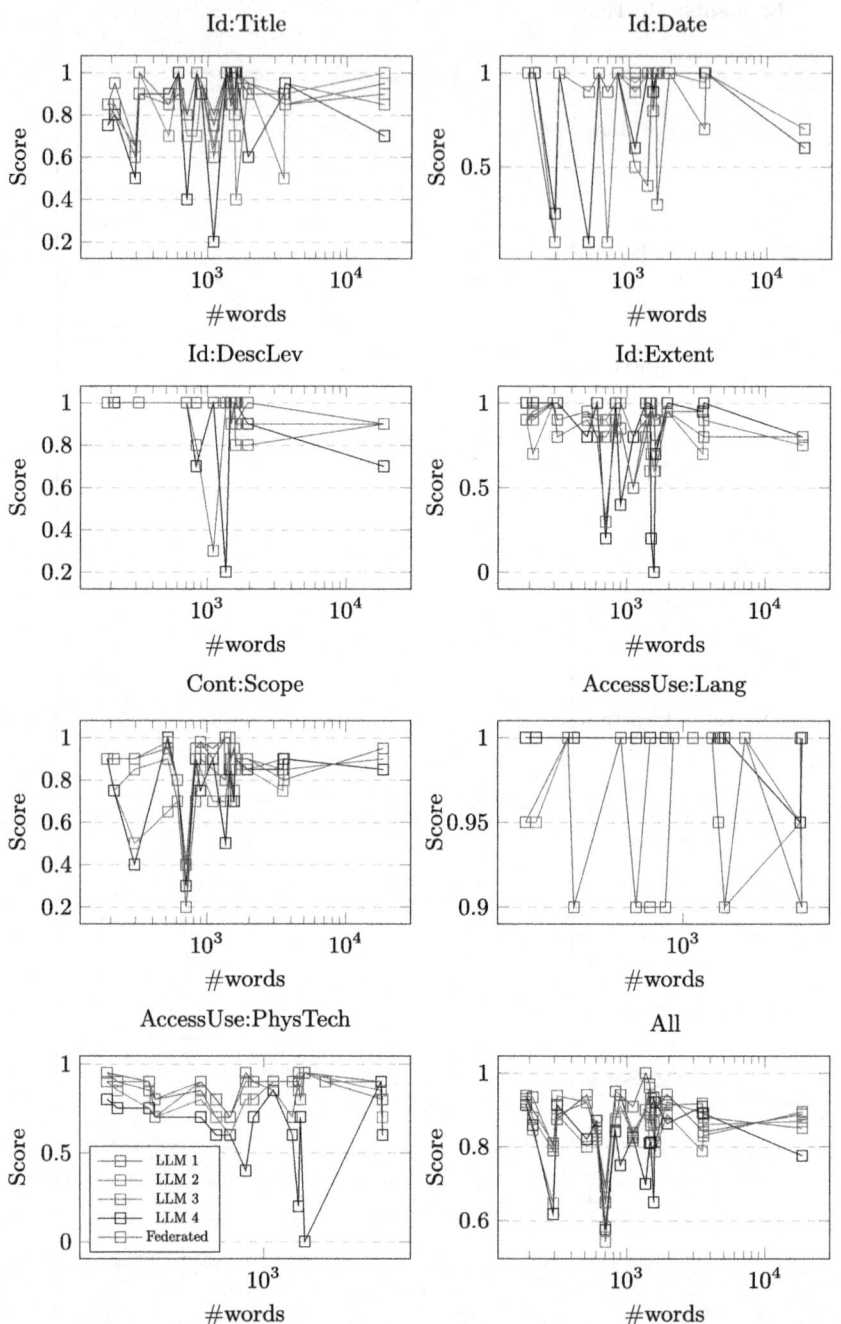

Fig. 5. Number of words versus scores for single elements of ISAD(G) metadata and overall score

the whole document. Looking at the averages of scores for each element and area, these average scores are never below 0.61. For the best LLM, these average scores are never below 0.78. Finally, the federated approach beats all single LLMs with an average of 0.90 for the whole metadata description, and its average scores for each element and area are never below 0.83.

Table 2 presents the performance enhancement of our federated approach over directly using single LLMs. For a few elements, such as Id:Extent and AccessUse:Lang, the federated approach does not outperform some individual models. The main reason for this is actually that the federated approach adheres to the archiving standards more strictly and produces more complete content. For example, according to the ISAD(G) specification, the element Id:Extent describes the physical or digital size and format of the unit. For the element, our federated approach creates its content as "3 pages, digital document (PDF)", while the ground truth is only "3 pages". Obviously, the content created by the federated approach is more complete than the ground truth. However, due to the difference between the generated value and the ground truth, the federated approach received a lower score than the LLMs, which extract only "3 pages" for this element, just like the ground truth. Looking at the performance improvements given by the federated approach, coupled with the fact that the federated approach can produce more complete content than the ground truth, we can say that the ability of the federated approach in generation of high-quality metadata is very impressive.

To further analyze the reasons of the superior performance of the federated approach, we also look at the box plots of scores in Figure 4, which shows the minimum, maximum, median, 25%, and 75% quartiles over all documents. We detect several cases: The federated approach takes over

- all (or almost all, respectively) the extracted information from the best single LLM (like for the elements Id:Title, Id:DescLev, and AccessUse:Lang (Cont:Scope, respectively)).
- the best values among the different LLMs (like for the element Id:Date and AccessUse:PhysTech)
- a mixture of good and bad values (like for the element Id:Extent). In this case the federated approach achieves an average score compared to the scores of the 4 LLMs.

Interestingly, we do *not* observe a case where the federated approach takes over the bad values in most cases, and the case, where a mixture of good values and bad values are taken over are rare (i.e., we only observed it for the element Id:Extent), and the reason for this, as discussed earlier, is the capabilities of the federated approach: it can adhere more strictly to the archiving standard and create more exact and complete content than human specialists do.

We present in Figure 5 the scores for each element of ISAD(G) metadata versus the number of words of the documents. Interestingly, we see that there are some outliers for documents with around 1,000 words, where the generated metadata descriptions are not similar to the ones of the ground truth. Looking at the overall score for the whole metadata descriptions, we see a tendency for slightly decreasing scores for larger documents (ignoring the outliers with low scores for smaller documents).

5 Conclusions

Large Language Models (LLMs) are advanced AI systems that are trained on large amounts of text data and thus acquire knowledge about the world and a remarkable ability to understand natural languages. They are being used to automate a variety of natural language processing tasks such as customer support, content creation, and question answering. This work explores their potential for automated archiving and introduces an LLM-agent-driven AI system for automatic generation of archival metadata. This system integrates LLM agents with validators, specialized context handling, and a federated optimization technique that unites the intelligence of individual LLMs to produce high-quality metadata descriptions for archival materials. We conducted an extensive experimental evaluation using real-world archival samples covering documents of various types and data formats. The evaluation results demonstrate the capability of the LLM-based approach in automatic archiving and the superior performance of the federated technique in generating high-quality archival metadata descriptions.

References

1. Barrueco, J.M., Termens, M.: Digital preservation in institutional repositories: a systematic literature review. Digit. Libr. Perspect. **38**(2), 161–174 (2022)
2. Brown, T., et al.: Language models are few-shot learners. In: Larochelle, H., Ranzato, M., Hadsell, R., Balcan, M., Lin, H. (eds.) Advances in Neural Information Processing Systems, vol. 33, pp. 1877–1901. Curran Associates, Inc. (2020)
3. Chen, Z., et al.: Harnessing multiple large language models: a survey on llm ensemble (2025). arXiv (arXiv:2502.18036), https://doi.org/10.48550/arXiv.2502.18036
4. Ezzabady, M., Ieng, F., Khorashadizadeh, H., Benamara, F., Groppe, S., Sahri, S.: Towards generating high-quality knowledge graphs by leveraging large language models. In: The 29th Annual International Conference on Natural Language & Information Systems (NLDB 2024), Turin, Italy (2024)
5. Gillner, B.: Abfragen statt anbieten. eine alternative Praxis im archivischen Umgang mit Dateisystemen. Archiv. Theorie und Praxis **4**, 317—-321 (2023)
6. Goel, A., et al.: Llms accelerate annotation for medical information extraction (2023). https://doi.org/10.48550/ARXIV.2312.02296
7. Groppe, S.: Emergent models, frameworks, and hardware technologies for Big data analytics. J. Supercomput. **76**(3), 1800–1827 (2018). https://doi.org/10.1007/s11227-018-2277-x
8. Jaillant, L., Aske, K., Goudarouli, E., Kitcher, N.: Introduction: challenges and prospects of born-digital and digitized archives in the digital humanities. Arch. Sci. **22**, 285–291 (2022)
9. Jaillant, L., Rees, A.: Applying AI to digital archives: trust, collaboration and shared professional ethics. Digit. Scholarsh. Humanit. **38**(2), 571–585 (2022)
10. Junior, S.B., et al.: Are large language models the new interface for data pipelines? In: Proceedings of the International Workshop on Big Data in Emergent Distributed Environments, Santiago, Chile (2024)

11. Kessel, A.L., Groppe, S., Röpert, D., Groppe, J.: AI-supported analysis and classification of digitized botanical collections. In: The 11th International Conference on machine Learning, Optimization and Data science (LOD), Tuscany, Italy (2025)
12. Kessel, A.L., et al.: Impact of chatbots on user experience and data quality on citizen science platforms. Computers **14**(1) (2025). https://doi.org/10.3390/computers14010021
13. Khorashadizadeh, H., et al.: Research trends for the interplay between large language models and knowledge graphs. In: VLDB 2024 Workshop: The International Workshop on Data Management Opportunities in Unifying Large Language Models + Knowledge Graphs (LLM+KG), Guangzhou, China (2024). https://vldb.org/workshops/2024/proceedings/LLM+KG/LLM+KG-9.pdf
14. Khorashadizadeh, H., et al.: Construction and canonicalization of economic knowledge graphs with llms. In: International Knowledge Graph and Semantic Web Conference (KGSWC) (2024)
15. Lenartz, S.: Digital ist besser? Möglichkeiten der automatisierten Aufbereitung und Bewertung von Fileablagen mit Python am Beispiel einer digitalen Fotosammlung. Wuerttembergische Landesbibliothek. Stuttgart (2022)
16. Naumann, K., Puchta, M.: Kreative digitale Ablagen und die Archive. Ergebnisse eines Workshops des KLA-Ausschusses Digitale Archive am 22./23 November 2016 in der Generaldirektion der Staatlichen Archive Bayerns. München (2017)
17. Parekh, T., Hsu, I.H., Huang, K.H., Chang, K.W., Peng, N.: Geneva: benchmarking generalizability for event argument extraction with hundreds of event types and argument roles. In: Proceedings of the 61st Annual Meeting of the Association for Computational Linguistics. Association for Computational Linguistics (2023)
18. Schimmenti, A., Pasqual, V., Tomasi, F., Vitali, F., van Erp, M.: Structuring authenticity assessments on historical documents using llms (2024). https://doi.org/10.48550/ARXIV.2407.09290
19. The National Archives: Using AI for Digital Records Selection in Government - Guidance for records managers based on an evaluation of current marketplace solutions (2021). https://cdn.nationalarchives.gov.uk/documents/using-ai-digital-selection-in-government.pdf
20. Wang, X., Li, S., Ji, H.: Code4struct: code generation for few-shot event structure prediction. In: Proceedings of the 61st Annual Meeting of the Association for Computational Linguistics. Association for Computational Linguistics (2023)
21. Wendt, G., Westphal, S.: Eine Herausforderung des Übergangs: Fileablagen als Quellen der digitalen Überlieferungsbildung. Transformation ins Digitale. 85. Deutscher Archivtag Karlsruhe, pp. 105–113 (2017)

Entropy-Guided Probing for Predicting LLM Hallucinations with Knowledge Graph Features

Ushtar Ali[1], Steven Lynden[2], Akiyoshi Matono[2], and Toshiyuki Amagasa[1](✉)

[1] University of Tsukuba, 1-1-1 Tennodai, Tsukuba, Ibaraki 305-8577, Japan
ushtar.ali@kde.cs.tsukuba.ac.jp, amagasa@cs.tsukuba.ac.jp
[2] National Institute of Advanced Industrial Science and Technology, 2-4-7 Aomi, Koto-ku, Tokyo 135-0064, Japan
{steven.lynden,a.matono}@aist.go.jp

Abstract. Large Language Models (LLMs) exhibit remarkable performance across a range of tasks, including question answering. However, their tendency to hallucinate, sometimes producing misleading or incorrect responses, remains a critical challenge. In this work, we investigate how LLM-generated answers align with structured relational knowledge from Knowledge Graphs (KGs). We analyze the relationship between KG-derived features and question-answering over various open-source LLMs, uncovering correlations between such features and LLM accuracy. Our study focuses on Wikidata, but is generalizable to other KGs. Leveraging these insights, we propose a feature-based probing strategy that identifies challenging questions likely to expose LLM limitations. We utilized these correlated features to perform an evaluation probing KG-LLM alignment to detect weak spots where KG-LLM alignment is poor. Experimental results show that KG features can effectively guide probing. This work highlights the value of using structured knowledge graphs in building more reliable and trustworthy generative AI systems.

Keywords: Large Language Models · Graph management and analytics · Data science techniques

1 Introduction

Large Language Models (LLMs) have demonstrated remarkable capabilities in tasks such as question answering, code generation, and text summarization. However, they are also prone to hallucinations–producing statements that are factually incorrect or unsupported by external knowledge sources. This undermines the reliability of LLMs, particularly in knowledge-intensive applications. To address this, recent approaches have investigated the use of structured external knowledge sources, such as knowledge graphs (KGs), to check the factual accuracy of LLM outputs [11,15]. However, existing evaluation methods often require exhaustive alignment between LLM outputs and KG tuples, which is computationally expensive and limits their scalability. In this paper, we propose a probing strategy guided by features extracted from a KG, particularly focusing on structural properties such as connectivity and specificity. Whereas LLMs are trained on text corpus usually obtained from the Internet, e.g. Wikipedia, knowledge graphs such as Wikidata contain linked graph-structured information – our hypothesis is that the graph structure can indicate how well an LLM can accurately answer questions. In the context of KGs such as Wikidata [12], entities refer to real-world items represented as nodes (e.g., Q76 for Barack Obama), while properties represent labeled edges (e.g., P39 for "position held") that connect entities to other entities or literal values. These form subject-predicate-object triples in the resource description framework (RDF) structure. We consider **entropy** as a measure of uncertainty or diversity in the distribution of linked KG elements, such as property values or outgoing relations. Additionally, we examine related structural features, such as the number of statements or references, which may reflect the *informativeness* or *coverage* of an entity in the KG. We hypothesize that such features are predictive of LLM performance on knowledge-based question answering tasks.

To quantify these relationships, we compute **correlation**, specifically the *Pearson correlation coefficient*, which measures the strength and direction of the linear relationship between a given KG-derived feature (e.g., property entropy, number of references) and answer correctness. We utilize the WikiWebQuestions [13] dataset, which provides natural language questions and answers (Q/As) and their corresponding SPARQL queries over Wikidata. This allows the relationship between KG features and the accuracy of a variety of open-source LLMs over the Q/As to be analyzed. Our key contributions are as follows:

- We perform a comprehensive correlation analysis between Wikidata-derived entropy-based and structural features and LLM question-answering performance, uncovering measurable signals that are predictive of hallucination likelihood.
- We introduce an efficient entropy-guided probing strategy that selects high-risk questions based on structural KG features, reducing the need for exhaustive alignment or querying.
- We compare our approach against a random baseline and a computationally simplified variant of the KGLens [15] method, showing that entropy-based selection effectively targets cases of factual error.

The remainder of this paper is structured as follows. Section 2 discusses related work on LLM hallucination detection and knowledge graph alignment. Section 3 presents our entropy-based feature extraction process from Wikidata and explores their correlation with LLM question answering accuracy. Section 4 introduces our entropy-guided probing methodology and compares it with random selection and a simplified version of the KGLens alignment technique. Finally, Sect. 5 summarizes our findings and outlines directions for future work.

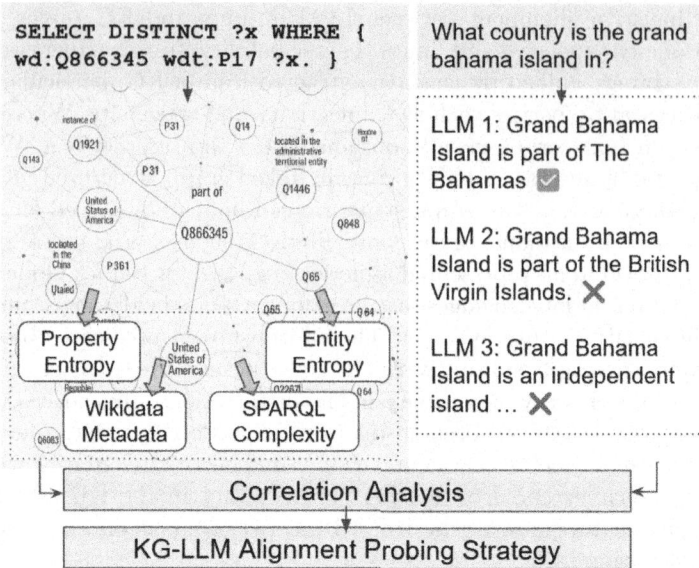

Fig. 1. Our approach takes SPARQL queries with their equivalent questions and analyzes the correlation between LLM accuracy and various KG features (some Wikidata-specific, some generalizable to most KGs) to guide the probing of LLMs to assess alignment.

2 Related Work

Large Language Models (LLMs) frequently produce unverifiable or hallucinated content [4]. A common mitigation strategy is internal verification, where models assess their own confidence scores [7,14]. However, due to constantly evolving information, LLMs often face unseen scenarios, leading to factual inconsistencies. Retrieval-Augmented Generation (RAG) has been proposed to ground responses in external documents [2,8,9], but retrieved text can be ambiguous or incomplete, resulting in residual hallucinations [3]. To improve grounding, structured resources such as Knowledge Graphs (KGs) offer a more consistent and interpretable format for verification.

Recent work has explored the alignment of LLM output with KGs [6] and shows that the use of organized structured resources, such as KGs, is closely aligned with the factual precision [1]. Guan et al. [5] use few-shot prompting over Wikidata to retrofit model responses, but the method still suffers from hallucinations due to limited reasoning and fact selection capabilities. The most closely related work to our contribution is KG-Lens [15], which identifies hallucination-prone regions in KGs by repeatedly probing the model with sampled triples. While effective, it is computationally expensive as extensive queries must be generated from the KG and evaluated against the LLM. In contrast, our approach utilizes a small number of queries to sample property/entity entropy and metadata from the Wikidata API. Other strategies include generating structured graphs from LLM outputs and aligning them with KGs [10], although such methods risk information loss due to parsing limitations. In contrast, Zhu et al. [16] take a reverse approach, generating false-premise questions from KGs to expose factual inconsistencies. We do not consider false premise questions in our approach, although there is potential to explore this in future work.

3 Entropy-Based KG Feature Correlation with LLM Question Answering

In this section, we analyze the correlation between Wikidata KG features and question answering (QA) performance using the following LLMs:

- Llama: `meta-llama/Llama-3.1-8B-Instruct`: 8B parameter model from Meta with a strong general-purpose performance profile.
- Alpha: `stabilityai/stablelm-tuned-alpha-7b`: A 7B parameter model optimized for dialogue and instruction following, developed by Stability AI.
- Mistral: `mistralai/Mistral-7B-Instruct-v0.3`: A dense, efficient model with strong reasoning capabilities and competitive performance.
- Qwen (7B/32B): `Qwen/Qwen2.5-7B-Instruct-1M` and `Qwen/QwQ-32B`: Multilingual instruction-tuned models developed by Alibaba for open-ended tasks.

All models are used as provided on Hugging Face[1], without additional fine-tuning/parameter modification except for `Llama-3.1-8B-Instruct`, where temperature = 0.07 and maximum_new_tokens=32 are set for more concise and deterministic responses. We selected a diverse set of instruction-tuned models with varying architectures, sizes, and training objectives to evaluate their performance on the same dataset. This allows us to assess generalization capabilities and reasoning robustness across different model families and parameter scales.

3.1 WikiWebQuestions QA Dataset

In this study, we exploit WikiWebQuestions [13] dataset, which serves as the question answering (QA) benchmark for evaluating LLM performance. WikiWebQuestions is a high-quality benchmark for QA over Wikidata. The dataset includes 3,665 questions in the training set and 2,032 in the test set, spanning a wide variety of entities and relations. We utilize WikiWebQuestions as a conventional QA benchmark for evaluating the accuracy of LLM's answers by a structural alignment with Wikidata's factual graph. The dataset is publicly available on Kaggle[2]. Originally, the WikiWebQuestions dataset was provided in the form of training and test sets, each containing both single-answer and multiple-answer questions intended for model training and evaluation. We merged the training

---

[1] https://huggingface.co.
[2] https://www.kaggle.com/datasets/paultimothymooney/stanford-oval-wikiweb-questions.

and test sets into a single dataset. To conduct a detailed analysis and assess the alignment of Wikidata features with LLMs for each question type separately, we split the unified dataset into two categories: single-answer questions and multiple-answer questions. As our approach requires a natural language question/answer and corresponding SPARQL query with at least one explicitly defined entity and property (concrete/bound internationalized resource identifiers (IRIs)) in the query to function, we removed any entries that did not match these constraints. We separated the resulting dataset into those with single and multiple answers, resulting in 941 single-answer Q/As and 835 multiple-answer Q/As.

3.2 Extracting Informational Entropy-Based Wikidata Features

We extract structural and entropy-based features from Wikidata to investigate their correlation with QA performance of LLMs. Each question in the WikiWebQuestions-based dataset is paired with a SPARQL query, from which we extract one or more entities and properties. Features are computed for each entity or property using SPARQL queries to the public endpoint and metadata from the Wikidata REST API. For features associated with multiple entities or properties in a single query, we compute their average. SPARQL complexity is computed directly from the number of triple patterns in the query. The following features are utilized:

- **Property Entropy** measures the diversity of values associated with a property type across a sample of linked entities. For a property p, let p_i be the count of value v_i, $P = \sum_i p_i$, and n the number of distinct values. We issue one SPARQL query per property (with a LIMIT clause introduced due to the fact that some properties link hundreds or thousands of entity/value pairs in Wikidata) as a sampling-based approach, for example for property P39 (position held), we execute SELECT ?value (COUNT(?value) AS ?count) WHERE { ?entity wdt:P39 ?value. } GROUP BY ?value ORDER BY DESC (?count) LIMIT 50). This query returns 50 samples, capturing a representative sample of the property's value distribution, enabling efficient entropy estimation. Property entropy is then computed as:

$$\frac{-\sum_i \frac{p_i}{P} \log_2 \left(\frac{p_i}{P}\right)}{\max(1, \log_2 n)}$$

The denominator uses $\max(1, \log_2 n)$ to avoid division by zero when $n = 1$. Property entropy is calculated as the average of the above over all properties extracted from the SPARQL query.
- **Entity Entropy** measures how evenly knowledge of an entity is distributed across its outgoing properties. This is also computed using a SPARQL query for each entity in the query, this time with the entity rather than the property bound. Unlike properties, entities typically have a manageable number of outgoing links, so no sampling (LIMIT) is applied when computing entity

entropy. For a given entity, let c_i be the count of outgoing links via property i, $C = \sum_i c_i$, and n the number of distinct property values. We compute:

$$\frac{-\sum_i \frac{c_i}{C} \log_2\left(\frac{c_i}{C}\right)}{\max(1, \log_2 n)}$$

The entity entropy is calculated as the average of this across all entities in the SPARQL query.
- **SPARQL Complexity** Number of triple patterns in the SPARQL query. Serves as a proxy for query difficulty. This is the only feature computed directly from the query structure.
- **Number of Statements** Total RDF triples (claims) associated with an entity, averaged if multiple entities are present. (This feature along with the subsequent features below are obtained from the Wikidata REST API).
- **Number of References** Number of sources cited for an entity's claims, averaged across entities.
- **Number of Properties** Number of distinct properties linked to an entity, averaged across entities.
- **Number of Sitelinks** Number of interwiki sitelinks (e.g., to Wikipedia pages), averaged across entities.
- **Last Modified Date** Timestamp of the entity's most recent edit, averaged as a numerical value if multiple entities are present.

These features are hypothesized to correlate with the difficulty or reliability of LLM responses, and a correlation analysis is subsequently presented. Refer to Appendix B for an example of features extracted for a WikiWebQuestions entry. To compute features, the following queries/calls are made: a single query is issued for all property types (e.g. P17) encountered in the dataset, and for each entity (e.g., Q937) one SPARQL query is issued to compute the entity entropy, and one call to the Wikidata REST API is made to retrieve the number of sitelinks, last modified date, and other information. During our experiments, we ensured responsible use of Wikidata's public infrastructure by rate-limiting our queries and following best practices when accessing both the SPARQL endpoint and REST API.

3.3 Experiments

In order to evaluate LLM responses, a classification of correctness and relevance of LLM answers is generated by prompting GPT-4o[3] with the LLM answer, the correct answer and asking GPT-4o to classify the question/answer into one of the following (see Appendix B for and example of the prompt):

1. **Perfectly accurate**: The response is fully correct, concise, and directly answers the question.

---
[3] https://platform.openai.com/docs/models/gpt-4o.

2. **Very accurate**: The response is mostly correct with minor nuances or wording differences but does not introduce errors.
3. **Accurate but with some redundant or irrelevant information**: The response contains correct information but includes unnecessary details or slight digressions.
4. **Inaccurate (probable intrinsic error, e.g., from outdated/inaccurate training data)**: The response is incorrect, likely due to limitations in the model's training data.
5. **Inaccurate (probable fabrication/hallucination)**: The response presents incorrect information that appears fabricated or unsubstantiated.
6. **Inaccurate (irrelevant)**: The response is incorrect because it answers a different question or is off-topic.
7. **Completely inaccurate**: The response is entirely incorrect and does not relate meaningfully to the question.

We analyze how these answer correctness categories correlate with entropy-based features derived from the Wikidata KG. Higher entropy values in Wikidata features may indicate increased uncertainty in LLM responses, particularly in cases where the model generates inaccurate or hallucinated answers. Conversely, lower entropy values may correspond to areas where LLMs perform more reliably. By studying these correlations, we aim to better understand how structured knowledge representations influence LLM response accuracy and identify potential predictors for hallucination likelihood.

3.4 Results

Answer distributions are shown in Fig. 2. We consider responses in categories 1–3 to be factually correct, while categories 4–7 are treated as factually incorrect and plot the Pearson correlation between features and accuracy in Fig. 3. An example of the evaluation prompt used is included in Appendix B. To evaluate the robustness of the observed Pearson correlations between dataset features and model performance, we analyzed the corresponding *p-values*. A p-value below 0.05 indicates that the correlation is statistically significant at the 95% confidence level. Table 1 summarizes the number of models (out of 5) for which each feature exhibits statistically significant correlation ($p < 0.05$), for both the multi-answer and single-answer settings. Based on these results, we conclude that–with the exception of SPARQL query complexity, which does not show a statistically significant correlation in the single-answer setting and only limited significance in the multi-answer case–all other features are viable candidates for probing LLM-KG alignment. Both property entropy and entity entropy exhibit negative correlations with model accuracy. This suggests that higher entropy, reflecting greater diversity or uncertainty in the distribution of linked values, makes it more difficult for models to generate correct answers. When a property maps to a wide range of objects, or an entity connects to many distinct properties, the resulting ambiguity and lack of specificity can hinder model performance and increase the likelihood of hallucinations.

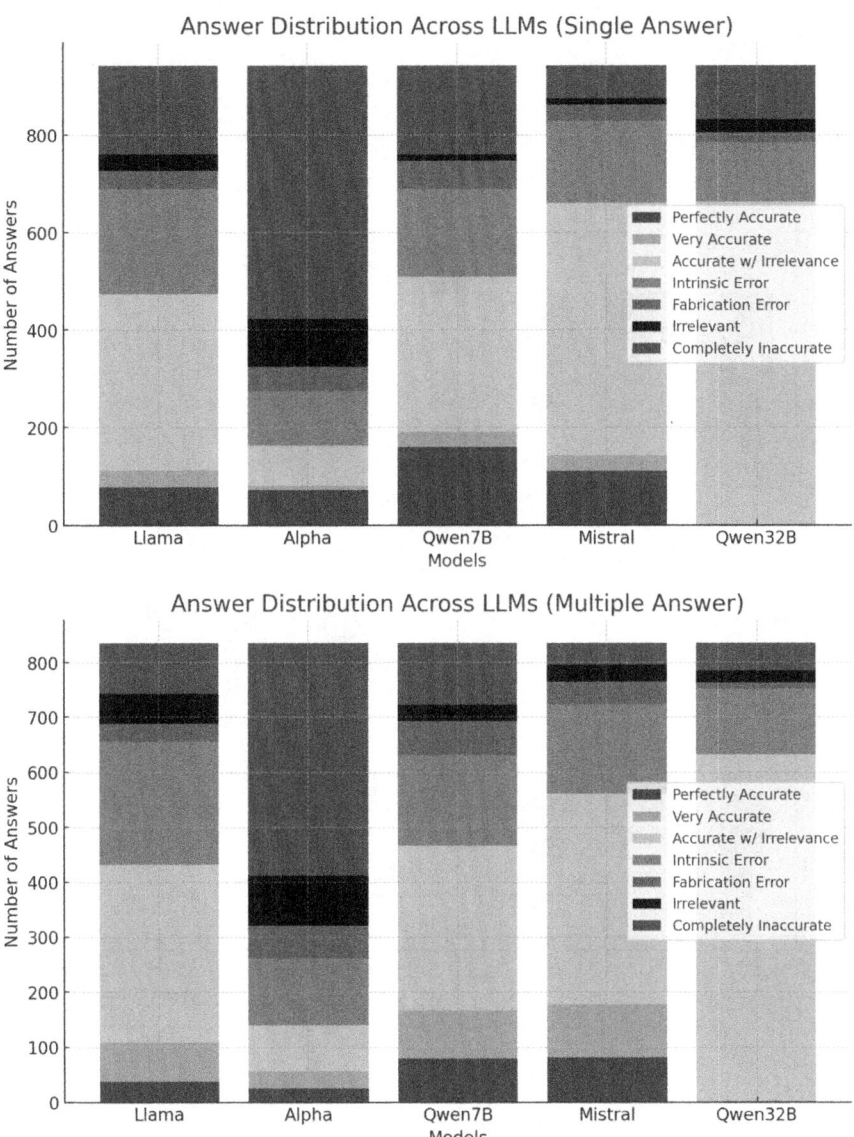

Fig. 2. Evaluated answer distributions over the LLMs investigated.

As perhaps expected, last modified date exhibits a slight positive correlation with model accuracy. This correlation may reflect that frequently edited or recently curated entities are more likely to be included in up-to-date training corpora. As such, entities with more recent modification dates may correspond to more prominent or well-documented topics, which in turn improves model output quality. For the number of statements/reference/properties/sitelinks, all

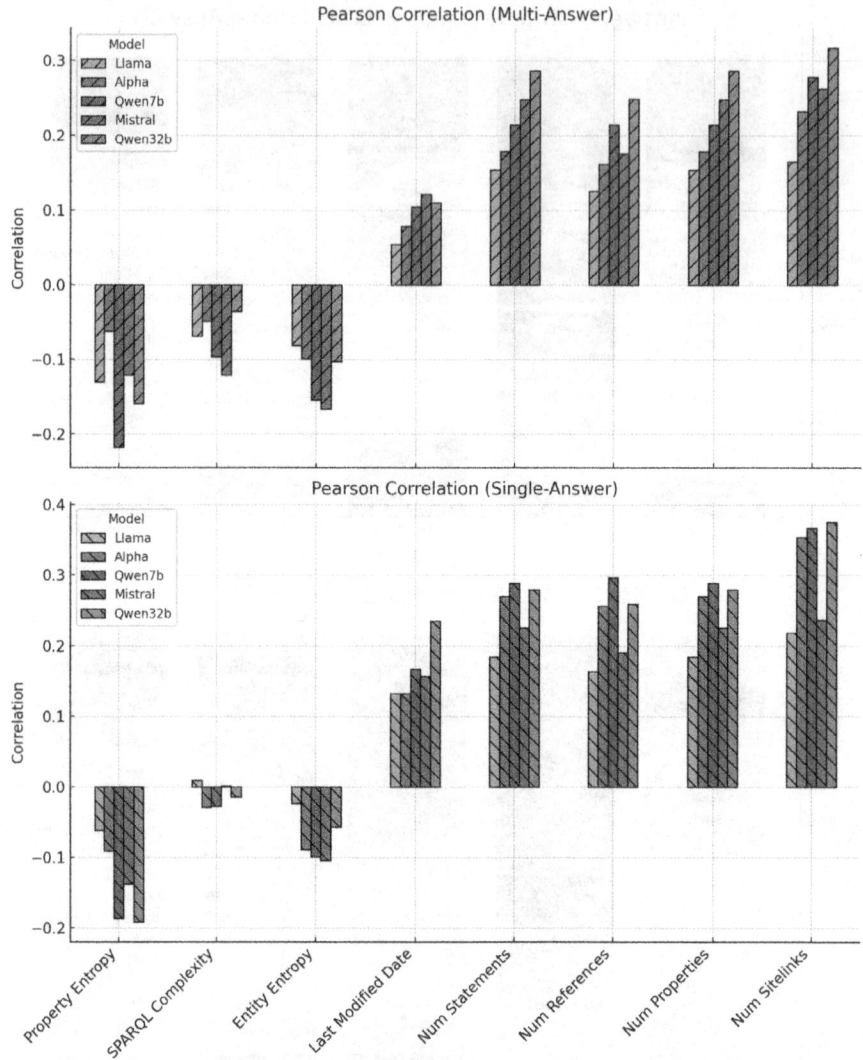

Fig. 3. Correlation of features with model accuracy.

show strong positive correlations with model responses. These features can be interpreted as proxies for how richly an entity is represented in Wikidata. Entities with greater structural detail and connectivity are more likely to appear in training data or be supported by redundant factual cues, enabling models to generate more accurate answers.

Table 1. Significance of correlated features (s/t) refers to significant (s) i.e. p < 0.05 out of the total (t).

Feature	Multi-Answer (s/t)	Single-Answer (s/t)
Property Entropy	4/5	4/5
SPARQL Complexity	3/5	0/5
Entity Entropy	5/5	5/5
Last Modified Date	5/5	5/5
Number of Statements	5/5	5/5
Number of References	5/5	5/5
Number of Properties	5/5	5/5
Number of Sitelinks	5/5	5/5

4 Probing KG-LLM Alignment Using Informational Entropy

Here we introduce our methodology for using informational entropy to assess the alignment between KGs and LLM-generated responses. We utilize our approach to select questions predicted to be less likely to be aligned with the KG and evaluate them against multiple LLMs, comparing our approach to two baselines. The two baselines we utilize to test against are random probing, where N question/answer pairs are randomly selected from the dataset and tested evaluated against each LLM, and a subgraph-based approximation of the parameterized KG-based approach introduced in KGLens [15] adapted as described below.

The original KGLens approach focuses on aligning subsets of the Wikidata KG with LLMs, whereas our WikiWebQuestions-based dataset covers arbitrary queries over the entirety of Wikidata. As of March 2025, Wikidata contains approximately 1.3 billion statements, making complete coverage of the dataset infeasible. Therefore, a simplified, approximated KGLens-based approach aimed at computational feasibility is employed, which is now described. Given a starting Wikidata entity the process begins by selecting all direct neighbors of the entity. Then, for each connected entity from this first hop, we recursively extract their direct neighbors. This expansion is limited to a maximum of 50 edges per hop to retain tractability. All triples are filtered to ensure they contain human-readable English labels and represent meaningful factual relations, i.e. links to external database etc. are discarded. The resulting subgraph is a localized two-hop portion of the KG centered on the entity, containing structured factual knowledge about e and its surrounding context. Each triple (subject/predicate/object) (s, p, o) in the subgraph is associated with a confidence estimate representing the model's likelihood of failing to correctly answer a question about that fact. This confidence is modeled as a Beta-distributed random variable, following [15], where $\theta_{s,p,o} \sim \text{Beta}(\alpha_{s,p,o}, \beta_{s,p,o})$ denotes the probability that the LLM does not know the fact. All triples are initialized with a uniform prior: $\alpha = 1$, $\beta = 1$. After querying the LLM, the parameters are updated based on correctness: If the model answers the question incorrectly: $\alpha \leftarrow \alpha + 1$; If the model answers the question correctly: $\beta \leftarrow \beta + 1$. These updates are also propagated to neighboring edges–i.e., all triples sharing the same subject or object–based on the assumption that gaps in knowledge are often correlated among related facts. At each iteration, the edge (s, p, o) with the highest expected failure probability is selected for querying. For simplicity, and diverging slightly from the approach in [15], questions are generated for each selected triple as follows: "What is the relation between s and o?". The generated question is posed to the LLM, and its answer is compared against the true predicate p from Wikidata. For evaluation, we use a second LLM (GPT 4o-mini) to judge the response. The evaluation prompt includes both the original question, the model's answer, and the actual Wikidata relation, and asks: "Does the answer match this information? The correct relation is: <relation>.". GPT's answer is used to update the (α, β) parameters as described. After a number of probing iterations, the LLM's alignment with the knowledge subgraph centered at the entity measured by the average of all current θ values for all triples in the subgraph. Lower values of $\bar{\theta}_e$ indicate better factual alignment between the LLM and the knowledge graph in the neighborhood of the starting entity. At each iteration up to a set number (chosen as 20 in our study), an edge (s, p, o) is selected greedily based on its expected failure probability $\theta = E[\theta_{s,p,o}]$. This constraint balances breadth (by exploring two-hop neighborhoods) and depth (by targeting the most uncertain facts), allowing the method to produce a meaningful alignment score without exhaustive querying.

4.1 Experiments

We choose Mistral7B as an LLM to evaluate LLM probing due to the fact that it was the best performing LLM in terms of correct answers without irrelevance. 250 Q/A pairs with single answers and 250 Q/A pairs with multiple answers are randomly selected from the WikiWebQuestions dataset, ensuring that each associated SPARQL query has exactly one bound entity around which the parameterized confidence graph can be constructed. Mistral7B is used as the LLM to test against. We compare the following approaches for selecting top-N Q/A pairs from the datasets and measure the overall accuracy (percentage of correct questions):

- Theta: selecting lowest N average theta values in the parameterized confidence subgraph generated from an entity. i.e. selecting Q/As predicted to be non-aligned with Wikidata by an approximation of the KGLens [15] approach.
- Random: selecting N Q/A pairs randomly.
- Entropy: combination of the correlated features presented in the previous section as described below.

Combining Correlated Entropy-Based Features: For each question–answer pair q_i, we compute a combined score $S(d_i)$ using six normalized features scaled to the range $[0, 1]$. Features negatively correlated with accuracy include entity entropy e_i and property entropy p_i. Features positively correlated with accuracy include the number of sitelinks s_i, number of statements t_i, number of references r_i, and last updated timestamp u_i. To ensure uniform directionality, the positively correlated features are inverted. The final score is computed as:

$$S(d_i) = \frac{1}{6} \left(e_i + p_i + (1 - s_i) + (1 - t_i) + (1 - r_i) + (1 - u_i) \right)$$

Lower values of $S(d_i)$ indicate greater accuracy. We select the top-N entries ranked by $S(d_i)$ in descending order to probe the LLM.

4.2 Results

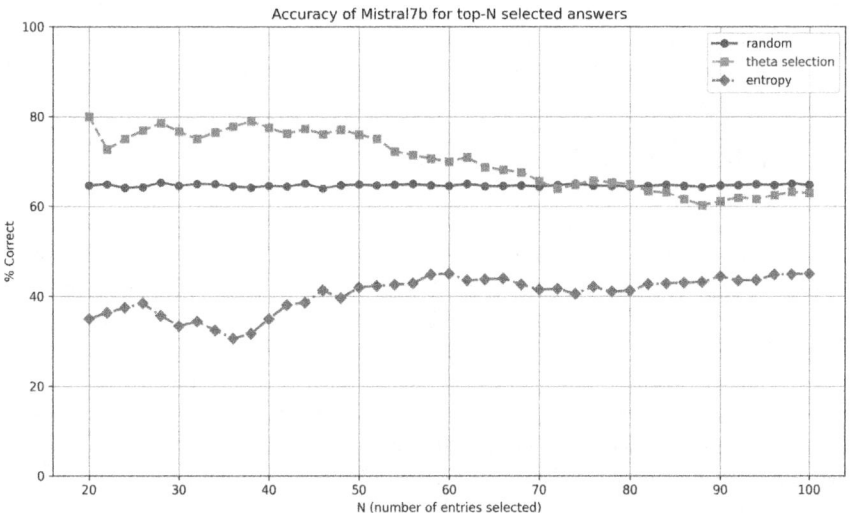

Fig. 4. LLM alignment scores for single-answer questions.

Figure 4 shows results for single-answer questions and Fig. 5 shows results for multiple-answer questions. We can see that for both question/answer sets, entropy-based selection proves more challenging for the Mistral LLM overall. For questions with multiple answers, entropy-based selection performs clearly better for all values of N and for single-answer questions theta-based selection and entropy-based selection perform more similarly (almost identical when N is low) and much better than random selection. Results show the effectiveness of the entropy-based selection approach, and although the caveat exists that we have used a limited approximation of the KGLens approach to achieve computational feasibility for the dataset used, entropy-based selection promise as a pragmatic KG-LLM alignment technique for assessing LLM "blind spots" efficiently.

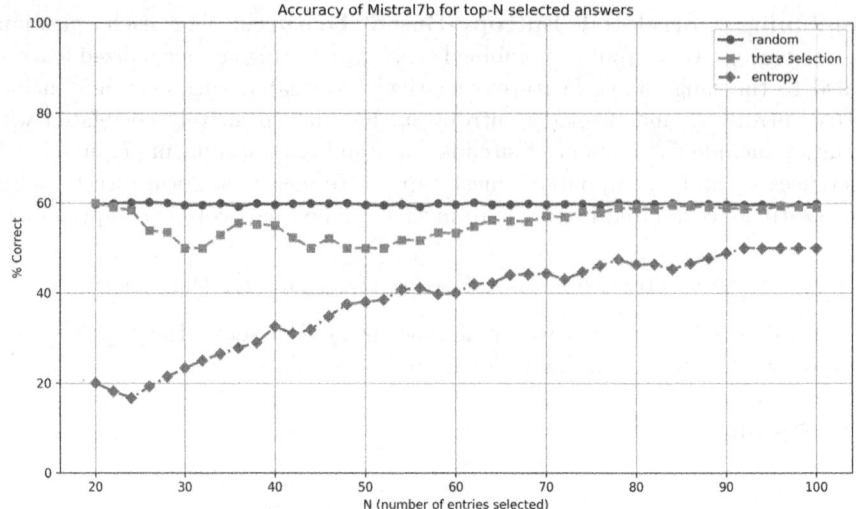

Fig. 5. LLM alignment scores for multiple-answer questions.

5 Conclusions and Future Work

In this paper, we investigated the relationship between entropy-based features extracted from Wikidata and LLM question-answering performance. We analyzed how informational entropy can serve as a predictor of LLM reliability and factual correctness. Our empirical findings suggest that certain entropy-derived features correlate with LLM performance. Furthermore, we demonstrated that entropy-guided probing can effectively reveal blind spots where LLMs/KGs are poorly aligned. Although this study is limited to Wikidata, future work will involve investigating whether the results hold for other KGs and LLMs. Furthermore, integrating these methods into active hallucination mitigation strategies has the potential to enhance the reliability of generative AI systems.

Acknowledgements. This paper is based on results obtained from the project, "Research and Development Project of the Enhanced infrastructures for Post-5G Information and Communication Systems" (JPNP20017), commissioned by the New Energy and Industrial Technology Development Organization (NEDO), JST CREST Grant Number JPMJCR22M2, and JSPS KAKENHI Grant Number JP23K24949.

Appendix

A Example: Entropy-Based Feature Extraction

```
Question: where are google headquarters located?
SPARQL Query: SELECT DISTINCT ?x WHERE { wd:Q694178 wdt:P131 ?x. }
```

```
Entity: Q694178 (Googleplex)

Computed Features:
  - Property Entropy: 0.797    - Entity Entropy: 0.9611
  - SPARQL Complexity: 1        - Number of Statements: 20
  - Number of References: 8     - Number of Properties: 20
  - Number of Sitelinks: 39     - Last Modified: 2025.08
```

B Evaluation Prompt Example (LLM Responses)

```
Given the following question and correct answer, evaluate the provided
model answer.
Question: what character did Natalie Portman play in star wars?
Correct Answer: Padme Amidala  Model Answer: Princess Leia.
Classify the model answer into one of the following categories:
1. Perfectly accurate. 2. Very accurate. 3. Accurate but with some
redundant or irrelevant information. 4. Inaccurate (probable intrinsic
error e.g from inaccurate/out-of-date training data). 5. Inaccurate
(probable fabrication/hallucination). 6. Inaccurate (irrelevant).
7. Completely inaccurate.
Respond with only the classification number and its label.
```

References

1. Agrawal, G., Kumarage, T., Alghamdi, Z., Liu, H.: Can knowledge graphs reduce hallucinations in LLMs? : a survey. In: Duh, K., Gomez, H., Bethard, S. (eds.) Proceedings of the 2024 Conference of the North American Chapter of the Association for Computational Linguistics: Human Language Technologies (Volume 1: Long Papers), pp. 3947–3960. Association for Computational Linguistics, Mexico City, Mexico (June 2024). https://doi.org/10.18653/v1/2024.naacl-long.219
2. Chen, J., Lin, H., Han, X., Sun, L.: Benchmarking large language models in retrieval-augmented generation. In: Proceedings of the AAAI Conference on Artificial Intelligence, vol. 38, no. 16, pp. 17754–17762 (2024)
3. Chitale, P., Gala, J., Dabre, R.: An empirical study of in-context learning in llms for machine translation. In: Findings of the Association for Computational Linguistics ACL,2024. pp. 7384–7406 (2024)
4. Farquhar, S., Kossen, J., Kuhn, L., Gal, Y.: Detecting hallucinations in large language models using semantic entropy. Nature **630**(8017), 625–630 (2024)
5. Guan, X., et al.: Mitigating large language model hallucinations via autonomous knowledge graph-based retrofitting. In: Proceedings of the AAAI Conference on Artificial Intelligence, vol. 38, no. 16, pp. 18126–18134 (2024)
6. Lavrinovics, E., Biswas, R., Bjerva, J., Hose, K.: Knowledge graphs, large language models, and hallucinations: an NLP perspective. J. Web Semant. **85**, 100844 (2025)
7. Manakul, P., Liusie, A., Gales, M.J.: Selfcheckgpt: zero-resource black-box hallucination detection for generative large language models. arXiv preprint arXiv:2303.08896 (2023)

8. Niu, C., et al.: Ragtruth: a hallucination corpus for developing trustworthy retrieval-augmented language models. arXiv preprint arXiv:2401.00396 (2023)
9. Radhakrishnan, P., et al.: Knowing when to ask–bridging large language models and data. arXiv preprint arXiv:2409.13741 (2024)
10. Rashad, M., Zahran, A., Amin, A., Abdelaal, A., Altantawy, M.: FactAlign: fact-level hallucination detection and classification through knowledge graph alignment. In: Ovalle, A., et al. (eds.) Proceedings of the 4th Workshop on Trustworthy Natural Language Processing (TrustNLP 2024), pp. 79–84. Association for Computational Linguistics, Mexico City, Mexico (June 2024). https://doi.org/10.18653/v1/2024.trustnlp-1.8, https://aclanthology.org/2024.trustnlp-1.8/
11. Sansford, H., Richardson, N., Maretic, H.P., Saada, J.N.: Grapheval: a knowledge-graph based llm hallucination evaluation framework. arXiv preprint arXiv:2407.10793 (2024)
12. Vrandečić, D., Krötzsch, M.: Wikidata: a free collaborative knowledgebase. Commun. ACM **57**(10), 78–85 (2014)
13. Xu, J., Lam, M.S.: Fine-tuned wikisp: a semantic parser for question answering over wikidata. In: Findings of the Association for Computational Linguistics: ACL 2023 (2023)
14. Yehuda, Y., Malkiel, I., Barkan, O., Weill, J., Ronen, R., Koenigstein, N.: Interrogatellm: zero-resource hallucination detection in llm-generated answers. In: Proceedings of the 62nd Annual Meeting of the Association for Computational Linguistics (Volume 1: Long Papers), pp. 9333–9347 (2024)
15. Zheng, S., Bai, H., Zhang, Y., Su, Y., Niu, X., Jaitly, N.: Kglens: a parameterized knowledge graph solution to assess what an llm does and doesn't know. arXiv preprint arXiv:2312.11539 (2023)
16. Zhu, Y., Xiao, J., Wang, Y., Sang, J.: Kg-fpq: evaluating factuality hallucination in llms with knowledge graph-based false premise questions. arXiv preprint arXiv:2407.05868 (2024)

Towards Automating RDF Extraction for Archaeological Knowledge Graphs with LLMs

Ali Hariri(✉), Stéphane Jean, and Mickaël Baron

ISAE-ENSMA, Université of Poitiers, LIAS, Poitiers, France
{ali.hariri,stephane.jean,mickael.baron}@ensma.fr

Abstract. Archaeological sites pose significant challenges in terms of accessibility and data integration. The Semantic Web, particularly through the CIDOC CRM ontology, enables structured data sharing. However, manually constructing knowledge graphs remains a complex and labor-intensive task. This study explores semi-automatic methods leveraging large language models (LLMs) to generate knowledge graphs for archaeological data. We propose different prompting strategies based on varying levels of information about CIDOC CRM and experiment with diverse prompt patterns. Experiments conducted on real datasets demonstrate that providing LLMs with a carefully selected subset of the CIDOC CRM ontology, combined with few-shot prompt patterns, enhances RDF extraction and improves performance in answering competency questions.

Keywords: Knowledge Extraction · LLMs · Ontology · CIDOC CRM

1 Introduction

The *Hypogeum of the Dunes* in Poitiers, France, is a significant Merovingian monument that exemplifies the complexity of cultural heritage research. Its study demands interdisciplinary collaboration across archaeology, geology, epigraphy, and other related fields. However, research on this site faces challenges, including developing a shared vocabulary for data integration, aligning discipline-specific data formats, and addressing restricted physical access due to the monument's fragility. These challenges are representative of broader issues in managing and preserving cultural heritage sites. Addressing them is a key objective of the French ANR project Digitalis (ANR-22-CE38-0011-01), which aims to develop a collaborative digital infrastructure to support interdisciplinary research through advanced data integration, semantic interoperability, and remote accessibility.

Semantic Web technologies, particularly the CIDOC Conceptual Reference Model (CIDOC CRM) ontology [3], offer a promising framework for representing and integrating heterogeneous data in a consistent and meaningful way. CIDOC CRM is designed to capture the complex relationships between cultural heritage objects, actors, and events, enabling researchers to map their data into a shared conceptual space and support remote collaboration. Moreover, CIDOC CRM supports the development of domain-specific extensions, enabling researchers to tailor the model to diverse research needs. A prominent example is CRMArcheo, an extension designed to represent archaeological excavation processes, stratigraphic information, and artifact management. This makes it particularly well-suited for documenting complex sites such as the Hypogeum of the Dunes. However, manually constructing knowledge graphs (KGs) that adhere to CIDOC CRM (and its extensions) remains a demanding task, requiring both deep domain knowledge and technical skills in data modeling and Semantic Web technologies. This challenge often limits the widespread adoption of semantic approaches in cultural heritage research.

In this paper, we investigate the potential of Large Language Models (LLMs) to assist researchers in automatically generating RDF triples that conform to the CIDOC CRM ontology. Given their training on extensive web data, including references to CIDOC CRM documentation and examples, we assume that LLMs are capable of generating relevant RDF triples from textual descriptions provided by experts. These automatically generated triples can subsequently be reviewed and refined by domain specialists to build high-quality knowledge graphs (KGs) that support advanced research tasks. We propose and evaluate three prompting strategies that vary in the level of guidance they provide to the LLM, using real archaeological datasets from the Hypogeum of the Dunes. Our results emphasize the importance of carefully selecting relevant CIDOC CRM classes and properties, designing effective prompts, and integrating expert validation to achieve accurate and meaningful RDF extraction.

2 Related Work

Ontology population has been extensively studied in various domains, with multiple approaches leveraging Natural Language Processing (NLP) [10] techniques, Machine Learning (ML) and Deep Learning (DL) algorithms. Several works [1,11] have explored methods for extracting entities and relationships from text, utilizing rule-based, statistical, and hybrid approaches. ML-based ontology population techniques have improved the automation of knowledge extraction [7], while deep learning models offer more advanced methods for semantic understanding [12]. However, these methods often require extensive human supervision, substantial annotated training data, and domain-specific adaptations, which limit their scalability and applicability in certain contexts [14]. In particular, the availability of high-quality training datasets remains a challenge for supervised ontology population approaches, especially in specialized domains where labeled data is scarce. In the context of our study on the *Hypogeum of the Dunes*, constructing such a training dataset is not feasible.

Recent advancements in LLMs have opened new possibilities for semantic knowledge representation and ontology population [15], particularly through tasks like entity recognition [4] and RDF triples generation [13]. Despite promising results in general-purpose domains, their use in cultural heritage is still in its early stages. Loffredo and De Santo [9] recently proposed a methodology that enhances LLM performance for cultural heritage tasks by combining ontologies with a Retrieval-Augmented Generation (RAG) framework. Yet research on automating the population of CIDOC CRM-based ontologies remains limited. To the best of our knowledge, the work by Ding et al. [2] is the closest to ours, proposing Knowledge Prompt Chaining, a framework that leverages prompt chaining to inject graph-structured knowledge into LLMs. However, their study primarily focuses on prompt design, rather than examining the level of CIDOC CRM information to include in the prompt for optimal results in a real-world setting. This is the gap we address in this paper.

3 Problem Statement

Let Ω be a set of ontologies, and $D = \{(X^i, S^i)\}_{i=1}^{n}$ be a set of structured data sources where $S^i \in \Omega$, and each structured data source is defined as:

$$X^i = \left\{ (x_{r,a_j})_{j=1}^{J_i} \right\}_{r=1}^{R_i}$$

where n is the number of structured data sources in D. Here, x_{r,a_j} represents the value of attribute a_j in row r, J_i is the number of attributes in the i-th data source, and R_i is the number of rows. The semantic modeling task consists of mapping structured data X to a semantic model S within the ontology Ω. This process involves two main components:

1. **Semantic Labeling**: Assigning ontology classes and properties to attributes of the structured data.
2. **Semantic Graph Construction**: Defining relationships between identified entities and attributes to build a semantic graph.

3.1 Semantic Modeling

Semantic Labeling is a function F that maps attributes of the structured data source X to ontology nodes in Ω. This process can be represented as a set of triples:

$$N^i = \{(a_j, p_j, n_j)\}_{j=1}^{J_i}$$

where each attribute a_j is associated with a property p_j and a node n_j in the ontology. The function F is defined as:

$$F : N^i = f(X^i, \Omega)$$

Semantic Graph Construction aims to establish meaningful links between semantic entities by defining relationships in accordance with a target ontology. This process results in a set of triples:

$$S^i = \{(e_g, p_g, e'_g)\}_{g=1}^{G_i}$$

where e_g and e'_g represent semantic entities, and p_g is the relationship connecting them.

In summary, the extraction process can be conceptualized as a two-step mapping:

$$X \xrightarrow{F} N \xrightarrow{G} S.$$

where, F performs the semantic labeling, while G constructs the semantic graph.

However, applying these functions to real-world archaeological tables presents significant challenges, as the effectiveness of F and G depends primarily on two factors: (i) the scope and specificity of CIDOC CRM, and (ii) the formulation of the prompt provided to the LLM. These practical constraints motivate the research questions considered in the following section.

However, applying these functions to real archaeological tables is challenging because the effectiveness of F and G depends on two main factors: (i) the extent of CIDOC CRM context, and (ii) the way the prompt is formulated for the LLM. These practical considerations motivate the research questions presented in the following section.

4 Our Approach to Automating RDF Extraction Aligned with CIDOC CRM

While LLMs demonstrate remarkable capabilities in generating structured RDF data from textual sources, their performance can be highly sensitive to two key factors: the explicit inclusion of ontology descriptions and the formulation of the prompt. Ontologies such as CIDOC CRM encode essential domain knowledge, but it remains an open question whether incorporating them directly into prompts improves extraction quality. Moreover, prompt engineering decisions may affect the model's ability to generalize from examples and correctly apply ontological structures. Gaining insight into these factors is critical for designing effective LLM-based knowledge extraction pipelines. To address these concerns, we formulate the following two research questions:

Ontology scope: *how does providing ontology description (O) impact the quality of RDF extraction?*
Prompt Pattern: *which prompt configuration (P) maximizes extraction accuracy and generalization?*

4.1 Ontology Scope

Our goal was to develop several Knowledge Extraction strategies, each varying in the level of information about CIDOC CRM provided in the prompt. To facilitate comparison, we ensured a consistent prompt template, based on the well-established role-based prompting method. This approach led to the definition of the following three strategies.

Unguided LLM Extraction. Since LLMs are trained on web data that includes the entire CIDOC CRM documentation, this approach evaluates whether prior knowledge is sufficient to extract meaningful RDF data. The first strategy, therefore, uses an LLM without any ontological guidance, relying solely on a generic instruction to extract RDF triples. The prompt used is presented in Fig. 1 (excluding "With Ontology Description" part for this strategy). It consists of the basic directives, i.e., the model's role, task definition and input/output format.

Full Ontology Prompting. Even if the LLM has prior knowledge of CIDOC CRM, including the description of the CIDOC CRM ontology in the prompt may help the model focus specifically on this ontology, leading to better RDF data extraction. Therefore, our second strategy injects the complete textual description of the CIDOC CRM ontology into the prompt. As shown in Fig. 1, the prompt includes both the model's role and task directive, along with the full set of ontology classes and properties. In addition to the ontology, the prompt contains an example transformation and the data chunk. This structure enables the LLM to map input content to the appropriate CIDOC CRM elements and generate syntactically valid RDF triples.

Ontology Subset Prompting. CIDOC CRM is a large ontology with hundreds of classes and properties, which introduces two potential challenges for the previously devised strategy. First, the prompt becomes excessively long due to the extensive nature of the ontology, potentially exceeding the model's context window limit. Second, the ontology includes class hierarchies with multiple levels, which may prevent the LLM from accurately identifying the specific class or subclass that needs to be instantiated.

To mitigate this, we introduce a third strategy that retains the original prompt structure but uses a curated ontology subset tailored to archaeology. It emphasizes precise and relevant classes, "E22 Human-Made Object" over the broader "E19 Physical Object", while excluding less relevant ones from biology or law. Leveraging the increased prompt space, we add detailed class and property descriptions to reduce model hallucinations. This approach aims to improve RDF granularity for tasks like artifact classification and provenance tracking. The experiments conducted to evaluate which strategy yields the best results are presented in the next section.

4.2 Prompt Pattern

Even when a suitable ontology is provided, the structure of the prompt plays a major role in shaping the model's interpretation of the task, its ability to generalize, and the consistency of its output. To investigate this, we complemented variations in ontology context with a systematic exploration of example-based prompt patterns, aiming to assess how different forms of exemplification influence the model's capacity to generate high-quality RDF triples.

Alternative approaches such as chain-of-thought prompting or full supervised fine-tuning are less suitable for our use case. These methods either require substantial amounts of training data or deviate from our focus on prompt-based

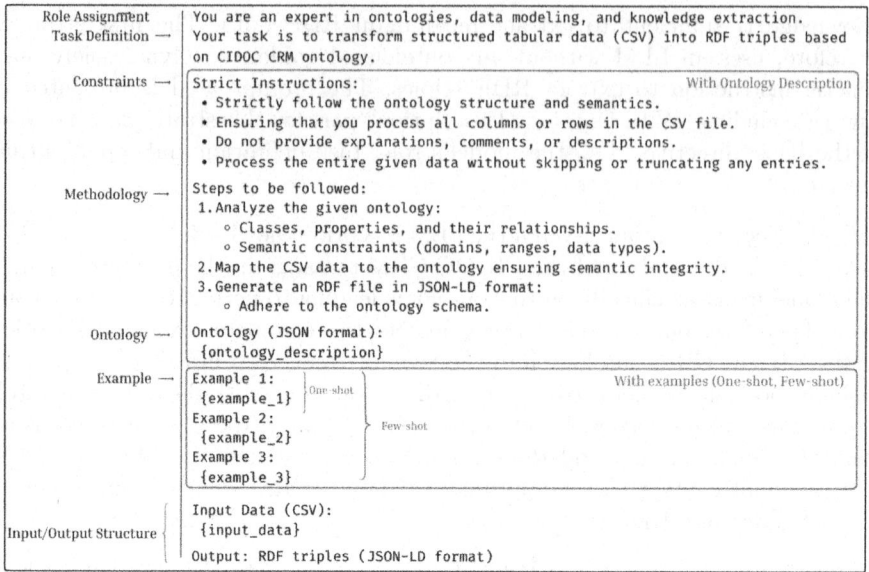

Fig. 1. Prompt Template Representation

knowledge extraction. Furthermore, techniques like chain-of-thought can introduce confounding variables (e.g., inconsistent reasoning paths), complicating controlled evaluations. By limiting our analysis to three well-established and directly comparable example-based prompt patterns, we aim to ensure that our findings remain interpretable and practically relevant for ontology-driven RDF extraction tasks.

Zero-shot Prompting. In this setting, the prompt follows the template illustrated in Fig. 1, but excludes any input–output examples. It includes only the essential role-based components: the model's assigned role, a definition of the task, specifications of the input and output formats, and, when applicable, the ontology description. Without explicit examples, the model relies solely on its pre-trained knowledge and its interpretation of the task description to generate RDF triples.

One-shot Prompting. This strategy extends the zero-shot prompt by incorporating a single, concrete example that illustrates the transformation of an input text into RDF triples compliant with CIDOC CRM. The goal is to guide the model by demonstrating the expected mapping process without overloading it with numerous examples. While this one-shot approach can improve performance, it still exhibits limitations when applied to real-world data. For instance, in cases involving incomplete information–such as "A fragmentary ceramic vessel with no discernible decoration, possibly from Chamber II, dating uncertain"–the model may struggle to appropriately handle missing or ambiguous attributes.

Similarly, when multiple instantiations are required, such as varying representations of locations ("Chamber I, south wall" versus "coordinates: 43.1234°N, 1.2345°E"), the model may find it difficult to select the correct CIDOC CRM classes, such as E53 Place or E47 Spatial Coordinates. These challenges motivated the adoption of a few-shot strategy, which introduces more diverse examples to support the model's ability to generalize across heterogeneous data inputs.

Few-Shot Prompting. To further enhance the model's ability to generalize and handle diverse data scenarios, the few-shot prompt includes multiple examples (three in our case), all drawn from the Hypogée des Dunes dataset. These examples are deliberately varied in terms of data source, the presence or absence of specific information, and the classification requirements imposed by different data types. Each example illustrates the transformation from raw text to RDF triples, showcasing a range of CIDOC CRM classes and properties relevant to the archaeological context of the Hypogée des Dunes. By exposing the model to these nuanced variations, the few-shot strategy aims to improve generalization, minimize hallucinations, and ensure consistent RDF generation across heterogeneous real-world inputs.

Example Format. For both one-shot and few-shot prompts, we employ a consistent and formalized example format that explicitly illustrates the transformation from unstructured text to structured RDF triples. Each example typically comprises an identifier, a class assignment, and a set of RDF properties linking the subject to relevant concepts or entities. The RDF data is represented using the machine-readable JSON-LD serialization, which ensures the output is both human-interpretable and compatible with downstream semantic web tools. This standardized format distinctly separates the original textual description from its corresponding RDF representation, facilitating the model's ability to learn the mapping between the two. By applying this format uniformly across all examples, regardless of dataset origin, we enhance reproducibility, support systematic evaluation, and enable seamless integration into diverse archaeological knowledge extraction workflows.

To clarify the propositions presented informally above, we introduce in the next section a formalization of these methods.

4.3 Formal Framework for Our Propositions

We define the prompt-based knowledge extraction task as a function:

$$f : (T, P, O) \longrightarrow R$$

where:

- T is the textual input chunk to be processed.
- P denotes the prompt configuration (zero-shot, one-shot, few-shot).
- O the ontology context, indicating the degree of ontological information in the prompt.

- R is the resulting RDF representation.

Prompt Pattern. We formalise the prompt pattern P as:

$$P \in \{P_0, P_1, P_n\}$$

where:

- P_0: zero-shot (prompt includes only task instructions and structure).
- P_1: one-shot (prompt includes one example transformation).
- P_n: few-shot (prompt includes n examples, with $n \geq 2$).

Ontology Context. We formalise the ontology context O as:

$$O \in \{O_\emptyset, O_{full}, O_{subset}\}$$

where:

- $O_\emptyset$: no ontology description provided in the prompt (unguided extraction).
- O_{full}: complete textual description of the CIDOC CRM ontology provided.
- O_{subset}: curated subset of the ontology tailored to archaeology.

Prompt Composition. Each prompt π is defined as the concatenation:

$$\pi = \text{[Task Instructions]} + O + P + T$$

where:

- [Task Instructions]: A role-based instruction describing the model's role (e.g., "You are an expert in RDF extraction...").
- O: the ontology context.
- P: one or more example transformations (for one-shot and few-shot).
- T: the textual input chunk.

We present our evaluation of the different prompts designed according to this formal framework in the next section.

5 Experimental Setup

5.1 Datasets

This study uses structured CSV data provided by archaeologists who researched the *Hypogeum of the Dunes*, including stratigraphic units, a lapidary inventory, and geological data. The dataset was converted into RDF triples for semantic representation. To ensure methodological reliability, unstructured textual data was excluded due to the absence of a validated reference RDF dataset. For consistency with other sections of this paper, we refer to the lapidary inventory dataset as "Inv. lapidary". Key dataset statistics are presented in Table 1.

Table 1. Statistics of the datasets used for RDF extraction.

Dataset	Nb. Rows	Nb. Columns	Nb. Annotations
Stratigraphic unit list	182	8	2,304
Facts list	80	10	1,186
Inv. lapidary	68	12	1,740
Stone and geology corresponding list	68	9	1,237

5.2 Models

To examine the influence of deployment strategy and model architecture on RDF extraction performance, we selected three representative LLMs spanning two deployment categories: online (cloud-based) and local (self-hosted). For the online model, we selected GPT-4o [6] (OpenAI), a state-of-the-art commercial language model recognized for its advanced reasoning capabilities, broad coverage of diverse training corpora, and widespread adoption in both academic research and industrial applications. For the local models, we selected Llama 3.1 [5] (Meta) and Mistral 7B [8] (Mistral AI), representing high-performance open-source models that can be hosted on local infrastructure, offering enhanced control, data privacy, and customization. These models differ in parameter size, context window, and training corpus composition, allowing us to systematically investigate how deployment strategy and model design influence the generation of CIDOC CRM-compliant RDF data. Table 2 summarises the key technical specifications of the selected models.

Table 2. Technical description of the selected LLMs.

Model	Provider	Parameter	Context Window	Deployment
GPT-4o[a]	OpenAI	~1.5T	128k tokens	Cloud
Llama 3.1[b]	Meta	8B	128k tokens	Local
Mistral[c]	Mistral AI	7B	32k tokens	Local

[a] https://openai.com/index/hello-gpt-4o/
[b] https://ai.meta.com/blog/meta-llama-3-1/
[c] https://mistral.ai/news/announcing-mistral-7b.

For the local deployment of Llama 3.1 and Mistral 7B, we utilized a high-performance local server to ensure efficient model inference and to enable precise performance measurements. The server specifications are as follows:

- *RAM*: 756 GB
- *CPU*: AMD EPYC 7543P, 32 cores, 64 threads
- *GPU*: 2 × NVIDIA A10 (24 GB)

5.3 Experimental Design

To facilitate the extraction of RDF triples from archaeological datasets, we developed a modular and scalable pipeline tailored to the challenges of processing structured CSV data with LLMs. Recognizing that LLMs often struggle with long inputs and that output truncation can degrade result quality, we divided the input CSV files into manageable 20-row chunks. This chunking strategy ensures that each prompt remains within the model's context window, while also preserving semantic coherence across rows.

Each chunk is then integrated into a carefully constructed prompt that includes: (1) the relevant CIDOC CRM ontology description (either the full ontology or a domain-specific subset selected by an expert); (2) a dataset-specific example that demonstrates how to map real archaeological data to ontology-aligned RDF triples; and (3) a prompt pattern (zero-shot, one-shot, or few-shot) that guides the model's reasoning process.

The output generated by the LLM is expected to be a structured list of RDF triples, which we serialize in JSON-LD format to enable seamless integration into knowledge bases. To enhance interoperability and facilitate downstream analysis, we also apply post-processing steps to ensure consistent syntax and alignment with CIDOC CRM standards.

The entire pipeline is implemented in Python using the LangChain framework, which provides flexible modules for prompt engineering, LLM interaction, and output handling. Figure 2 illustrates the complete architecture, highlighting how each component contributes to transforming raw archaeological data into structured knowledge graphs.

Code is available at https://forge.lias-lab.fr/cidoccrm-llm-extractor.

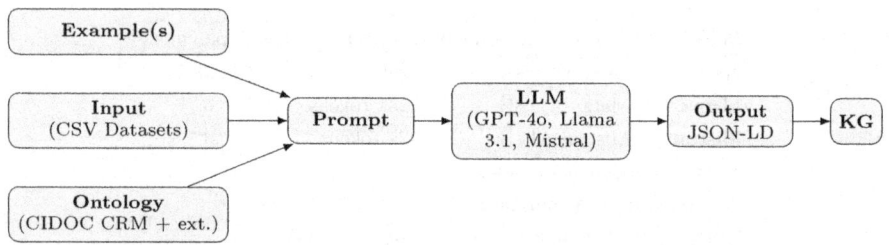

Fig. 2. Ontology-guided RDF extraction pipeline using LLMs

5.4 Evaluation Metrics

We evaluate ontology population using precision and recall, comparing extracted triples to an expert-curated gold standard. Precision measures the proportion of correct triples, while recall assesses how comprehensively relevant semantic relations are captured. The evaluation also uses domain-specific competency questions to test the relevance of the extracted triples, focusing on capturing interdisciplinary links (e.g., archaeology and geology) and identifying key domain concepts. For example, the question "Which stratigraphic units contain mortar, and where are they located within the site?" evaluates the model's ability to link material use with spatial and contextual information.

6 Results

In this section, we present the results of our experiments. Our analysis addresses the following key questions:

1. *Ontology scope:* how does the inclusion of an explicit ontology description affect the quality of RDF extraction?
2. *Prompt Pattern:* which prompt configuration (zero-shot, one-shot or few-shot) maximizes extraction accuracy and generalization?
3. *Environmental Impact:* what is the carbon footprint associated with using these LLMs, considering their deployment type and configuration?

6.1 Ontology Scope

The *Unguided LLM Extraction* strategy led to unsatisfactory results: the LLM, trained for general-purpose tasks, often failed to align with the CIDOC CRM ontology. It frequently used overly generic labels (e.g., "object" instead of "E22 Human-Made Object" or "E78 Collection"), causing semantic errors.

The other two strategies yielded better results, highlighting the importance of including the CIDOC CRM description in the prompt. Table 3 presents precision and recall results on the most representative datasets, selected due to space constraints, while Table 4 shows the CQ Scores.

Our results show that using a subset of the CIDOC CRM leads to improved precision and recall compared to the full ontology (see Table 3). In the *Full Ontology Prompting* strategy, the model occasionally defaulted to overly general classes, for example labeling artifacts as "E19 Physical Object" instead of the more specific "E22 Human-Made Object," which limited the granularity and precision of the extracted knowledge.

By restricting the ontology to archaeology-specific entities, the *Ontology Subset Prompting* strategy produced more accurate RDF triples and fewer errors from ambiguity or irrelevant mappings. It favors precise classes like "E22 Human-Made Object" over generic ones like "E19 Physical Object," aligning better with expert expectations. This also improved competency question success, as shown in Table 4, since the model was guided toward relevant semantic structures.

Analyzing the results from our best strategy, *Ontology Subset Prompting*, revealed that some errors were not due to the extracted triples but stemmed from inconsistencies in the source data. Since our datasets were derived from working study documents, they often included heterogeneous formatting, such as inconsistent delimiters, irregular punctuation, or ambiguous notations. In these cases,

Table 3. Precision and recall by prompting strategy, model, and dataset

Strategy	Dataset	Model	Precision (%)			Recall (%)		
			Zero	One	Few	Zero	One	Few
Full Ontology Prompting	Facts List	GPT-4o	17.0	76.3	88.1	11.5	63.2	76.8
		Llama 3	14.0	60.2	68.4	9.0	49.5	58.2
		Mistral	11.5	55.8	63.1	7.0	45.3	52.0
	Inv. Lapidary	GPT-4o	16.5	77.2	89.3	10.8	64.0	78.0
		Llama 3	13.2	61.7	70.0	8.7	50.4	59.3
		Mistral	11.0	57.3	64.5	7.5	46.1	53.2
Ontology Subset Prompting	Facts List	GPT-4o	19.0	85.0	90.2	12.5	71.5	79.1
		Llama 3	15.5	66.5	74.2	10.0	55.7	63.0
		Mistral	13.0	61.9	68.3	8.2	51.0	58.6
	Inv. Lapidary	GPT-4o	18.0	84.8	91.0	11.8	70.7	80.5
		Llama 3	14.5	65.1	73.0	9.5	54.4	61.5
		Mistral	12.0	60.0	66.1	7.8	49.3	56.0

Table 4. CQ score by prompting strategy, model, and prompt type

Strategy	Model	CQ Score (%)		
		Zero-shot	One-shot	Few-shot
Full Ontology Prompting	GPT-4o	18.0	72.4	79.3
	Llama 3	15.3	45.8	52.6
	Mistral	13.4	41.3	47.0
Ontology Subset Prompting	GPT-4o	21.1	85.2	88.5
	Llama 3	17.4	54.6	60.1
	Mistral	14.8	49.2	53.5

the model attempted to correct the issues, leading to semantic confusion. For example, hyphenated entries like "US12-US14" were sometimes misinterpreted as a single unit rather than distinct ones.

The analysis also showed that further work is needed to define the optimal CIDOC CRM subset to include in the prompt. In the *Ontology Subset Prompting* strategy, we observed that the model might miss useful information if their corresponding classes were excluded from the subset. While this improves overall precision, it limits the breadth of the extracted knowledge. Among all tested models, GPT-4o achieved the best results across all configurations.

6.2 Prompt Pattern

The *Zero-shot Prompting* configuration yielded the lowest performance, with the model often producing incomplete or inconsistent triples. Without any guidance from examples, the LLM relied solely on its pre-trained knowledge, leading to variability in both structure and content. This often manifested as missing relationships or inconsistent application of CIDOC CRM classes.

In contrast, the *One-shot Prompting* configuration showed noticeable improvements in both precision and recall. The inclusion of a single example provided a concrete template for the LLM to emulate, leading to a more consistent output structure and a moderate reduction in semantic errors.

The *Few-shot Prompting* configuration outperformed both zero-shot and one-shot configurations. By providing multiple, diverse examples, the LLM was better able to generalize across different archaeological contexts and handle variations in data presentation. This configuration produced the highest precision and recall scores (see Table 3), as well as the most consistent CQ score structure across different models (see Table 4).

Interestingly, we observed a trade-off between example diversity and extraction flexibility. One-shot prompting, despite providing only a single transformation case, allowed the model to generalize more freely and propose novel RDF structures not explicitly shown in the example. In contrast, few-shot prompting exposed the model to a wider range of typical cases, improving consistency and alignment with expected outputs but also reducing its ability to generate original or unexpected triples.

6.3 Environmental Impact

Beyond extraction performance, Table 5 presents the estimated carbon footprint (in gCO_2eq) for each prompting strategy across all evaluated models and prompt configurations. Emissions were computed using CodeCarbon[1] for local inference and EcoLogits[2] for cloud-based API calls, ensuring consistent measurement across deployment environments.

The results show that GPT-4o produces substantially higher emissions particularly under few-shot settings that involve longer prompts and more extensive inference. In contrast, local models such as Llama 3.1 and Mistral 7B exhibit significantly lower carbon footprints.

This difference stems from several contributing factors, notably the number of model parameters, the length of prompts, and the size of the generated output. Larger models inherently require more computational power, and their energy consumption increases further with complex prompting schemes. While few-shot prompting consistently improves RDF extraction quality, it also demands greater resources, resulting in higher environmental impact. These observations highlight the importance of smaller, task-oriented models as a more sustainable alternative to large, general-purpose systems, whose broad capabilities often come at the cost of significant and sometimes unnecessary computational overhead in domain-specific scenarios.

---

[1] https://codecarbon.io/.
[2] https://ecologits.ai/.

Table 5. Carbon Footprint (gCO$_2$eq) by Model, Prompting Strategy, and Prompt Type

Strategy	Model	Prompt Type		
		Zero-shot	One-shot	Few-shot
Unguided Prompt	GPT-4o	13.72	17.94	20.44
	Llama 3.1	0.16	0.22	0.25
	Mistral	0.06	0.08	0.13
Full Ontology Prompt	GPT-4o	23.76	23.85	27.18
	Llama 3.1	0.12	0.11	0.15
	Mistral	0.06	0.08	0.09
Ontology Subset Prompt	GPT-4o	26.03	26.67	27.63
	Llama 3.1	0.13	0.18	0.19
	Mistral	0.06	0.07	0.10

7 Conclusion and Future Work

In this paper, we examined the ability of LLMs to extract CIDOC CRM compliant RDF triples from real-world archaeological datasets. Our results indicate that although LLMs are trained on web data that includes the CIDOC CRM documentation, this prior knowledge alone is insufficient to achieve accurate and consistent RDF extraction. Simply providing the full CIDOC CRM description in the prompt also has drawbacks: it leads to long prompts that may exceed some LLMs' context window limitations, and it often causes the models to default to overly general classes due to the extensive class hierarchies within CIDOC CRM.

To address these challenges, we introduced a new approach based on a carefully selected subset of CIDOC CRM classes and properties, curated by a domain expert and embedded in the prompt. This focused approach improved performance, yielding precision and competency question success rates exceeding 80%. Additionally, our experiments highlighted the significant impact of prompt patterning (zero-shot, one-shot, and few-shot configurations) on RDF extraction quality. For example, in the *Ontology Subset Prompting* configuration, the best-performing model, GPT-4o, achieved a precision of 90.6% and a CQ Score of 88.5% in the few-shot setting. These results demonstrate that few-shot prompting consistently produced the most accurate and consistent outputs, thanks to its inclusion of multiple, diverse examples that improved generalization across different archaeological contexts.

However, the subset approach also restricts the scope of the extracted knowledge, occasionally omitting relevant triples outside the selected classes. This limitation highlights the need for more adaptable ontology integration to balance precision with broader coverage.

Future work should explore both the automatic selection of relevant classes and properties, reducing the manual effort required to curate ontology subsets, and the use of smaller, locally deployed models as a viable alternative to large scale cloud based systems. Fine-tuning these models with domain specific examples and structured RDF knowledge can significantly enhance their extraction capabilities, enabling them to handle a broader range of archaeological data while maintaining high-quality output. This approach not only improves precision but also addresses key concerns related to data security and environmental sustainability, as smaller models typically require fewer resources and produce lower carbon emissions than large, general-purpose systems.

References

1. Biemann, C.: Ontology learning from text: a survey of methods. J. Lang. Technol. Comput. Linguist. **20**(2), 75–93 (2005)
2. Ding, N.P., Du, J., Feng, Z.: Knowledge prompt chaining for semantic modeling. arXiv preprint arXiv:2501.08540 (2025)
3. Doerr, M.: The cidoc conceptual reference module: an ontological approach to semantic interoperability of metadata. AI Mag. **24**(3), 75–75 (2003)
4. Freund, M., Dorsch, R., Schmid, S., Wehr, T., Harth, A.: Enriching rdf data with LLM based named entity recognition and linking on embedded natural language annotations. In: Tiwari, S., Villazón-Terrazas, B., Ortiz-Rodríguez, F., Sahri, S. (eds.) Knowledge Graphs and Semantic Web. KGSWC 2024. LNCS, vol. 15459, pp. 109–122. Springer, Cham (2024). https://doi.org/10.1007/978-3-031-81221-7_8
5. Grattafiori, A., et al.: The llama 3 herd of models. arXiv preprint arXiv:2407.21783 (2024)
6. Hurst, A., et al.: Gpt-4o system card. arXiv preprint arXiv:2410.21276 (2024)
7. Imsombut, A., Sirikayon, C.: An alternative technique for populating Thai tourism ontology from texts based on machine learning. In: 2016 IEEE/ACIS 15th International Conference on Computer and Information Science, pp. 1–4. IEEE (2016)
8. Jiang, F.: Identifying and mitigating vulnerabilities in llm-integrated applications. Master's thesis, University of Washington (2024)
9. Loffredo, R., De Santo, M.: Using ontologies for llm applications in cultural heritage (2024)
10. Maynard, D., Li, Y., Peters, W.: NLP techniques for term extraction and ontology population (2008)
11. Petasis, G., Karkaletsis, V., Paliouras, G., Krithara, A., Zavitsanos, E.: Ontology population and enrichment: state of the art. Knowledge-driven multimedia information extraction and ontology evolution: bridging the semantic gap, pp. 134–166 (2011)
12. Sambandam, P., Yuvaraj, D., Padmakumari, P., Swaminathan, S.: Spiking equilibrium convolutional neural network for spatial urban ontology. Neural Process. Lett. **55**(6), 7583–7602 (2023)
13. de Souza, R.R., Pinheiro, T.L., Oliveira, J.C.B., dos Reis, J.C.: Knowledge graphs extracted from medical appointment transcriptions: Results generating triples relying on llms. In: KEOD, pp. 129–139 (2023)
14. Suchanek, F., Ifrim, G., Weikum, G.: Leila: learning to extract information by linguistic analysis. In: Proceedings of the 2nd Workshop on Ontology Learning and Population: Bridging the Gap between Text and Knowledge, pp. 18–25 (2006)
15. Trajanoska, M., Stojanov, R., Trajanov, D.: Enhancing knowledge graph construction using large language models. arXiv preprint arXiv:2305.04676 (2023)

Ontology-Based Forest Fire Management Using Complex Event Processing and Large Language Models

Ritesh Chandra(✉) [iD], Sonali Agarwal, and Sadhana Tiwari

Indian Institute of Information Technology Allahabad, Prayagraj,
Uttar Pradesh, India
{rsi2022001,Sonali,rsi2018507}@iiita.ac.in

Abstract. Forests sustain ecology, but wildfires are a significant threat. Fire weather indices help assess hazards but require constant real-time processing. To address this, we developed a Semantic Sensor Network (SSN) ontology model using data from Monesterial Natural Park, enhanced with Semantic Web Rules Language (SWRL). The system integrates Large Language Models (LLMs) and Complex Event Processing (CEP) engines for real-time fire detection. Sensor networks collect climate data (humidity, temperature, wind speed, etc.), which is processed via Spark Streaming and CEP to identify fire-related events. LLMs analyze detected events, while SPARQL queries retrieve relevant insights from the ontology. The results are combined to estimate the overall risk, allowing informed decision-making within a comprehensive Decision Support System (DSS) framework. It makes it easier to understand and deal with the risks of wildfires, as shown by tests that use ontology metrics, query-based testing, event alerts, and LLM performance (F1 score, precision, and recall).

Keywords: LLMs · SSN · Spark · CEP

1 Introduction

Forest fires are a significant global issue, threatening ecological integrity and economic stability. In 2025, 5,245 wildfires burned 108,535 acres across the United States [1], showcasing the scale of destruction. Each year, vast forest areas are lost, destroying habitats, reducing biodiversity, and releasing carbon dioxide, worsening climate change. While natural causes like lightning can ignite fires, human activities such as agricultural burning, land clearing, and accidental ignition remain the primary contributors.

India is highly prone to forest fires due to human encroachment, unsustainable land use, and poor fire management. Most fires stem from human activity [2], impacting nearby communities, destroying ecosystems, and worsening air pollution. Climate change intensifies fire risks by extending fire seasons. Existing global fire risk models, such as the Forest Fire Danger Index and the Canadian Fire Weather Index (CFWI) [3], may not be suitable for India's diverse forest conditions. A real-time DSS using Spark, Kafka, Esper, LLMs, and ontology enhances fire risk assessment. Kafka streams data to Spark for analytics, while the Esper CEP engine detects fire-related events. Ontology structures domain knowledge, and LLMs automate reasoning, improving decision-making and response efficiency [4].

Ontologies help create a consistent data model by organizing information across different sources and infrastructure services. They provide domain-specific information and establish a common language for seamless interaction between systems. Ontologies also make it easier to discover, access, and integrate data, ensuring efficient knowledge representation. The SSN ontology gathers data on multiple aspects of interest, specifically representing sensor networks. Descriptive Ontology for Linguistic and Cognitive Engineering (DOLCE) + Description-and-Situation (DnS) UltraLite is the upper-level ontology in this framework, showing the resources of the infrastructure and the freely available datasets [5]. This study covers multiple challenges.

1. Determining appropriate data sources based on their content, such as notable features, observations, sensors, and the defined scope of the dataset.
2. Addresses concerns related to data models, interfaces, mobility, and the heterogeneity of data sources, ensuring semantic interoperability between systems that incorporate multiple data models.
3. Integrating a CEP engine with LLMs will enable the real-time identification of important patterns and complex events in data streams while ensuring decision-making accuracy and facilitating rapid responses.

By integrating forest ontology, LLMs, and CEP engines, we create a comprehensive Decision Support System (DSS) for forest management.

The created system accomplishes the following goals and addresses particular user needs:

- To propose a real-time alert generation model based on LLMs.
- We developed a forest ontology using the Ontology Web Language (OWL), following the forest guidelines.
- We developed SWRL rules to improve the reliability of the ontology and utilized these rule parameters for event detection.
- We convert raw data into Resource Description Framework (RDF) triples, a structured format that enhances representation, querying, and interoperability across diverse systems.
- Merging real-time events with static data sources enhances query efficiency and enables more accurate, context-aware decision-making.
- The system integrates Kafka, Spark, and Esper to simulate real-time events.

We structure the remaining sections of this research work as follows: Sect. 2 discusses related work. Section 3 provides details about the specifics of the experiment. Section 4 presents the results and discussion of the experiment, while Sect. 5 describes the conclusion and future work to be undertaken.

2 Related Works

This section reviews recent work to highlight gaps requiring further exploration to implement the proposed methodology.

Ginkal et al. [6] proposed a cloud-based Wireless Sensor Network (WSN) for forest fire detection, integrating the WSN with the Internet of Things (IoT), Unmanned Aerial Vehicles (UAVs), computer vision, You Only Look Once (YOLO) object detection algorithm, and Moderate Resolution Imaging Spectroradiometer (MODIS). Their approach improves real-time monitoring and detection efficiency over traditional methods.

Ibraheem et al. [7] developed a forest fire detection system using the Intermediate Fusion Visual Geometry Group 16 (VGG16) model and Energy Consumption Prediction Low Energy Adaptive Clustering Hierarchy (ECP-LEACH), achieving 99.86% accuracy while optimizing energy consumption. Huiyi et al. [8] analyzed fire causes using Jenks Natural BreaksâĂŞDensity-Based Spatial Clustering of Applications with Noise (Jenk-DBSCAN), MODIS MCD64A1 burned area product, and machine learning (ML) techniques including Support Vector Machine (SVM), Random Forest (RF), and Logistic Regression (LR), identifying meteorological variables as key contributors.

A study by Hamed et al. [9] presents the Forest Observatory Ontology-based Decision Support system (FOODS), an ontology-based knowledge graph for forest observatories that enhances data integration and supports sustainable forest management through improved analytical capabilities. Noroozi et al. [10] explore the use of LLMs for ontology construction in fields with limited expert availability, leveraging prompt engineering. Fine-tuned LLMs–Meta AI's LLaMA-7B (7 billion parameters) and BigScience's BLOOM-3B (3 billion parameters)–significantly improved learning and theory construction.

Existing research highlights the need for a model that integrates ontology-based methods, Spark streaming, LLM, Esper CEP[1], and Kafka to enhance accuracy and reduce costs. Current systems, human-based, optical camera-based, and satellite-based, lack this unified approach. Ontology-driven systems like FWI [5], Geo-ontologies[2], and Ackhaul-Aware Channel Allocation for Reliable Emergency X-network (BACAREX) aid monitoring and tracking, but none combine LLMs, Spark, CEP, and Kafka in one framework.

The purpose of this study is to create a framework within the SSN ontology that makes it easier for both automated systems and human users to connect with observation systems through SPARQL. This framework aims to give more complete information by making it easier to understand and use data and surveillance tools about forest fires. It utilizes Spark streaming, LLMs, and CEP tools (Esper and Kafka) to enhance the system's strength and expedite and improve decision-making. The proposed structure also outlines the key components essential for monitoring and controlling forest fires.

---

[1] https://www.espertech.com/esper/.
[2] https://www.ebi.ac.uk/ols4/ontologies/geo.

3 Experiment Details

This section includes complete experimental details of the proposed methodology, covering the proposed model architecture, dataset description, preprocessing, RDF conversion, ontology development, streaming generation using Spark, complex event processing, query processing through LLMs, and alert generation using SPARQL.

3.1 Proposed Model

The working process of the proposed system, shown in Fig. 1, integrates Spark Streaming, Ontology, Esper CEP Engine, Kafka, and LLMs. Kafka efficiently ingests and processes real-time data streams, which Spark Streaming and Esper CEP then analyze to identify complex patterns and anomalies. SPARQL queries on LLM and ontology facilitate advanced decision-making in forest fire management. The system can make better predictions and decisions by seamlessly combining real-time data processing, event correlation, and semantic reasoning.

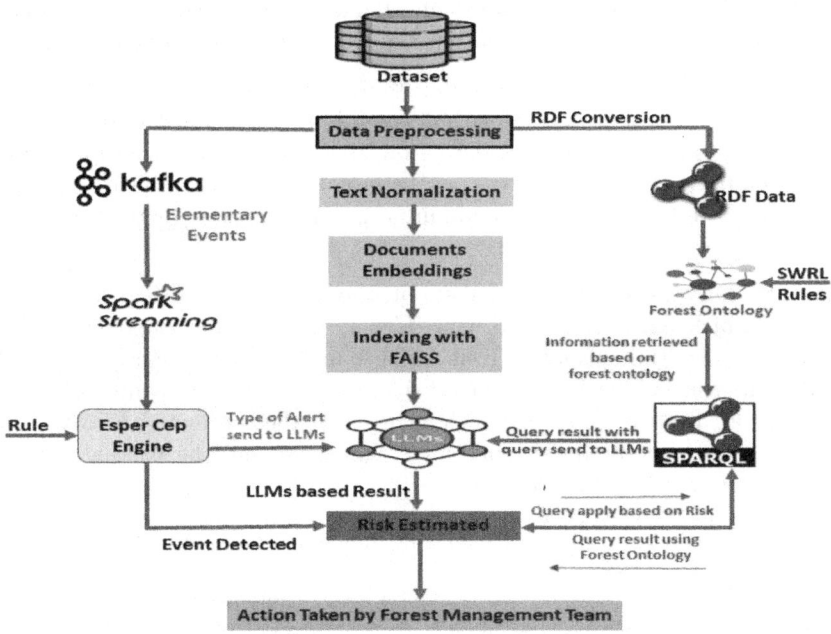

Fig. 1. Complete working process of model architecture.

Dataset. The Montesinho Park data set [11] provides geographical, temporal, and meteorological data for the analysis of forest fires in Portugal. It includes spatial coordinates, seasonal fire occurrences, and weather factors like temperature, humidity, wind, and rainfall. The burned area, a key variable, is skewed toward lower values, benefiting from logarithmic transformation for better modeling. This dataset helps in fire prevention and management.

Preprocessing. Data preprocessing involves several key steps to prepare the dataset for effective modeling. First, the "area" variable, which is often highly skewed, is logarithmically transformed to reduce skewness and improve model performance. Numerical features, including the Fine Fuel Moisture Code (FFMC), the Duff Moisture Code (DMC), the Drought Code (DC), the Initial Spread Index (ISI), the temperature, the relative humidity (RH), the wind, and the rain, are normalized using Z-score standardization to ensure it is a comparable scale. Categorical variables such as "month" and "day" are converted into a numerical format through one-hot encoding, enabling their use in ML algorithms. Outliers within the data set are identified and addressed using Z scores to prevent them from skewing the analysis. Additionally, correlation analysis is conducted to detect and eliminate redundant features, reducing dimensionality and improving model efficiency. We apply oversampling and undersampling techniques to address class imbalance in the "area" variable, particularly the disproportion between low and high fire-affected areas. These adjustments enhance the model's ability to recognize patterns across all classes more effectively.

RDF Conversion. The W3C standardized RDF stores information as triples subject-predicate-object. This makes it possible to manage the data in a structured and flexible way. Absorbs data from its structure, allowing for seamless integration between several sources. This study uses Python (rdflib)[3] to change a CSV data set to RDF format and add information to make it more interoperable and easier to query and link [12].

Ontology Development. The forest ontology is based on the upper-level SSN ontology, which was first created using Canadian forest fire guidelines [13] and then expanded with information from India to make it more accurate around the world. Enhancements include policies, fire prevention measures, and inter-agency coordination. The SWRL rules, which incorporate the parameters FFMC, DMC, and DC, create a comprehensive ontology. Table 1 presents the existing rules, and Fig. 2 illustrates the FWI structure.

Table 1. SWRL guidelines for managing human resources in firefighting [14].

SWRL Rules	Explanation
Specialized_Personnel(?p) ˆhas_Training(?p, "Firefighting") → deploy(?p)	Deploy trained firefighters for fire control.
Firefighter(?p) ˆhasIndividual Protective Equipment(?p, "Gear") → ensureSafety(?p)	Ensure firefighters wear protective gear.

Kafka to Spark Streaming. After preprocessing, the publish-subscribe module in Apache Kafka ingests the data as events. The event producer collects the

---
[3] https://rdflib.readthedocs.io/.

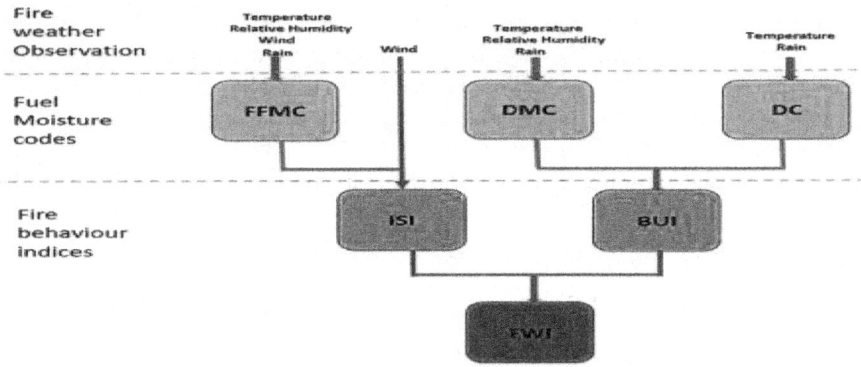

Fig. 2. Structure of the FWI System [5]

preprocessed data and posts it to Kafka topics. These topics assign the data to several brokers, including A, B, and C, who manage replication, distribution, and event scheduling. Kafka maintains fault tolerance by replicating the data across several brokers to guarantee dependability and reduce the possibility of data loss from broker failures. After the brokers safely hold the data, they send it to Apache Spark for streaming processing [15]. Processing data in small batches, as shown in Fig. 3, lets Spark do real-time analytics and make further adjustments to continuously stream data.

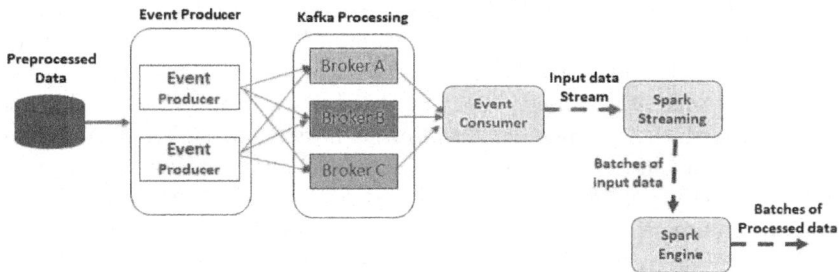

Fig. 3. Kafka to Spark Streaming model architecture

CEP Using Esper Engine. We use the Esper CEP engine to ingest Spark events or through streaming mode. This enables real-time processing to identify complex events based on predefined rules, which were developed in previous work [5] based on FWI. Esper is an open-source analytics platform that facilitates real-time data stream processing and pattern detection for deployment and application development.

In this case, the events correspond to metrics such as FFMC, DMC, etc. We analyze these parameters to identify complex events using risk estimation criteria derived from the Canadian Forest System guidelines. These guidelines help standardize this complex event detection based on predefined parameters. Next, we forward these complex events to the individuals listed below: 1) The LLMs receive the complex events for further interpretation and analysis. 2) The forest management team conducts an actionable risk assessment and decision-making, as shown in Fig. 4.

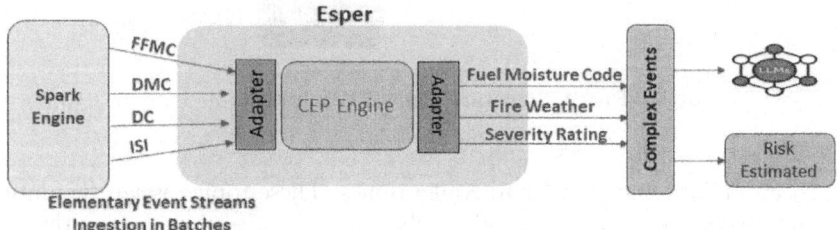

Fig. 4. Esper CEP-based stream processing

Query on LLMs Based on Complex Events. This study uses advanced approaches to process and understand forest fire data, resulting in a structured text format. The method begins with a thorough preprocessing stage that aims to remove errors and noise from the dataset, ensuring data quality and reliability. The cleaned data is text-normalized to standardize its format, resulting in a consistent dataset ready for subsequent analytical operations.

To extract semantic insights, the normalized text is translated into document embeddings with the cutting-edge all-MiniLM-L6-v2 model[4]. These embeddings capture the semantic essence of the text, which allows accurate information extraction based on similarity in context and relevance. This embedding process ensures that even seemingly insignificant information is retained for meaningful interpretation and retrieval.

These embeddings are organized by Facebook AI Similarity Search (FAISS) to allow for efficient retrieval. This powerful indexing structure enables quick access to semantically rich content, which is critical for facilitating rapid query processing.

A question-and-answer module, based on the Intel/dynamic_tinybert[5], forms the system's core and is integrated into a retrieval-based framework. When a query is submitted, the retriever finds the two most relevant documents in the index. The QA model uses these documents to generate exact replies. To improve

---

[4] https://huggingface.co/sentence-transformers/all-MiniLM-L6-v2.
[5] https://huggingface.co/Intel/dynamic_tinybert.

continuity and relevance, the system includes a conversation buffer memory, which allows it to keep context across many conversations.

The system provides real-time responsiveness, managing requests concerning ambient conditions, sensor data, or alarm occurrences quickly and precisely. The system uses Apache Spark and Esper to handle large-scale datasets and concurrent queries seamlessly. This design ensures robust scalability, allowing for consistent performance even during moments of high demand.

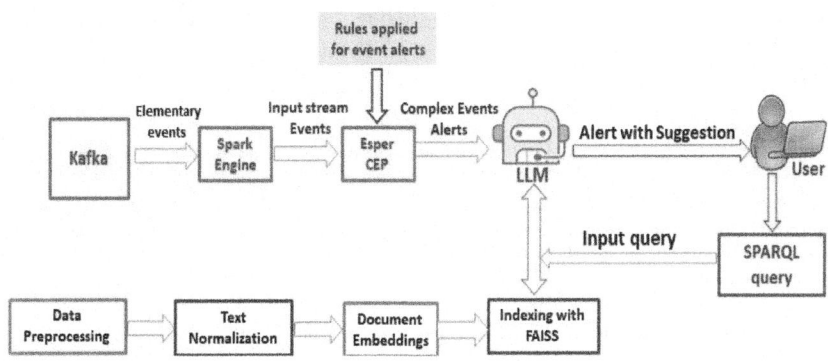

Fig. 5. Integration of Eper CEP engine framework with LLMs [14]

Integrating pre-trained LLMs with Apache Spark improves system efficiency, speed, adaptability, and scalability. As seen in Fig. 5, this architecture ensures that the system stays adaptable to changing fire safety requirements, advances in sensor technology, and environmental changes. By retaining its dependability and cost-effectiveness, the system continues to deliver accurate fire alarm detection, providing better protection for both lives and property.

Alerts Generation Using SPARQL Query on LLMs and Forest Ontology: The LLM and ontology-based alert systems produce nearly identical responses when performing the same SPARQL query. Due to its strong learning capacity, the LLM system provides additional insights and recommendations. The ontology system then uses its huge knowledge base to check the LLM outputs. This knowledge base has important things like geographic information, temperature data, and instructions for setting up sensors. This procedure guarantees that the LLM's recommendations are contextually appropriate and relevant. Results are shown in Table 2.

Table 2. SPARQL Query results on Forest Ontology and LLMs.

SPARQL Query
SELECT ?location ?windSpeed ?fireRisk WHERE { ?location rdf:type ex:ForestZone . ?location ex:hasWindSpeed ?windSpeed . ?location ex:hasFireRisk ?fireRisk . FILTER (?windSpeed > 20 && ?fireRisk = "High") }

Forest Ontology-based Results

Location	WindSpeed	FireRisk
ex:RegionA	25	High
ex:RegionB	22	High
ex:RegionC	24	High

Predictive Insight (LLMs)

South Forest, currently at a wind speed of 18 km/h and moderate fire risk, is expected to transition into the high-risk category within the next 24â€“48 h due to forecasted increases in wind speed and drier conditions.

Precautionary Measures (LLMs)

Resource Allocation Suggestion: North Forest has already surpassed critical risk thresholds, with a wind speed of 25 km/h and high fire risk. It is recommended to:
1) Deploy additional fire watch personnel.
2) Increase aerial surveillance using drones.
3) Set up rapid response teams in nearby regions to mitigate potential fire outbreaks.

Figure 6 shows how questions can be asked of both, i.e., ontology and LLM. It illustrates how LLMs help to locate information from an FAISS system. It provides benefits in understanding the workings of the proposed system to obtain the desired information. The Esper CEP engine, [16], also boosts the LLM model to give accurate results based on complex events sent to the LLMs.

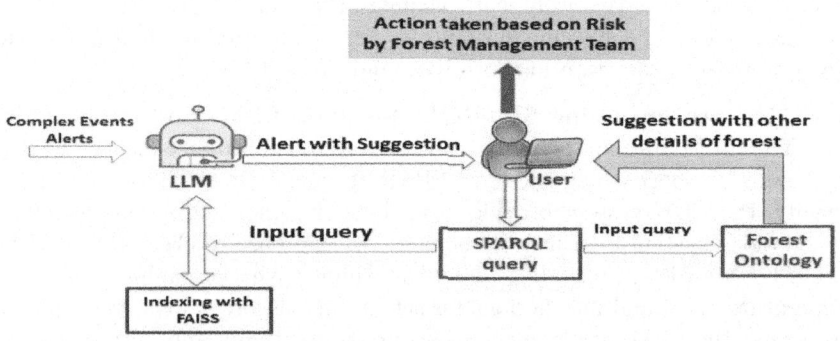

Fig. 6. Integration of Eper CEP engine framework with LLMs and Forest ontology

4 Results and Discussion

This research focuses on forest fire prevention and management by developing a DSS. The previously developed system uses sensor data generated from Montesinho Natural Park, which includes four weather parameters: temperature, humidity, wind speed, and precipitation. Survey reports are supplied to improve the model and offer context to sensor location mapping within the ontology. When examined with a reasoner in the Protege environment, the ontology provides exact recommendations based on SWRL rules. The model, which was constructed using the SSN ontology framework, included 43 rules in all. SPARQL queries were used to obtain extensive insights, with 97 queries being conducted to produce accurate results [5].

We expanded this work by enhancing the ontology, including forest management system documents of Australia, Canada, and India, resulting in a more accurate and comprehensive forest ontology. We integrate the Spark Streaming-based alert mechanisms and LLMs to automate the system and minimize the need for human intervention. Many conversations with different forest documentation sources helped create the LLMs, which give results and personalized advice on how to avoid certain situations. We have added more than 200 SWRL rules to the ontology to enhance Spark's streaming alert detection function. The LLMs receive these alarms and provide context-specific recommendations, thereby enhancing the system's accuracy. After that, the warnings are looked at in the ontology, which gives us useful information that helps the forest department make a decision [14].

This work integrates Esper CEP-based alert systems and LLMs to automate forest fire risk detection and response, reducing human intervention. Esper CEP continuously processes real-time data, detecting patterns like extreme weather conditions to trigger alerts. LLMs, trained in forest documentation, refine these alerts by suggesting preventive measures such as evacuations and resource allocation. The CEP engine, using SWRL rules, triggers alerts based on predefined conditions, such as high temperatures and strong winds. The forest department applies these alerts to an ontology-based system, ensuring rapid assessment and action. Over time, integrating more documentation enhances the forest ontology, improving predictive accuracy.

Forest Ontology Evaluation Based on Metrics Count. Several essential metrics are created using the ontology metrics count in Table 3, such as inheritance richness, relationship richness, class/relation ratio, axiom/class ratio, and inheritance richness, as mentioned in Table 4. An online tool is used to validate[6] ontologies based on these specified parameters. Attribute richness is the average number of attributes allocated to each class, indicating the effectiveness of the ontology's design and the importance of the data associated with each instance. Inheritance richness is defined as the average number of subclasses per class. It is a valuable metric for assessing how well the ontology organizes knowledge through its classification into classes and subclasses.

We can describe the ontology using a five-tuple model $O = (C, Dr, Sc, Re, Ind)$. In this model, C stands for the classes, Dr for the data properties or attributes, Sc for the subclasses, Re for the relationships between classes, and Ind for the individuals. Key evaluation metrics include attribute richness, which measures attribute descriptive detail; relationship richness, which assesses

---

[6] https://ontometrics.informatik.uni-rostock.de/ontologymetrics/.

Table 3. Metrics count value of Forest ontology and comparison with existing work.

Metrics	Value [5]	Value [14]	Proposed Value
Axiom Count	3813	5897	6875
Logical Axiom Count	1504	2129	2895
Object Property Count	143	454	546
Class Count	407	1007	1211
Data Property Count	142	198	265
Individual Count	287	587	625
Annotation Property Count	25	53	76
Object Property Domain	158	483	675
Object Property Range	280	281	298

the diversity of connections between classes; class richness, which indicates class distribution balance; and average population, which represents the number of individuals per class [14,17].

RR quantifies the depth of connections between concepts in an ontology. Equation 1 is used to determine this.

$$RR = \frac{|Prop|}{|Subclass| + |Prop|} \quad (1)$$

$|Prop|$ represents the number of properties. $|Subclass|$ represents the number of subclasses.

Equation 2 demonstrates how AR is determined by averaging the number of characteristics throughout the whole class.

$$AR = \frac{|Attribute|}{|Class|} \quad (2)$$

$|Attribute|$ represents the number of attributes. $|Class|$ represents the number of classes.

CR denotes the amount of real-world knowledge represented by the ontology. Equation 3 divides the number of classes having instances by the total number of classes:

$$CR = \frac{|Class\ with\_instance|}{|Class|} \quad (3)$$

$|Class\ with\_instance|$ represents the number of classes that have instances. $|Class|$ represents the total number of classes.

Equation 4 expresses how AP determines the average number of persons in each class.

$$AP = \frac{|Individual|}{|Class|} \quad (4)$$

$|Individual|$ represents the number of individuals. $|Class|$ represents the total number of classes.

Table 4 presents the computed values of several assessment metrics for the proposed Forest ontology, along with a comparison to existing work. The proposed ontology outperforms existing forest ontologies in terms of knowledge coverage.

Table 4. Comparison of the proposed forest ontology metric values with existing ontologies.

Schema Metrics	Proposed Value	Value [5]	Value [14]
Relationship Richness	0.9102	0.8093	0.8703
Class/Relation Ratio	0.9961	0.7948	0.9781
Inheritance Richness	0.9081	0.8587	0.8961
Axiom/Class Ratio	57.6612	42.1742	67.2622
Attribute Richness	0.9712	0.8289	0.9019

Query-Based Evaluation of LLM Model. Before judging how well LLMs work, we use the input query to get useful context from the FAISS database. The ordered context is based on cosine similarity, a measure of how relevant the context is to the question. The cosine similarity values range from -1 to +1, with values closer to +1 indicating stronger importance. After receiving the context, it is given to the LLMs, which construct a response using this information. We next assess the LLM's output by computing precision, recall, and F-measures and comparing the produced answer to a specified test response. Precision represents the proportion of relevant information in the response; recall assesses how much important information from the context was acquired; and the F-measure strikes a balance between precision and recall. High values for these indicators indicate that the LLM's answer is correct and contextually appropriate, hence verifying its performance shown in Table 5 *(Prec. = Precision, Rec. = Recall, and F-m. = F. measure)*.

Table 5. LLMs score based on queries.

Query	Score Similarity (FAISS)	Prec.	Rec.	F-m.
What steps should be made to minimize forest fires if Ignition Potential is exceptionally high and DMC Rules are impossible to follow?	0.8460	0.713	0.615	0.701
What precautions should be implemented when the ignition potential is extremely low, the DMC and DC rules are stringent and complex, the rate of fire spread is rapid, and both precipitation and wind speed are uncertain?	0.8651	0.670	0.740	0.710
What measures should be implemented when the ignition potential is moderate, the DMC and DC rules are stringent and extensive, the rate of fire spread is slow, and weather conditions such as rain and wind speed remain uncertain?	0.8217	0.847	0.614	0.747

Query Performance Analysis on Different Models. The query runs over the LLMs, and the ontology is shown in Fig. 7 as a separate part of the suggested work. However, we ran comparable queries using the CEP engine and the LLM's ontology. The processing times for Q2 and Q5 are very different because the questions are more complex. It takes longer for the combined models to finish the same queries than it does for separate models.

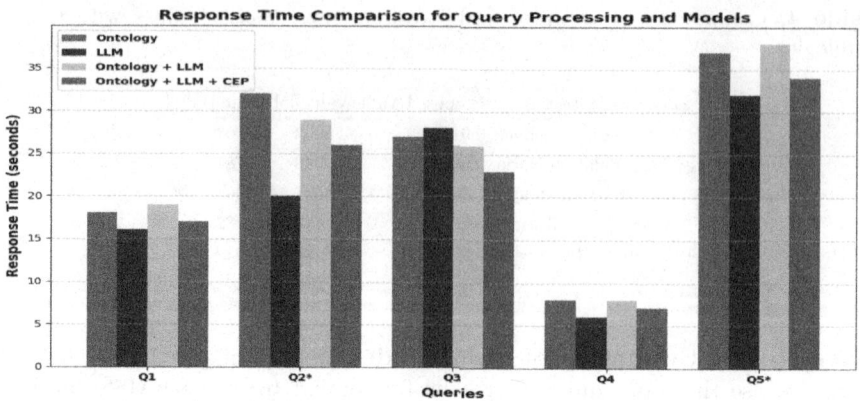

Fig. 7. Query performance analysis on different models with combination without combination

CEP Analysis Based on Event Processing Time. Table 6 displays the execution timings for processing events using the Esper CEP Engine, considering various criteria and event numbers. We assess each rule's performance using an increasing number of occurrences, ranging from 10,000 to 35,000. The data show variances in execution time, reflecting changes in computational complexity or resource use between the rules. For example, Rule 5 exhibits considerable oscillations, such as a large decrease in execution time at 25,000 events, but other rules execute pretty consistently over the dataset. This research helps identify potential enhancements for event processing procedures.

Table 6. Event execution time in Esper CEP Engine, representing the time taken to process and detect patterns in real-time data.

Events	Rule 1	Rule 2	Rule 3	Rule 4	Rule 5
10000	1.65 s	1.66 s	1.75 s	1.84 s	1.74 s
15000	2.58 s	2.25 s	2.54 s	2.24 s	2.36 s
20000	2.76 s	2.87 s	2.84 s	2.62 s	3.58 s
25000	3.36 s	2.52 s	3.12 s	3.10 s	2.19 s
30000	2.87 s	3.43 s	3.68 s	2.77 s	2.63 s
35000	2.75 s	2.82 s	2.72 s	2.61 s	2.72 s

Comparative Analysis with Existing Work. Table 7 highlights key studies using semantic technologies like OWL, SPARQL, and SSN ontology to improve fire emergency management. Methods range from ontology-based reasoning and knowledge fusion to real-time sensor data processing. The proposed system advances this by integrating streaming, CEP, and AI for real-time, explainable fire risk detection and alerts.

Table 7. Comparison with existing work.

Reference	Objective	Approach	Outcome
Masa et al. [18]	Semantic reasoning for fire emergencies	OWL, SPARQL, Pellet rules	Situational awareness
Marotta et al. [19]	Unified wildfire knowledge base	RDF fusion, API integration	Improved analytics
Kirubakaran et al. [20]	Real-time DSS from sensors	SSN, SWRL, Jena reasoning	Real-time alerts
Proposed	Intelligent DSS with CEP	SSN, CEP, LLMs, FAISS, Spark	Automated fire risk detection

5 Conclusion and Future Scope

This study proposes an integrated DSS for forest fire management by combining SSN ontology, CEP, LLMs, and big data tools like Kafka and Spark Streaming. Kafka enables distributed data ingestion, Spark performs real-time analytics, and the CEP engine detects fire-related events using SWRL rules. LLMs and SPARQL queries over the ontology support automated reasoning, real-time alerts, and context-aware recommendations, reducing the need for manual intervention.

Evaluation was performed using ontology metrics, SPARQL query response time, LLM scores (precision, recall, and F1), and CEP event execution time. The results indicate enhanced knowledge coverage, system scalability, and accurate fire risk detection. However, the system's effectiveness is limited by the availability and reliability of WSN data in remote areas.

In future work, we will add fire protocols specific to different regions, adjust SWRL rules, and include advanced ML models to make detection more accurate, adaptable, and capable of real-time decision-making.

References

1. National Interagency Fire Center. https://www.nifc.gov/. Accessed 28 Mar 2025
2. Sagar, N., et al.: Forest fire dynamics in India (2005–2022): unveiling climatic impacts, spatial patterns, and interface with anthrax incidence. Ecol. Ind. **166**, 112454 (2024). https://doi.org/10.1016/j.ecolind.2024.112454
3. Barber, Q.E., et al.: The Canadian fire spread dataset. Sci. Data **11**(1), 764 (2024). https://doi.org/10.1038/s41597-024-03436-4
4. Delouee, M. L., Pernes, D. G., Degeler, V., Koldehofe, B.: Towards federated LLM-powered cep rule generation and refinement. In: Proceedings of the 18th ACM International Conference on Distributed and Event-based Systems, pp. 185–186 (2024). https://doi.org/10.1145/3629104.3672429
5. Chandra, R., Agarwal, S., Singh, N.: Semantic sensor network ontology-based decision support system for forest fire management. Ecol. Inf. **72**, 101821 (2022). https://doi.org/10.1016/j.ecoinf.2022.101821
6. Ginkal, P.M., et al.: Forest fire detection using AI. Grenze Int. J. Eng. Technol. (GIJET) **10** (2024)

7. Ibraheem, M.K.I., Mohamed, M.B., Fakhfakh, A.: Forest defender fusion system for early detection of forest fires. Computers **13**(2), 36 (2024). https://doi.org/10.3390/computers13020036
8. Su, H., et al.: A novel framework for identifying causes of forest fire events using environmental and temporal characteristics of the ignition point in fire footprint. Ecol. Ind. **160**, 111899 (2024). https://doi.org/10.1016/j.ecolind.2024.111899
9. Hamed, N., Rana, O., Orozco Ter Wengel, P., Goossens, B., Perera, C.: FOODS: ontology-based knowledge graphs for forest observatories. ACM J. Comput. Sustain. Soc. **3**(1), 1–42 (2025). https://doi.org/10.1145/37076
10. Noroozi, F., Ghanbarian, G., Safaeian, R., Pourghasemi, H.R.: Forest fire mapping: a comparison between GIS-based random forest and Bayesian models. Nat. Hazards, 1–24 (2024). https://doi.org/10.1007/s11069-024-06457-9
11. "Forest Fires Data Set" https://archive.ics.uci.edu/dataset/162/forest+fires. Accessed 26 Mar 2025
12. Haider, N., Hossain, F., et al.: CSV2RDF: generating RDF data from CSV file using semantic web technologies. J. Theor. Appl. Inf. Technol. **96**(20), 6889–6902 (2018)
13. "Canadian Wildland Fire Information System: FWI System Summary" https://cwfis.cfs.nrcan.gc.ca/background/summary/fwi. Accessed 15 Mar 2025
14. Chandra, R., Kumar, S.S., Patra, R., Agarwal, S.: Decision support system for forest fire management using ontology with big data and LLMS. arXiv preprint: arXiv:2405.11346 (2024)
15. Ichinose, A., Takefusa, A., Nakada, H., Oguchi, M.: A study of a video analysis framework using kafka and spark streaming. In: 2017 IEEE International Conference on Big Data (Big Data), pp. 2396–240. IEEE (2017). https://doi.org/10.1109/BigData.2017.8258195
16. Ortiz, G., Bazan-Muñoz, A., Lamersdorf, W., Garcia-de Prado, A.: Evaluating the integration of Esper complex event processing engine and message brokers. PeerJ Comput. Sci. **9**, e1437 (2023). https://doi.org/10.7717/peerj-cs.1437
17. Lourdusamy, R., John, A.: A review on metrics for ontology evaluation. In: 2018 2nd International Conference on Inventive Systems and Control (ICISC), pp. 1415–142 (2018). https://doi.org/10.1109/ICISC.2018.8399041
18. Masa, P., Meditskos, G., Kintzios, S., Vrochidis, S., Kompatsiaris, I.: Ontology-based modelling and reasoning for forest fire emergencies in resilient societies. In: Proceedings of the 12th Hellenic Conference on Artificial Intelligence, pp. 1–9 (2022). https://doi.org/10.1145/3549737.354976
19. Marotta, S.M., Masucci, V., Kontogiannis, S., Avgerinakis, K.: From unified ontology to knowledge base: data fusion for enhanced wildfire management. In: International Conference on WorldS4, pp. 131–150. Springer Nature Singapore, Singapore (2024). https://doi.org/10.1007/978-981-97-8695-4_13
20. Kirubakaran, S., Hussein, R.R., Ramakrishna, T.V., Reddy, G.V., CS, S.: Semantic sensor network ontology for forest fire management decision support. In: 2025 International Conference on Intelligent Control, Computing and Communications (IC3), pp. 1287–1292. IEEE (2025). https://doi.org/10.1109/IC363308.2025.10957683

Table Annotation Utilizing Large Language Model and Knowledge Graph

Ying Zhang[1,2] and Mizuho Iwaihara[1(✉)]

[1] Graduate School of Information, Production, and Systems, Waseda University, Kitakyushu 808-0135, Japan
yingzhang@asagi.waseda.jp, iwaihara@waseda.jp
[2] School of Mechatronic Engineering and Automation, Shanghai University, 333 Nanchen, Dachang, Baoshan, Shanghai, China

Abstract. Tabular data are organized in a structured format that contains rich information. However, challenges exist in interpreting this style of data in an easily understandable way. The annotation of semantic and atomic types of the columns, as well as the relationships between different columns, can assist both users and machines in comprehending tabular data across various scenarios. Existing deep learning methods rely on large amounts of training samples per type and suffer from long running times. In this paper, we explore the utilization of large language models (LLMs) and Knowledge Graph (KG) for column type annotation (CTA) and column property annotation (CPA). External knowledge resources, such as DBpedia and CaliGraph, are used to preliminarily refine the label set for selection by the LLMs, while noisy labels in the candidates are removed. A hierarchical search is performed by retrieving concrete entities matched to resources from the subgraph of the golden labels. Moreover, we evaluate both zero-shot and few-shot scenarios using a well-crafted prompt. Specifically, we compare the approaches of selecting few-shot demonstrations either randomly selected from the training set or by leveraging vector similarities between test and few-shot samples. As the experimental results show, our proposed method outperforms the baseline on the SOTABv2 benchmark for both CTA and CPA tasks.

Keywords: Table Annotation · Tabular data · Knowledge Graph · Large Language Model

1 Introduction

Tabular data are organized in rows and columns with semantic information in a structured format. Precise annotation of tabular data can help both humans and machines interpret the complex structure and comprehend the overall content of the tables in web search, question answering and many other processes. Table Annotation (TA) includes embedding comments, descriptions, labels, or annotations in various sections of the table (such as the title, rows, columns, or cells), which help identify the real-world concepts that capture the semantics of the data [7]. One important subtask of TA is Column

Type Annotation (CTA), which focuses on identifying the most appropriate semantic type that characterizes the majority of entities within a column [1]. Another subtask is Column Property Annotation (CPA), which infers to describing the semantic relationships between columns (between the leftmost column and other associated columns) to capture their dependencies [4].

Knowledge graph-based methods leverage knowledge graphs (KGs) to annotate table columns and often relies on rule-based or statistical approaches like TF-IDF and the semantic alignment between tables and KGs. A second line of work has employed learning-based techniques for TA tasks. These techniques utilize a large corpus of columns annotated with column types to either train models from scratch or fine-tune pre-trained language models (PLMs). However, most methods need large amounts of task-specific training data to achieve good performance, and costs for manually labeled data are often intensive. The advances in generative large language models (LLMs) provide an opportunity to address the aforementioned problems. They are effective in handling missing values or irregular phrases in tabular data. This paper investigates novel use of LLMs for TA tasks. In-context learning with demonstrations is used to provide few-shot examples to guide LLMs in more precise type/property prediction. External knowledge resources like DBpedia [10] and CaliGraph [11] are searched to retrieve candidate classes for the label set in prompts. Ablation experiments are conducted to validate the effectiveness of each component, and the final performance is compared with the baselines on both CTA and CPA tasks.

The paper is organized as follows: Sect. 2 reviews the related work. Assumptions and definitions are formulated in Sect. 3, and the details of the proposed method are described in Sect. 4. The method is evaluated with the SOTABv2 dataset, and experimental results are presented in Sect. 5. Finally, conclusions are drawn in Sect. 6.

2 Related Work

KG Lookup-Based Methods. TA task is transformed into a matching task between table cells and knowledge graph (KG) resources. KG lookup-based methods start by generating an initial set of candidate entities using a SPARQL endpoint lookup service. These candidates are then evaluated based on metrics associated with table elements, including cell values. Feature engineering-based approaches use statistical and lexical features from table rows and columns to train machine learning models such as random forest (RF) and K-nearest neighbors (KNN). The performance of these models is heavily dependent on the quality and quantity of labeled training data. Neumaier et al. [5] enhance numerical CTA by labeling columns with contextual information, utilizing a hierarchical KG from DBpedia. They employ KNN for predictions based on statistical features derived from this background KG.

Deep Learning-Based Method. Deep learning has advanced significantly across many fields, thanks to large datasets and powerful computing. TURL [8] uses BERT to encode table elements and employs a visibility matrix to refine interactions between cells, columns, and rows. Both open-source and closed-source LLMs are applied to TA tasks through fine-tuning and prompt engineering. TableLlama [6] introduces an open-source LLM fine-tuned for table-based tasks using the TableInstruct dataset, which includes diverse real-world tables and tasks for instruction tuning. The method in [2] employs in-context learning in the prompt to improve LLM prediction accuracy by providing task-specific examples during inference.

3 Problem Definition and Background

Problem Formulation. The table annotation (TA) task is divided into two subtasks: column type annotation (CTA) and column property annotation (CPA). Notations are listed in Table 1. Assume that a column ki ($1 < i \leq m$) in a given table T is given. Vocabularies C* and P*, respectively, represent the column type set for column ki and the property set for column k1 and ki, respectively. CTA refers to the task of determining an ontology type M(ki, ci) ∈ C* that represents the relationship between column k1 and ki.

Table 1. Notation

Symbol	Description
Knowledge graph	
$E = \{e_1, e_2, ..., e_w\}$	A set of resources in a KG
$P = \{p_1, p_2, ..., p_x\}$	A set of ontology properties in a KG
$C = \{a_1, a_2, ..., a_y\}$	A set of ontology classes in a KG
$L = \{l_1, l_2, ..., l_z\}$	A set of resource labels in a KG
$\{(e, \textit{rdfs:label}, l) | e \in E, l \in L\} \in K$	A set of resource-label pairs in a KG
Table	
$T = \{T_1, T_2, ..., T_n\}$	A set of tables
$T^t = \{c_1, c_2, ..., c_m\}$	Columns in a table
$c_i^t = \{c_{i,1}^t, c_{i,2}^t, ..., c_{i,3}^t\}$	Cell values in a column
$S = \{s_1, s_2, ..., s_k\}$	A set of labels for possible types
$s^t = \{s_1^t, s_2^t, ..., s_m^t\}$	Column types in a table
M	Large Language Model

4 Methodology

As shown in Fig. 1, the methodology leverages both KG-retrieved information and in-context learning demonstrations to guide the LLM in generating accurate answers.

Retrieving Candidate Classes From KG. Given the table cell entities and the original golden label set, the process begins with a fuzzy matching step that restricts the

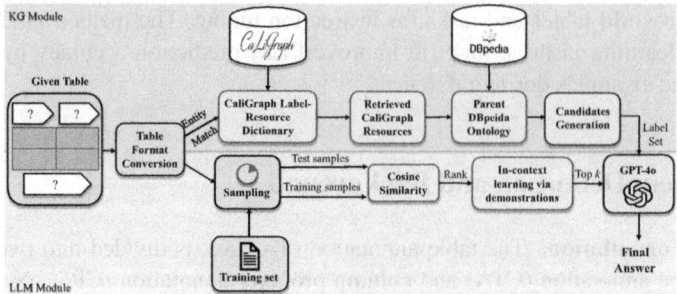

Fig. 1. Proposed Framework.

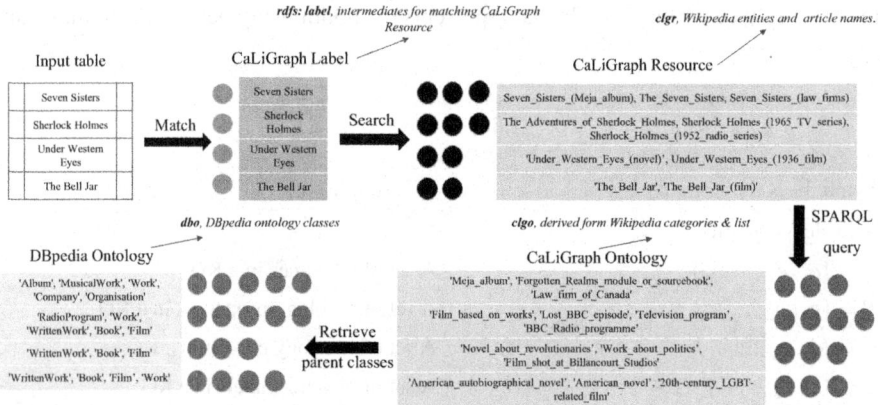

Fig. 2. Workflow of KG Module

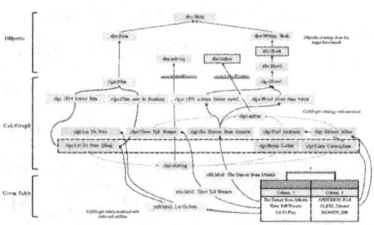

Fig. 3. An example of a tree structure of DBpedia and CaLiGraph. The ontology classes of DBpedia are at the top of the tree structure, representing the classes to be classified. In the middle are the resources (clgr) and ontologies (clgo) from CaLiGraph. The relationships between nodes and their hierarchical levels are illustrated. The rdfs:label serves as an intermediary to match table entities with CaLiGraph resources.

search to paths reachable from the golden labels within the DBpedia graph and, where applicable, the broader Wikipedia dataset. For each cell entity, candidate resources are retrieved using rdfs:label mappings from the CaLiGraph Label-Resource Dictionary [13]. To mitigate the issue of ambiguous results, the retrieved resources along with the table cell entities, are converted into dense vector representations for capturing semantic similarity at the sentence or phrase level. As depicted in Fig. 2 and Fig. 3, the process of type/property-value pairs lookup begins with retrieving parent ontology classes and relevant property-value pairs for the retained resources through SPARQL queries.

In the CPA tasks, object column entities, such as "ANDERSON, Poul" in Fig. 3, are matched against these retrieved property-value pairs. For example, for the entity "The Dancer from Atlantis," the algorithm retrieves its dbo:author property linking it to "Poul Anderson" from the CaLiGraph ontology. To ensure semantic alignment, cosine similarity filtering is applied between the object column values (e.g., "ANDERSON, Poul") and the retrieved property values.

This iterative refinement step addresses challenges related to entity ambiguity and helps reduce the candidate space to the most relevant classes or properties. In Fig. 2, the input table entities, such as "Seven Sisters" or "Sherlock Holmes," are initially matched to their corresponding resources in the CaLiGraph Label-Resource Dictionary [13], followed by retrieval of their parent classes (e.g., dbo:Work, dbo:WrittenWork, or dbo:Book) from DBpedia. Additionally, Fig. 3 illustrates how parent classes, such as dbo:Book or dbo:Film, are derived from hierarchical relationships and property mappings in the KG, which enables more granular refinement of candidates. The whole KG retrieval algorithm is illustrated in Algorithm 1.

Algorithm 1: KG retrieval algorithm for CTA and CPA tasks

Input: $DBO\_C_{golden\ labels} = \{dbo_{c1}, dbo_{c2}, ..., dbo_{cn}\}$, $DBO\_P_{golden\ labels} = \{dbo_{p1}, dbo_{p2}, ..., dbo_{pn}\}$, Cell entities $c_{i,k} = \{c_{i,1}, c_{i,2}, ..., c_{i,k}\}$, $c_{j,k} = \{c_{j,1}, c_{j,2}, ..., c_{j,k}\}$;
  Subgraph $SG \leftarrow$ the paths reachable from graph of the $DBO_{golden\ labels}$
Output: $DBO_{candidates} = \{dbo_{c1}, dbo_{c2}, ..., dbo_{cm}\}$

CPA:
1. for each $c_{i,k} \in \{c_{i,1}, c_{i,2}, ..., c_{i,k}\}$
2.   match *label* and retrieve *clgr* from *SG*
3.   if retrieved *clgr* set $= \emptyset$
4.   retrieve resources from Wikipeida
5.   Convert $c_{i,k}$, *clgr* into embeddings
6.   if $cosSim(\vec{c_{i,k}}, \vec{clgr}) > \theta$
7.   *clgr* set $\leftarrow$ *clgr*
8.   for *clgr* in *clgr* set
9.   retrieve parent *dbo_p* - *clgr* (*value*) pairs
10.   for each $c_{j,k} \in \{c_{j,1}, c_{j,2}, ..., c_{j,k}\}$
11.   convert $c_{j,k}$, *clgr* (*value*) into embeddings
12.   if $cosSim(\vec{c_{i,k}}, clgr\ (value)) > \theta$
13.   *dbo_p* set $\leftarrow$ *dbo_p*
14. $DBO_{candidates} \leftarrow \{dbo\_p\} \cap DBO\_P_{golden\ labels}$
15. return $DBO\_P_{candidates}$

CTA:
1. for each $c_{i,k} \in \{c_{i,1}, c_{i,2}, ..., c_{i,k}\}$
2.   match *label* and retrieve *clgr* from *SG*
3.   if retrieved *clgr* set $= \emptyset$
4.   retrieve resources from Wikipeida
5.   convert $c_{i,k}$, *clgr* into embeddings
6.   if $cosSim(\vec{c_{i,k}}, \vec{clgr}) > \theta$
7.   *clgr* set $\leftarrow$ *clgr*
8.   for *clgr* in *clgr* set
9.   retrieve parent *dbo_c*
10. $DBO_{candidates} \leftarrow \{dbo\_c\} \cap DBO\_C_{golden\ labels}$
11. return $DBO\_C_{candidates}$

In Context Learning via Demonstrations. Unlike traditional fine-tuning, which involves training the model with new data, ICL allows the model to adapt on the fly by simply observing task demonstrations. The test samples without labels, along with the available training samples, are embedded for selecting few-shot examples by comparing the similarity between vectors. The converted vectors capture semantic information from the text. For each test sample, we compute the cosine similarity between the embedded test and training samples, then select the top-k most similar training samples to construct the demonstration set.

Final Answer Generation by LLM. The structured prompts for CTA and CPA tasks are organized into three roles: System, User, and AI. The System message defines the task (e.g., annotating column types or relationships) and provides instructions. The User message issues specific annotation commands, while the AI message either gives correct labels for training samples or predicts labels for test samples. The structured prompts enforce a restriction where answers must be strictly derived from the predefined label set in the dataset, which ensure the adherence to task constraints.

5 Dataset and Evaluations

The SOTAB_v2_DBpedia dataset [12] leverages DBpedia types and properties to provide semantic labels for columns. We compare our model with the baselines using the SOTABV2 dataset: DODUO [9] and model in [2] for CPA. For the comparison model, we fine-tuned the model DODUO for 30 epochs with a learning rate of 5e-5, a batch size of 16, and a maximum sequence length of 32. For the proposed method, the temperature was fixed at 0, and the top-k parameter was set to 5.

As shown in Table 2 and Table 3, for CTA task, the best-performing model uses five-shot ICL learning with candidates selected from KGs. When using randomly selected one-shot examples to guide the annotation, precision consistently decreases compared to the zero-shot setting for both tasks, which infers that highly random examples not only fail to enhance performance but may also introduce noise, which actually give negative impact to the annotation process. Consequently, the selection of examples must be carried out carefully. When similarity-based examples are employed, the Micro-F1 scores for CTA and CPA tasks increase by approximately 5% and 3%, respectively. Furthermore, incorporating more examples, such as five-shot, results in a 7% improvement in Micro-F1 scores for both tasks compared to the zero-shot condition. At this stage, the performance on the CPA task becomes comparable to the baseline. Moreover, by retrieving external knowledge from KGs to generate candidates, rather than utilizing all golden types and properties from the dataset as input, the label set is further refined. For the CPA task, simply using column-wise formatting with Markdown still does not yield optimal results compared to the model in [2]. Therefore, an additional set of comparative experiments was conducted to validate the input formatting.

As shown in Table 4, the shift from a column-wise to a table-wise approach, combined with HTML-based table rendering, delivers measurable improvements. The initial column-wise Markdown implementation surpassed baseline performance, but these refinements achieve even greater gains. The 5-shot setting achieves the best performance across all metrics for both CTA (90.46 Micro-F1) and CPA (96.02 Micro-F1), consistently outperforming both the DODUO baseline and lower-shot configurations. By retrieving external knowledge from KGs to generate candidates (instead of using all golden types/properties from the dataset as input), the label set becomes more precise. In the 5-shot scenario, this refinement helps the proposed method outperform the best CPA baseline.

Table 2. Ablation Study on CTA (column-wise) Task with Markdown Input Format.

Model	Recall	Precision	Micro-F1	Macro-F1
DODUO (Baseline)	87.81	89.33	87.81	87.82
GPT-4o + Zero-shot	71.28	74.53	74.45	70.40
GPT-4o + One-shot (random examples)	72.84	73.10	74.74	71.34
LLaMA 3.1_8B + Five shot (ICL via demonstrations) + Candidates retrieved from KG	74.90	77.22	84.43	74.95
GPT-4o + One-shot (ICL via demonstrations) + 46 DBpedia ontology types	77.36	78.36	79.15	76.13
GPT-4o + Five-shot (ICL via demonstrations) + 46 DBpedia ontology types	77.45	78.14	81.35	76.37
GPT-4o + Five-shot (ICL via demonstrations) + Candidates retrieved from KG	**91.31**	**91.71**	**89.13**	**90.46**

Table 3. Ablation Study on CPA (column-pair) Task with Markdown Input Format.

Model	Recall	Precision	Micro-F1	Macro-F1
DODUO (Baseline)	90.38	91.07	90.38	89.12
Model in [2] / 5-shot	92.56	93.16	**93.73**	92.24
GPT-4o + Zero-shot	79.77	81.83	84.18	79.18
GPT-4o + One-shot (random examples)	78.28	80.13	87.17	77.86
GPT-4o + One-shot (ICL via demonstrations) + 49 DBpedia ontology properties	84.11	85.43	90.27	83.85
GPT-4o + Five-shot (ICL via demonstrations) + 49 DBpedia ontology properties	89.69	91.68	90.38	89.66
GPT-4o + Five-shot (ICL via demonstrations) + Candidates retrieved from KG	**93.19**	**94.55**	93.58	**93.23**

Table 4. Comparison of Table-wise Input Formats (HTML) Across Different Shot Settings (0-shot, 1-shot, 5-shot) for CTA and CPA Tasks Using GPT-4o.

CTA task / x-shot	Recall	Precision	Micro-F1	Macro-F1
0-shot	89.01	90.37	88.84	89.66
1-shot	90.57	91.66	89.43	90.56
5-shot	**92.01**	**91.74**	**90.46**	**91.49**
CPA task / x-shot	Recall	Precision	Micro-F1	Macro-F1
0-shot	89.98	91.14	91.59	90.10
1-shot	94.22	95.53	95.46	94.53
5-shot	**94.88**	**96.10**	**96.02**	**95.17**

6 Conclusion

In this paper, we discussed LLM prediction on table annotation tasks of column type annotation and column property annotation. In-context learning with few-shot examples shows improvement on LLM prediction accuracy. External knowledge from DBpedia and CaliGraph is incorporated to generate refined few-shot candidates. Compared to the strongest baseline, our model demonstrates 2.65% (CTA) and 2.29% (CPA) improvements in Micro-F1 scores.

Acknowledgement. This work was in part supported by JSPS KAKENHI Grants Number 22K12044 and 25K03230.

References

1. Chen, J., Jiménez-Ruiz, E., Horrocks, I., Sutton, C.: ColNet: embedding the semantics of web tables for column type prediction. Proc. AAAI Conf. Artif. Intell. **33**(01), 29–36 (2019)
2. Korini, K., Bizer, C.: Column property annotation using large language models. In: European Semantic Web Conference, Cham: Springer Nature Switzerland, pp. 61–70(2024)
3. Pham, M., Alse, S., Knoblock, C.A., Szekely, P.: Semantic Labeling: A Domain-Independent Approach. In: Groth, P., et al. (eds.) ISWC 2016. LNCS, vol. 9981, pp. 446–462. Springer, Cham (2016). https://doi.org/10.1007/978-3-319-46523-4_27
4. Nguyen, P., Yamada, I., Kertkeidkachorn, N., Ichise, R., Takeda, H.: SemTab 2021: tabular data annotation with MTab tool. In: SemTab@ ISWC, pp. 92–101 (2021)
5. Neumaier, S., Umbrich, J., Parreira, J.X., Polleres, A.: Multi-level semantic labelling of numerical values. In: 15th International Semantic Web Conference (ISWC), pp. 428–445 (2016)
6. Zhang, T., Yue, X., Li, Y., Sun,H.: TableLlama: towards open large generalist models for tables. arXiv:2311.09206 (2024)
7. Khurana, U., Galhotra,S.: Semantic Annotation for Tabular Data (2019)
8. Deng, X., Sun, H., Lees, A., Wu, Y., Yu, C.: TURL: table understanding through representation learning. Proc. VLDB Endowment **14**(3), 307–319 (2020)

9. Suhara, Y., et al.: annotating columns with pre-trained language models. In: Proceedings of the 2022 International Conference on Management of Data (SIGMOD '22), pp. 1493–1503 (2022)
10. http://dbpedia.org/sparql
11. http://caligraph.org/sparql
12. https://webdatacommons.org/structureddata/sotab/v2/
13. https://zenodo.org/record/4662515/files/caligraph-instances_labels.nt.bz2?download=1

Improving Software Security Through a LLM-Based Vulnerability Detection Model

Syeda Sadia Alam[1,2,3,4], Mst Shapna Akter[1,2,3,4], and Alfredo Cuzzocrea[1,2,3,4](✉)

[1] Department of CS and Engineering, Metropolitan University, Sylhet, Bangladesh
[2] Department of CS and Engineering, Oakland University, Rochester, MI, USA
akter@oakland.edu
[3] iDEA Lab, University of Calabria, Rende, Italy
alfredo.cuzzocrea@unical.it
[4] Department of CS, University of Paris City, Paris, France

Abstract. The significance of early vulnerability identification in ensuring security during software development cannot be denied. In this research, we introduce *CWEpredBELL*, a unique automated vulnerability prediction method that makes use of a modified pre-trained language model derived from *CodeBERT*. With a binary classification layer, an improved optimizer, and a fine-tuned loss function to boost model performance, our method is especially tailored for identifying vulnerabilities in source code. We used cross-validation techniques and the *Local Interpretable Model-Agnostic Explanations* (LIME) approach to identify particular lines of error in the source code. The experimental comparison demonstrates that CWEpredBELL is an effective method of automatically identifying vulnerabilities.

Keywords: Vulnerability Identification · CodeBert · Cybersecurity · SWE · LLM · NLP · LIME

1 Introduction

A *vulnerability* is recognized as a deficiency in an information system, security protocols, internal controls, or execution that could be exploited by a threat source [1]. A vulnerability constitutes a substantial threat in software. Vulnerabilities can result in data breaches, malware infestations, and the disruption of essential services. Identifying and eliminating vulnerabilities early in the *Software Development LifeCycle* (SDLC) reduces organizational expenses [2]. Additionally, effective techniques are necessary to identify and fix security vulnerabilities in the development life cycle, when remediation is comparatively inexpensive and easy to understand. The approach known as *Vulnerability Prediction* (VP) facilitates the early identification and mitigation of security vulnerabilities, primarily aiming to identify possible vulnerable software components [3]. The numerous

---

This research has been done in the context of the Excellence Chair in Big Data Management and Engineering at University of Paris City, Paris, France.

© The Author(s), under exclusive license to Springer Nature Switzerland AG 2026
R. Wrembel et al. (Eds.): DEXA 2025, LNCS 16046, pp. 122–129, 2026.
https://doi.org/10.1007/978-3-032-02049-9_9

types of vulnerability make it difficult to eliminate, compared to other bugs in code. Each *Common Weakness Enumeration* (CWE) denotes a unique group of vulnerabilities characterized by specific features, code semantics, and patterns. Addressing all vulnerabilities under just one label through a binary classification method, might excessively simplify the issue since it fails the particulars and contextual specifics inherent to each CWE. Consequently, a singular binary classifier depends solely on superficial textual patterns instead of comprehending the complexities of each vulnerability type [4]. Recent advancements in computing capacity, data accessibility, and novel algorithms have resulted in significant progress in *Artificial Intelligence* (AI) and *Machine Learning* (ML) over the past decade. Several techniques of AI or ML have grown prominent in daily life, including automation, *Natural Language Processing* (NLP), predicting, and security flaws [5,6]. Leveraging data-driven approaches such as AI and ML for automated, intelligent source code analysis falls into several challenges. Key among these are accurately representing the code for effective processing by ML algorithms and identifying vulnerabilities within the code itself [7]. In this research, we have proposed a comprehensive LLM based model termed *CWEpredBELL* to precisely identify the CWE vulnerability from the source code. The CWE vulnerability contains various kinds of vulnerabilities, such as buffer errors, buffer overflow, Pointer miscalculation, null pointers etc. [7]. Our model demonstrated outstanding performance across all metrics, accurately identifying vulnerabilities within the source code.

Our methodology demonstrates a significant contribution and optimal efficiency in vulnerability prediction, outperforming the performance of existing approaches.

2 CWEpredBELL: Models, Methods, Methodologies

2.1 Dataset Description

In order to train our customized model, we downloaded the dataset from the publicly accessible link https://osf.io/ d45bw/. The dataset is CWE and it widely utilized in the research of vulnerability prediction. It contains several categories of vulnerability such as CWE-119,CWE-120,CWE-476, CWE-469 and CWE-others. In our experiments, we have analyzed the CWE-other category which contains various types of vulnerabilities that are not specifically addressed by another group of CWE dataset. Besides, the CWE-120 and CWE-469 have also been applied to validate the generalizability of the proposed architecture. The CWE-120 indicates the vulnerability of buffer overflows and CWE-469 identify the vulnerabilities that are produced from the subtraction of pointers to determine size.

The dataset is divided into two levels: TRUE, indicating available vulnerability in source code, and FALSE indicating non-vulnerability in the source code. In our research, we have used 19,500 samples, with 10,500 allocated for training and 9,000 designated for testing.

2.2 Overview of Proposed Methodology

We began by gathering a suitable dataset, as outlined in [8], and applied essential pre-processing techniques to address the imbalance in the data. The majority of the source code samples in the dataset were classified as non-vulnerable (FALSE), necessitating specific adjustments to ensure more balanced and accurate model training. To construct an effective vulnerability detection model, we applied several data balancing techniques to balance the True and False classes of our dataset. Furthermore, we executed type conversion on the Boolean data within our target class, transforming it into numeric values. We have employed a pre-trained *Large Language Model* (LLM) to extract contextual information from our source code snippets and convert them into vector forms. Subsequently, we loaded the pre-processed data to train nine LLM models, including our developed CWEpredBELL model. We performed all crucial modifications in implementation to improve the model's performance. Through using the *AdamW optimizer* the model was trained in three epochs. We assessed the model's performance by accuracy, precision, sensitivity, and F1-score. Based on several metrics, our findings show that the proposed model CWEpredBELL consistently performs better than others. Figure 1 shows the overall methodology.

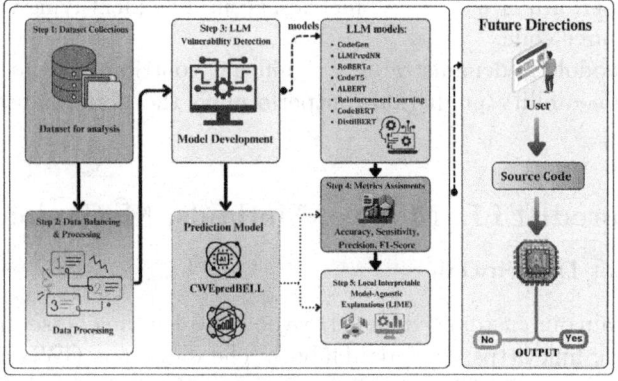

Fig. 1. Overall Methodology of the Proposed Approach

2.3 Dataset Pre-processing

We began by collecting our dataset, which then underwent several pre-processing steps to ensure it was well prepared and optimized for the models we applied. These pre-processing steps were crucial to make the dataset suitable for robust and accurate analysis. Initially, we converted the Boolean values to numeric format with the help of the type conversion function from pandas in our target class. To address the imbalance in our dataset, we applied oversampling [9] and undersampling [10] techniques.

2.4 Model Training

We utilize the pre-trained LLM model *CodeBERT*, which is intended for recognizing both code snippets and regular syntax. CodeBERT was developed on the principle of *Bidirectional Encoder Representations from Transformers* (BERT) [11]. In order to tokenize the code snippets and turn them into a format that is suitable for the transformer model, the *AutoTokenizer* is implemented. The tokenizer makes sure that the sequences are standardized by padding and reduces them to a maximum length of 512 tokens. Additionally, there are twelve transformer encoder layers in CodeBERT, which are based on the techniques developed by BERT. For the purpose of transforming the input embeddings into contextual representations, each of these layers makes use of a self-attention mechanism combined with a feed-forward network approach. Moreover, the feed-forward network plays the crucial role for modifying non-linearity, while the self-attention mechanism is required for shedding light on the links between tokens.

2.5 Model Description

In this research, we employed ML models with distinct hyper-parameters. In machine learning, hyper-parameters are the configuration settings that are not learnt from the data but are set before the training process begins and which can significantly influence the model's performances [12]. Examples include evaluation strategy, batch size, number of hidden layers, and dropout rates. The models analyzed in this article are given as follows: (*i*) CodeBERT [13]; (*ii*) LLMPredNN [14]; (*iii*) RoBERTa [15]; (*iv*) Reinforcement Learning [16,17]; (*v*) CodeGen [18]; (*vi*) CodeT5 [19]; (*vii*) ALBERT [20]; (*viii*) DistilBERT [21].

2.6 The Proposed CWEpredBELL Model

In our proposed model, the development of a binary classification model utilizing CodeBERT, leveraging the *PyTorch* framework and enhanced by the *Transformers* library. CWEPred-BELL model initializes with the pre-trained CodeBERT model and incorporates a linear classifier to generate logits for two output classes.

Initially, the tokenizer make tokens of code snippets and passed to the BERT model and prepares them for model ingestion. The dataset is split into training and validation sets, with *DataLoader* objects facilitating batch processing. An AdamW optimizer and *CrossEntropyLoss* function are defined for training. The epoch function iterates through the training *DataLoader*, computing the loss and updating model weights via backpropagation.

The loss function is defined as follows:

$$L = -\frac{1}{N} \sum_{j=1}^{N} \log P(Z = t_c \mid x_j) \qquad (1)$$

where: (*i*) N is the batch size; (*ii*) t_c is the true class; (*iii*) x_j is the sample number; (*iv*) $P(Z = t_c \mid x_j)$ is the predicted probability. The weight update rule

is given by:
$$\omega_1 = \omega - \alpha \cdot \nabla L(\omega) \tag{2}$$
where: (i) ω is the weight; (ii) α is the learning rate; (iii) $\nabla L(\omega)$ is the gradient of the loss function with respect to the weights.

The workflow of the model is shown in Fig. 4.

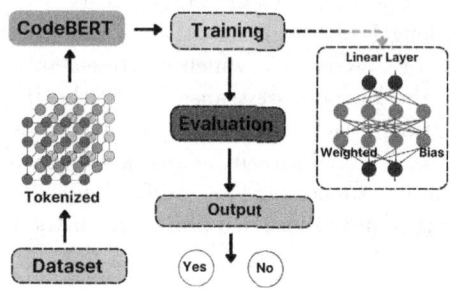

Fig. 2. Workflow of CWEpredBELL Model

2.7 Evaluation Metrics

We employed the accuracy, precision, sensitivity, and F1-score validation metrics. These metrics provide a comprehensive assessment of the model's efficacy in detecting vulnerabilities [22,23].

3 Experimental Results

In our research, we have implemented multiple LLM models. Furthermore, we applied the *Local Interpretable Model-Agnostic Explanations* (LIME) technique to enhance the specificity of vulnerability detection [24]. It identifies the accurate location of vulnerability by highlighting the vulnerable line.

An ablation study has been executed to justify the robustness of our proposed model. It included the customization of feed-forward network configuration, training the model with different parameters (learning rate and batch size) and including an additional linear layer. The accuracy of these models are given in Table 1.

We have implemented eight LLM models. The performance of these models is illustrated in Table 2, and the evaluation of our proposed CWEpredBELL model is shown in Table 3.

The results of the accuracy metrics shows that the Reinforcement Learners outperforms the other models by obtaining the highest accuracy, precision, sensitivity, and F1-score. RoBERTa and ALBERT model also demonstrated a good accuracy but the precision and Sensitivity score indicate RoBERTa performs

Table 1. Ablation Study

Ablation Details	Accuracy
Customized feed-forward network	0.81
Variation in learning rate	0.71
Additional linear layer	0.77

Table 2. Experimental Results of Comparison Models

Models	Accuracy	Precision	Sensitivity	F1-Score
CodeGen	0.77	0.61	0.61	0.60
LLMpredNN	0.80	0.68	0.67	0.66
RoBERTa	0.87	0.87	0.87	0.86
CodeT5	0.82	0.71	0.67	0.65
CodeBERT	0.69	0.69	0.69	0.68
Reinforcement Learners	0.93	0.93	0.93	0.93
ALBERT	0.86	0.61	0.61	0.60
DistilBERT	0.84	0.67	0.64	0.63

better than ALBERT. This analysis indicates that both of the models, RL and RoBERTa performed very well in vulnerability detection with minimal of misclassification. The DistilBERT and LLMpredNN models achieved the accuracy of 80% and 84%, Despite having lower accuracy the CodeT5 performed better than the ALBERT and DistilBERT due to the higher precision, sensitivity, and F1-score.

Table 5 describes the performance of the CWEpredBELL model gained in direct implementation and including cross- validation. In this case, steps stands for the number of model iterations or epochs and direct process signifies using a train-test-based dataset splitting method. The CWEpredBELL executed excellent results in both of the processes. In the direct implementation, the accuracy score along with the precision, sensitivity, and F1-score were increasing consistently with the steps. The third step generated the most satisfactory output with an accuracy of 96%, precision and F1-score of 97% along with 98% sensitivity.

Table 3. Experimental Results of the CWEpredBELL Model

Process	Steps	Accuracy	Precision	Sensitivity	F1-Score
3*Direct	1	0.90	0.92	0.93	0.93
3*Direct	2	0.93	0.93	0.98	0.96
3*Direct	3	**0.96**	**0.97**	**0.98**	**0.97**
5*Cross Validation	1	**0.94**	**0.91**	**0.94**	**0.94**
5*Cross Validation	2	0.92	0.92	0.92	0.92
5*Cross Validation	3	0.93	0.91	0.93	0.91
5*Cross Validation	4	0.93	0.93	0.91	0.93
5*Cross Validation	5	0.92	0.90	0.99	0.99

In cross-validation, the model also showed a remarkable performance. At the first cross-validation, CWEpredBELL acquired the highest 0.94 accuracy. The other three accuracy metrics precision, sensitivity, and F1-score were the same as the accuracy. It denotes the model's thoroughness in vulnerability prediction. The accuracy fluctuated slightly from the first to fifth cross-validation about 92%. The model also showed an accuracy of 92% in utilizing CWE-120 and 93% in the CWE-467dataset.

4 Conclusions

The study's findings demonstrate the various ways that LLMs model particularly, CodeBERT identify source code vulnerabilities. By leveraging CodeBERT for vulnerable code detection, we obtain a deeper understanding of learning methods can effectively identify vulnerabilities in code syntax. When combined with specialized neural networks, the multilayer architecture significantly boosts this capability. The resulting model demonstrates both accuracy and stability, underscoring its potential as a valuable tool in safe software development.

Acknowledgments. This work was partially supported by project SERICS (PE00000014) under the MUR National Recovery and Resilience Plan funded by the European Union - NextGenerationEU and by the National Aeronautics and Space Administration (NASA), under award number 80NSSC20M0124, Michigan Space Grant Consortium (MSGC).

References

1. Ross, R.S.: Information security. Joint Task Force Transformation Initiative, Guide for Conducting Risk Assessments, NIST Special Publication
2. "Top 8 cyber security vulnerabilities," Check Point Software. https://www.checkpoint.com/cyber-hub/cyber-security/top-8-cyber-security-vulnerabilities/. Accessed 11 Jan 2023
3. Kalouptsoglou, I., Siavvas, M., Ampatzoglou, A., Kehagias, D., Chatzigeorgiou, A.: Software vulnerability prediction: a systematic mapping study. Inf. Softw. Technol. **164**, 107303 (2023). https://doi.org/10.1016/j.infsof.2023.107303
4. "From Generalist to Specialist: Exploring CWE-Specific Vulnerability Detection," ResearchGate. https://www.researchgate.net/publication/382884480_From_Generalist_to_Specialist_Exploring_CWE-Specific_Vulnerability_Detection. Accessed 21 Sep 2024
5. Soykan, E.U., Bilgin, Z., Ersoy, M.A., Tomur, E.: Differentially private deep learning for load forecasting on smart grid. In: Proc. IEEE Globecom Workshops (GC Wkshps), pp. 1–6 (2019)
6. Bilgin, Z., Tomur, E., Ersoy, M.A., Soykan, E.U.: Statistical appliance inference in the smart grid by machine learning. In: Proc. IEEE 30th Int. Symp. Pers., Indoor Mobile Radio Commun. (PIMRC Workshops), pp. 1–7 (2019)
7. Bilgin, Z., et al.: Vulnerability prediction from source code using machine learning. IEEE Access **8**, 150672–150684 (2020). https://doi.org/10.1109/access.2020.3016774

8. Draper VDISC Dataset - Vulnerability Detection in Source Code. https://osf.io/d45bw/. Accessed 10 Apr 2024
9. Wongvorachan, T., He, S., Bulut, O.: A comparison of undersampling, oversampling, and SMOTE methods for dealing with imbalanced classification in educational data mining. Information **14**(1), 54 (2023). https://doi.org/10.3390/info14010054
10. Soltanzadeh, P., Feizi-Derakhshi, M.R., Hashemzadeh, M.: Addressing the class-imbalance and class-overlap problems by a metaheuristic-based undersampling approach. Pattern Recogn. **143**, 109721 (2023). https://doi.org/10.1016/j.patcog.2023.109721
11. Feng, Z., et al.: CodeBERT: a pre-trained model for programming and natural languages. In: Findings of the Association for Computational Linguistics: EMNLP 2020, pp. 1536–1547 (2020). https://doi.org/10.18653/v1/2020.findings-emnlp.139
12. Bacanin, N., et al.: On the benefits of using metaheuristics in the hyperparameter tuning of deep learning models for energy load forecasting. Energies **16**(3), 1434 (2023). https://doi.org/10.3390/en16031434
13. Accubits Technologies Inc., "CodeBERT." https://accubits.com/open-source-program-synthesis-models-leaderboardcodebert/. Accessed 27 Mar 2023
14. Liu, K., et al.: EL-CodeBert: better exploiting CodeBert to support source code-related classification tasks. In: Proc. 13th Asia-Pacific Symp. Internetware (Internetware'22), pp. 147–155 (2022). https://doi.org/10.1145/3545258.3545260
15. "Overview of ROBERTa model," GeeksforGeeks. https://www.geeksforgeeks.org/overview-of-roberta-model/. Accessed 24 Nov 2020
16. Venkatesh, L.: 3 LLM: reinforcement learning – GPT, Medium. https://luxananda.medium.com/reinforcement-learning-gpt-742016025359. Accessed 4 Mar 2024
17. Zhang, S., et al.: How can LLM guide RL? A value-based approach. arXiv [cs.LG] (2024). http://arxiv.org/abs/2402.16181
18. Liu, Z., et al.: Enhancing code summarization with deep learning: a survey. ACM Comput. Surv. **56**(1), 1–39 (2024). https://doi.org/10.1145/3557987
19. Choi, S.R., et al.: Survey on large-scale pre-trained models for software engineering tasks. IEEE Trans. Software Eng. (2024). https://doi.org/10.1109/TSE.2024.3078496
20. Nijkamp, E., et al.: CodeGen: an open large language model for code with multi-turn program synthesis (2022). arXiv: https://arxiv.org/abs/2203.13474
21. "CodeT5," Serp.Ai. https://serp.ai/codet5/. Accessed 29 Jan 2025
22. "ALBERT," Huggingface.Co. https://huggingface.co/docs/transformers/model_doc/albert. Accessed 29 Jan 2025
23. "distilbert-base-uncased," Huggingface.Co. https://huggingface.co/distilbert/distilbert-base-uncased. Accessed 29 Jan 2025
24. "LIME: Local interpretable model-agnostic explanations," C3 AI. https://c3.ai/glossary/data-science/lime-local-interpretable-model-agnostic-explanations/. Accessed Oct 19 2020

SysResolve: Study on In-Context LLM Generation of Resolution Scripts

Harsh Borse[✉], Utkalika Satpathy, Mainack Mondal, and Bivas Mitra

IIT Kharagpur, Kharagpur, India
harshzf2@gmail.com, utkalikasatapathy01@gmail.com
{mainack,bivas}@cse.iitkgp.ac.in

Abstract. Microservices pose challenges for automated fault resolution due to their distributed and complex nature. We present *SysResolve*, a framework that automates the entire resolution pipeline by combining multi-modal Root Cause Analysis (RCA) with Large Language Models (LLMs). RCA outputs are converted to natural language and passed through a Retrieval-Augmented Generation (RAG) pipeline to produce executable scripts. We evaluated and experimented on two microservices applications with three LLM (LlaMa3-70B, GPT-4, Claude 3.7). Our analysis highlights significant gains of current LLMs generation power from few-shot learning, with SysResolve achieving expert-level remediation while reducing recovery time.

1 Introduction

Modern software systems increasingly rely on microservices for scalability and flexibility, but this architecture complicates fault detection and resolution. Failures often span multiple services, producing heterogeneous signals across monitoring tools. Traditional resolution approaches depend on manual signal correlation and expert intervention, resulting in longer downtimes and higher operational costs [11].

Traditional fault management frameworks face key limitations in modern microservices settings. They often rely on static anomaly-action mappings, requiring prior exposure to every fault scenario, which restricts adaptability to emerging failures [12]. Some focus only on resource-level fixes, overlooking complex issues needing deeper intervention [13]. Existing solutions also neglect container orchestration complexities in platforms like Docker and Kubernetes [14]. Even recent LLM-based approaches retain manual steps, as human operators must interpret outputs and craft follow-up queries, reintroducing human dependence into the resolution loop [10].

In this paper, we present a novel automated fault resolution framework that leverages distinct monitoring modalities for RCA. We aim to develop a framework that does not require natural language solution step guidance from Site Reliability Engineers (SREs) to generate the resolution scripts, but rather directly utilizes the anomaly signature output from the RCA model, which is converted into a natural language sentence of an anomaly. Our approach uniquely combines multi-modal online RCA with a natural language builder to query a vector database to retrieve similar solutions/resolutions from past resolution steps to augment a pretrained LLM to generate remediation scripts/steps to address the faults. To enable this automation, we created a specialized dataset mapping RCA outputs to corresponding resolution commands in microservices environments, effectively capturing expert knowledge that was previously confined to human operators [11,15].

This end-to-end automation represents a significant advancement over existing approaches that terminate at fault diagnosis, requiring manual interpretation of results and implementation of solutions. By eliminating the human-in-the-loop requirement for resolution implementation, our framework substantially reduces the time from fault detection to resolution while maintaining the accuracy and reliability of expert-driven solutions [11].

2 Related Work

Automated fault resolution in microservices systems has advanced through machine learning, graph-based models, and large language models (LLMs).

LLM-Driven Fault Resolution Frameworks. Rahman et al. combined knowledge graphs with LLMs to enhance reasoning about system dependencies during fault diagnosis and resolution [5], generating actionable commands from natural language fault descriptions. Chen et al.'s AutoResolver uses fine-tuned LLMs to translate fault descriptions into executable remediation scripts [6].

Hybrid Approaches Integrating Multi-modal Data. Ma et al. proposed a topology-aware framework combining metrics anomaly detection with dependency graphs to identify performance bottlenecks and automate resolution [7]. These systems leverage diverse data modalities for comprehensive fault analysis.

Challenges in Automated Fault Resolution. Current systems struggle with novel failure scenarios due to reliance on predefined mappings or supervised datasets [8]. Maintaining up-to-date service dependency graphs in dynamic environments remains computationally expensive [9]. LLM-based systems frequently require human validation before executing remediation steps, creating bottlenecks in autonomous pipelines [10]. Even recent generative AI-based frameworks require human operators to interpret RCA model outputs and manually formulate queries or scripts for remediation, undermining automation by reintroducing human-in-the-loop bottlenecks.

3 System Design and Data Collection

3.1 Application Architecture

We deploy two microservice-based applications: *Library* (Lib) with 5 microservices, and *TrainTicket* (TT) with 27 interacting microservices. These applications are deployed using Docker Swarm across 6 distributed hosts in a worker-manager setup. We implement a state-of-the-art RCA framework that combines heterogeneous modalities (traces, logs, metrics, system calls) to detect anomalies and their types, providing unique signatures for different anomalies [16].

3.2 Anomaly Injection

We introduce non-functional like High-resource utilization due to zombie process, high latency and functional anomalies like upscaling issues etc. Non-functional faults impact resource utilization, while functional faults stem from coding errors or increased complexity. These anomalies require expert SRE invlocement for resolution.

3.3 Resolution Generation

Our automated fault resolution approach uses a supervised dataset created with expert SREs, establishing one-to-one mappings between anomalies and resolution commands. We collected over 5,000 distinct resolution patterns per anomaly, encompassing various fault types with different metrics, files, and configurations. Each resolution script incorporates sophisticated operational logic including accurate service identification, secure access procedures, process prioritization, protection mechanisms, and threshold-based decisions. This dataset serves as the foundation for our LLM-based system.

Resolutions: The identified anomalies require informed and automated remediation strategies typically performed by Site Reliability Engineers (SREs). For example, resolving resource leaks from zombie processes involves identifying the affected microservice and host, accessing the node via SSH, analyzing resource consumption, filtering out critical system processes, and terminating the offending process using predefined thresholds and heuristics. Latency-related issues may necessitate resource scaling, which includes pinpointing the bottlenecked service via the manager node and applying vertical (increasing allocated resources) or horizontal scaling (spawning additional instances) based on workload demands and service-level objectives.

3.4 Research Questions

We investigate the effectiveness of LLMs (GPT-4, LLaMA-3-70B and Claude 3.7) in generating automated resolution scripts with minimal human involvement, utilizing a Retrieval-Augmented Generation (RAG) pipeline. Our research questions:

- **RQ1:** How accurately can state-of-the-art LLMs generate resolution commands given a natural language description of an anomaly?
- **RQ2:** What impact does the volume of training data have on the performance of LLMs in generating fault resolution scripts?

4 Methodology

We divide automated resolution into 3 steps: converting RCA output signatures into natural language anomaly descriptions, using a RAG pipeline to generate resolution scripts, and executing and validating the solutions (Fig. 1).

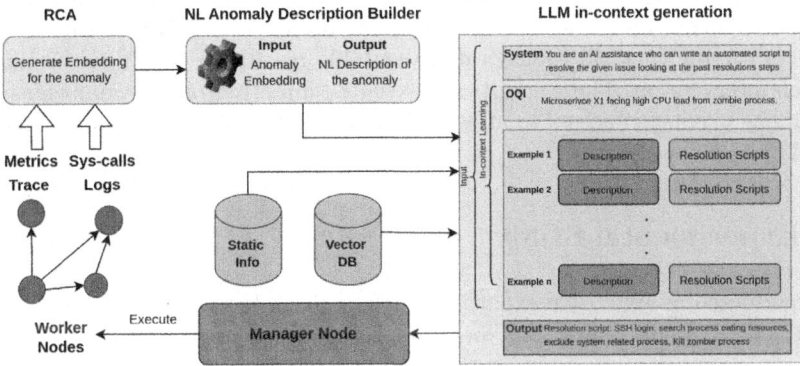

Fig. 1. SysResolve Pipeline

4.1 NL Anomaly Description Builder

Since LLMs cannot directly utilize the signatures (embeddings) from the RCA model, we develop a supervised model to convert these into natural language descriptions. Our approach uses a fine-tuned pre-trained language model with an embedding projection layer that maps system health embeddings to the language model's hidden representation space. The architecture employs cross-modal attention to focus on relevant aspects when generating descriptions. Training followed a multi-stage process: first establishing embedding-text alignment through contrastive learning, then fine-tuning with conditional language modeling based on expert annotations.

4.2 RAG Resolution Scripts

When an anomaly is detected, we retrieve similar descriptions and their resolution scripts to inform generation of a new resolution script.

RAG Database Architecture: We implement a vector database storing associations between anomaly descriptions and resolution scripts with three primary fields: (1) natural language *description*, (2) executable *resolution script*, and (3) dense *vector embeddings* from a pre-trained transformer model. For novel anomalies, we compute vector representations and perform approximate nearest neighbor search to retrieve the top-kk k most similar entries.

Prompt Template for Auto-resolution: Our prompt architecture integrates three components for in-context learning, as shown in Fig. 1. The *System* component establishes the model's operational identity and provides contextual directives that constrain the LLM to generate Ansible playbooks within specific requirements. The Operator Query Input (*OQI*) incorporates comprehensive anomaly descriptions, specific conditions, and static information about the targeted microservice. The prompt structure implements few-shot learning with in-context examples retrieved from the RAG database. The LLM uses the provided static information to generate executable resolution scripts with minimal human intervention. These scripts are executed directly from the manager node, which has privileged access to all worker nodes and Docker Swarm management capabilities.

5 Experimental Study

We conducted an extensive experimental evaluation to assess the feasibility and effectiveness of adapting LLMs for resolution code generation.

5.1 Evaluation Metrics

We employed three key metrics: **Functional Correctness (FC)**, measuring the proportion of script lines requiring no manual correction; **Average Correctness (AC)**, a custom metric assessing the correctness of individual function task blocks within a script; and **Successful Resolution Count (SRC)**, indicating whether the generated script successfully resolved the anomaly upon execution.

In Ansible, a task refers to a single unit of execution, such as service configuration or microservice scaling. Each block generated by an LLM was tested for correctness and success.

Table 1. Comparative analysis of LLMs for resolution script generation

LLMs	FC	AC	SRC
LLaMa-3-70B	88%	80%	75%
GPT-4	93%	91%	83%
Claude 3.7	95%	95%	89%

(a) Performance with few-shot learning

LLM	Method	FC	AC	SRC
LLaMa-3-70B	Zero	81%	30%	0%
	One	86%	65%	30%
	Few	88%	80%	75%
GPT-4	Zero	85%	40%	0%
	One	92%	70%	45%
	Few	93%	91%	83%
Claude 3.7	Zero	88%	45%	0%
	One	92%	80%	55%
	Few	95%	95%	89%

(b) Impact of learning methods

5.2 RQ1: Accuracy of LLMs for Script Generation

We evaluated three LLMs–LLaMa-3-70B, GPT-4, and Claude 3.7–under a uniform few-shot learning setting (10 examples). Table 1(a) shows, Claude 3.7 achieved the highest accuracy across all metrics, outperforming GPT-4 and LLaMa-3-70B. While all models produced functionally correct scripts, Claude 3.7 better adhered to syntactic and semantic conventions. The performance gap widened with increasing complexity of anomaly descriptions, where Claude 3.7 showed higher robustness, indicating its suitability for automated remediation with minimal human oversight.

5.3 RQ2: Effect of In-Context Learning Volume

We compared performance under zero-shot, one-shot, and few-shot learning. All models showed adequate FC even in zero-shot settings, but both AC and SSC improved markedly with more examples.

Claude 3.7 outperformed others consistently. In Table 1(b), we can see that, SSC was 0% for all models in zero-shot, highlighting the need for examples to ensure syntactic correctness. Transitioning to one-shot significantly boosted AC and SSC, especially for Claude 3.7. Few-shot learning yielded further, though marginal, gains, with Claude 3.7 reaching 95% in both FC and AC, and 89% in SSC. These results underline the critical role of example volume in LLM performance. Claude 3.7's ability to generalize from limited data makes it especially effective for resolution script generation.

6 Conclusion

We presented *SysResolve*, a fully automated framework that converts RCA outputs into executable scripts for microservices. Claude 3.7 achieved 95% correctness with few-shot learning, enabling expert-level remediation with reduced recovery time. Future work includes broader anomaly coverage and continuous learning.

Acknowledgment. This work has been partially supported by the DST-SERB funded project with grant CRG/2021/005316

References

1. Wu, L., Bogatinovski, J., Nedelkoski, S., Tordsson, J., Kao, O.: MicroRCA: root cause localization of performance issues in microservices. In: NOMS, pp. 1–9 (2020)
2. Lin, X., Chen, P., Zheng, Z.: MicroDiag: fine-grained performance diagnosis for microservice systems. In: ICSE, pp. 1–12 (2022)
3. Liu, Y., Jafarpour, B.: Graph attention networks for fault diagnosis in distributed systems. Comput. Chem. Eng. **175**, 108271 (2021)

4. Soldani, J., Tamburri, D.A., Van Den Heuvel, W.J.: Incremental graph update mechanism for fault diagnosis. ACM Comput. Surv. **55**(3), 1–39 (2022)
5. Rahman, M., et al.: Knowledge graph-enhanced LLM framework for automated fault resolution. IEEE Trans. Netw. Serv. Manag. **20**(2), 1–14 (2023)
6. Chen, Y., et al.: AutoResolver: fine-tuned LLMs for fault remediation. In: ICSE, pp. 1–11 (2023)
7. Ma, M., et al.: Topology-aware analysis framework for performance bottlenecks. IEEE Trans. Parallel Distrib. Syst. **33**(8), 1–12 (2022)
8. Zhou, X., et al.: Challenges in novel failure scenarios for automated systems. In: ICSE, pp. 1–10 (2021)
9. Zhao, Z., et al.: Computational bottlenecks in dynamic dependency graph maintenance. ACM Trans. Softw. Eng. Methodol. **31**(4), 1–25 (2022)
10. Kumar, S., et al.: Limitations of LLM-based systems in fully autonomous pipelines. In: NeurIPS, pp. 1–12 (2023)
11. Zhang, K., et al.: Failure diagnosis in microservice systems: a comprehensive survey and analysis. ACM Comput. Surv. **57**(1), 1–45 (2024)
12. Ikram, A., et al.: Root cause analysis of failures in microservices through causal discovery. In: NeurIPS (2022)
13. Adeoye, O.: Scalable fault tolerance for microservices-based systems. Master's thesis, University of Toronto (2022)
14. Sheriffdeen, S., Ade, R.: Achieving fault isolation in microservices architectures. In: ICDCS, pp. 1–8 (2019)
15. Sarda, K., et al.: Leveraging large language models for the auto-remediation of microservice applications: an experimental study. In: FSE Companion, pp. 358–369 (2024)
16. Borse, H., Satapathy, U., Mandal, M., Mitra, B.: URCD: unsupervised root cause detection in microservices (ICDCS) (2024)

Data Quality

A Novel Unsupervised Anomaly Detection Method Based on TCN-LSTM-CMA Autoencoder

Jiaji Feng[1](✉), Yongpan Zhang[2], Cheng Ding[2], and Su Pan[1]

[1] School of Internet of Thing, Nanjing University of Posts and Telecommunications, Nanjing 210003, China
jiajifeng200012@163.com
[2] China Telecom Jiangsu Branch, Nanjing 210003, China

Abstract. This paper proposes an unsupervised deep machine learning (DML) method for anomaly detection in unlabeled time series data. The method uses autoencoder for data reconstruction, combined with temporal convolutional networks (TCN), long short-term memory networks (LSTM), and causal masking attention mechanisms (CMA) to model short, medium, and long-term dependencies. This method effi-ciently models time series data while fully exploring short, medium, and long-term dependencies under the premise of preserving the causal structure of the data. Additionally, the method directly uses the output of the causal attention mechanism for parallel decoding, which effectively reduces the model's training time cost. After data reconstruction, Isolation Forest is used to analyze the reconstruction errors for anomaly detection. Experiments on multiple datasets of various sizes and types demonstrate that the proposed method performs better in anomaly detection.

Keywords: Time Series · Anomaly Detection · Deep Learning · Unsupervised Learning · Autoencoder

1 Introduction

With the rapid development of the Internet of Things (IoT) and big data technologies, the generation of massive time series data has simultaneously increased the demand for processing these data. Accurately detecting anomalous events within these data to ensure system stability and reliability has become an important and urgent challenge. Anomaly detection refers to the problem of identifying patterns in data that deviate from expected behavior [1]. Traditional anomaly detection methods, such as statistical-based ESD [2] and autoregressive models like ARIMA [3] and GARCH [4], perform excellently when dealing with simple time series data. However, as the dimensionality and complexity of data increase, the limitations of these methods become more apparent. They often suffer from poor generalization ability and insufficient detection accuracy to meet the requirements.

In recent years, with the rapid advancement of deep learning technologies, deep learning-based anomaly detection algorithms have been applied to various tasks [5], and research has shown that deep learning has significantly outperformed traditional methods in anomaly detection [6].

Based on whether data labels are available during training, anomaly detection can be divided into three categories: supervised learning, which uses labeled data for model training; unsupervised learning, which relies entirely on unlabeled data for training [7,8]; and semi-supervised learning, which jointly trains the model with a small amount of labeled data and a large amount of unlabeled data [9]. Due to the massive volume of data, manual labeling of data is challenging, so, like existing studies, we focus on unsupervised anomaly detection. In unsupervised anomaly detection, autoencoders are used for feature learning and data reconstruction, and anomalies are detected by comparing the actual data with the reconstructed data. For example, in [10–12], LSTM-based autoencoders are used to extract temporal features and reconstruct sequences, which are then used for anomaly detection. In [13], the authors introduce CNN combined with LSTM. CNN is first used to extract temporal features from time series data, and LSTM captures time dependencies, which further improves the performance of anomaly detection.

However, in [13], the convolution operation of CNN is limited by the fixed size of the receptive field, making feature extraction less effective. To address this issue, in [14], the authors use Temporal Convolutional Networks (TCN) to build autoencoders. By introducing dilated convolutions, they capture dependencies over longer time series. In the decoder, they introduce a fully connected layer to map the decoded output back to the same dimension as the input. However, the fully connected layer connects output nodes to all neurons, leading to information leakage, which causes overfitting and weakens the model's ability to reconstruct time series patterns. Additionally, this could result in the loss of critical temporal information, affecting the accuracy of anomaly detection.

To address the potential information leakage issue in [14] and fully preserve the causality of the data while balancing the global feature learning, this paper introduces a Causal Masked Attention (CMA) mechanism to globally model temporal features. To expand the receptive field of the model, we use TCN for feature extraction and combine it with LSTM to capture long-term dependencies. Based on this, we propose an autoencoder architecture that integrates TCN, LSTM, and CMA for time series reconstruction. During the reconstruction process, the model learns and recovers the inherent features of time series data and finally identifies anomalies by calculating the reconstruction error. TCN uses dilated convolutions to expand the receptive field, allowing the model to extract features from more distant time steps. Then, LSTM is used to model the nonlinear dependencies in long time series, enhancing the model's expressive power. On this basis, we incorporate CMA to enhance the model's ability to capture critical information. CMA dynamically weights the hidden states of each time step using a self-attention mechanism, ensuring that each time step only associates with past information, thus preventing the leakage of future information and addressing the potential issue of information dilution when capturing long-distance dependencies. When comparing the reconstructed data with actual data, anomaly detection also requires a classification method. For example, One-Class Support Vector Machine (OCSVM) [15] uses a Gaussian kernel function to map the data into a high-dimensional space and finds a hyperplane that separates normal data from anomalous data. By comparing the position of actual data points relative to the hyperplane, it determines whether the data is anomalous. Isolation Forest [16,17] constructs a set of randomly partitioned decision trees to isolate data points and uses anomaly scores to identify anomalies. Compared to other methods, Isolation Forest does not rely on data density or distance metrics, which gives it lower computational complexity. Therefore, we propose using Isolation Forest for large-scale anomaly detection tasks.

The main contributions of this paper can be summarized as follows:

- Proposing an unsupervised autoencoder architecture that integrates TCN, LSTM, and Causal Masked Attention mechanism, effectively combining the strengths of these modules to capture short-term, medium-term, and long-term dependencies in time series data.
- Combining the efficient reconstruction capabilities of the autoencoder with the rapid anomaly detection capability of Isolation Forest, significantly reducing computational complexity, thus enhancing efficiency in practical scenarios.
- Conducting extensive experiments on the Yahoo Webscope and China Telecom real-world datasets, comprehensively evaluating performance metrics such as accuracy, recall, and F1-score, clearly demonstrating the advantages of the proposed approach, along with detailed comparisons of training efficiency.

2 Proposed Method

The anomaly detection method proposed in this paper consists of two stages: data reconstruction and anomaly detection. The overall implementation process is shown in Fig. 1.

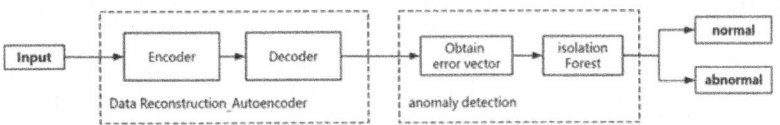

Fig. 1. Overall implementation flow of the proposed method.

2.1 Data Preprocessing

In the data preprocessing phase, the raw data undergoes missing value imputation to ensure data integrity. Next, the data is normalized. In this model, we use a sliding window approach to slice the raw data and generate our dataset.

The dataset is constructed using the sliding window approach. Suppose the original data st at each time step has n dimensions, $s_t = [data_1, data_2, data_3, ..., data_n]$, where $data_n$ represents the value of the n-th dimension at time t. The time series obtained after sliding window processing is denoted accordingly, $x_t = [s_{t+1}, s_{t+2}, s_{t+3}, ..., s_{t+timestep}]$, where timestep is the parameter of the win-dow size.

2.2 Data Reconstruction

Autoencoders are unsupervised learning models primarily used for dimensionality reduction, feature learning, and data reconstruction. An autoencoder consists of an encoder and a decoder. The core idea is to map the input data to a hidden layer using the encoder, and then reconstruct the original input through the decoder. Figure 2 illustrates the general structure of an autoencoder [18].

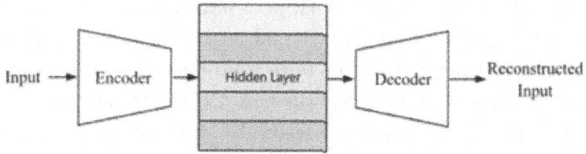

Fig. 2. Autoencoder architecture.

The encoder maps the input x to a hidden layer representation h, and the decoder maps the hidden layer representation h back to the original data space. This mapping and reconstruction output process can be represented as follows.

$$h = f(x) \tag{1}$$

$$x' = g(h) = g(f(x)) \tag{2}$$

where, the function f represents the encoding process, and the function g represents the decoding process.

Based on the general structure of the autoencoder above, we propose a novel autoencoder architecture that combines Temporal Convolutional Networks (TCN), Long Short-Term Memory (LSTM) networks, and Causal Masked Attention (CMA). This architecture fully leverages the strengths of TCN and LSTM in capturing temporal information, and enhances the model's learning ability during the reconstruction process through the CMA, effectively capturing features of time series data and per-forming reconstruction, ultimately improving the accuracy of anomaly detection. Specifically, we use a TCN-LSTM-CMA module as the encoder of the autoencoder and an LSTM-TCN module as the decoder. Figure 3 illustrates the proposed autoencoder architecture.

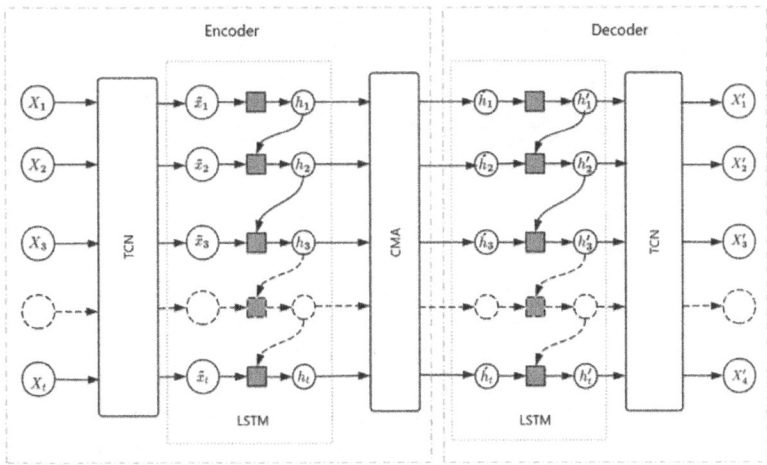

Fig. 3. structure of the TCN-LSTM-CMA autoencoder.

Encoding. In the model design presented in this paper, the encoder section is composed of TCN, LSTM, and CMA, which work together to efficiently extract the complex dependencies in time series data. Figure 3 shows the specific structure of the encoder.

TCN are based on CNN and are specifically designed for time series data, addressing the limitations of traditional CNN in capturing temporal information and dependencies. TCN employs causal convolutions to ensure that the output at the current time step depends solely on the current and previous inputs, thereby preventing future information leakage. Additionally, dilated convolutions are used to expand the receptive field, enabling the model to capture long-range dependencies. The one-dimensional convolution operation of TCN is given by the following equation:

$$\tilde{x}_t = \sum_{K}^{K-1} x_{t-d \cdot K} \cdot \omega_K \tag{3}$$

where x_{t-dK} is the input from the sliding window, $\tilde{x}_t$ is the convolution result, K is the kernel size, d is the dilation factor, and ω_K is the kernel weights.

In practice, multiple TCN layers can be stacked, with different dilation factors as-signed to each layer, allowing the model to extract features over larger time spans (Fig. 4).

The main task of the TCN layer is to capture the temporal information in the input time series and map the feature dimensions of each time step from input dimension to hidden dimension. The time series extracted by the TCN will then serve as the input to the LSTM to further model the long-range dependencies in the time series data.

LSTM networks efficiently manage the storage and transmission of information by introducing three key gating mechanisms: the forget gate, input gate, and output gate. These gating mechanisms allow the network to selectively retain or forget in-formation at each time step, ensuring that the model can capture

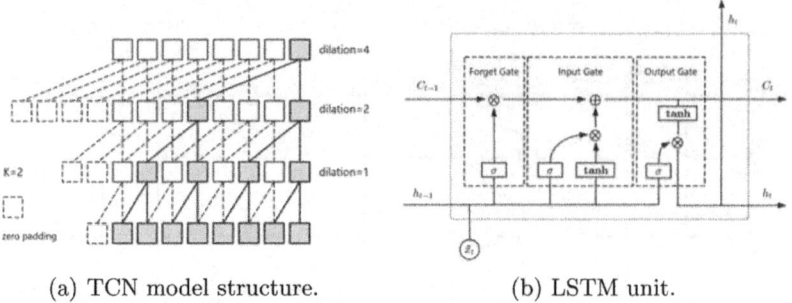

(a) TCN model structure. (b) LSTM unit.

Fig. 4. xxxxxxxx

important patterns even in long time sequences. Additionally, the cell state is used to store long-term information and runs throughout the entire time series. It is regulated by the gating mechanisms, which determine the flow and update of information.

The forget gate decides what information from the previous cell state c_{t-1} should be discarded. It performs a linear transformation followed by a Sigmoid activation function, producing an output between 0 and 1 to selectively discard irrelevant information.

$$f_t = \sigma\left(W_f \cdot [h_{t-1}, \tilde{x}_t] + b_f\right) \tag{4}$$

where, f_t is the output of the forget gate, W_f and b_f are the weight and bias, h_{t-1} is the previous hidden state, x_t is the TCN output used as LSTM input. σ is the Sigmoid activation function.

The input gate selects the information to be added to the cell state using a Sigmoid activation function and generates candidate information using a tanh function to update the cell state.

$$i_t = \sigma\left(W_i \cdot [h_{t-1}, \tilde{x}_t] + b_i\right) \tag{5}$$

$$\tilde{C}_t = \tanh\left(W_c \cdot [h_{t-1}, \tilde{x}_t] + b_c\right) \tag{6}$$

where it is the output of the input gate, $\tilde{C}_t$ is the candidate cell state. W_i and W_C is the corresponding weight matrix. b_i and b_C is the corresponding bias.

The cell state C_t is updated by the combined control of the forget gate and the input gate. The forget gate controls the retention of information from the previous state, and the input gate controls the incorporation of new information at the current time step.

$$C_t = f_t \bullet C_{t-1} + i_t \bullet \tilde{C}_t \tag{7}$$

where, C_t and C_{t-1} is the cell state, $\bullet$ is element-wise multiplication.

The output gate controls the information that is output from the cell state, and it determines the subsequent hidden state calculation.

$$o_t = \sigma\left(W_o \cdot [h_{t-1}, \tilde{x}_t] + b_o\right) \tag{8}$$

where, O_t is the output of the output gate, W_o is the weight matrix, b_o is the bias.

The final hidden state h_t is determined by the output gate O_t and the current cell state C_t after a tanh nonlinear transformation.

$$h_t = o_t \bullet tanh(C_t) \tag{9}$$

where, h_t is the hidden state at the current time step.

The final hidden state h_t is then passed to the next LSTM unit.

Integrating CMA into the encoder can enhance the output of the LSTM encoder by dynamically adjusting the influence of different time steps in the sequence through the self-attention mechanism. This process helps the model better capture long-term dependencies, thus improving the accuracy and effectiveness of reconstructing time series data. The benefit of adding CMA is that it forces the model to learn the underlying structure and patterns of the data during reconstruction by masking part of the input data. This causal masking mechanism ensures that when generating the current time step, the model only relies on past inputs, thereby preventing the leakage of future information. This strategy not only enhances the model's generalization ability but also improves its sensitivity to anomaly detection.

We map the input sequence H to query vectors Q, key vectors K, and value vec-tors V through a linear transformation. The query vector Q represents the demand for the current time step and is used to find information from past time steps relevant to the current time step; the key vector K encodes information for each time step, and together with Q, it calculates the attention weights; the value vector V stores the specific content of each time step, which is weighted and summed according to the attention weights to obtain the output H'.

$$Q^{(t \times d)}, K^{(t \times d)}, V^{(t \times d)} = H^{(t \times d)}(W_Q^{(h \times d)}, W_K^{(h \times d)}, W_V^{(h \times d)}) \tag{10}$$

$$d = h/nums\_head \tag{11}$$

where h is the dimension of the hidden state, d is the embedding dimension of each attention head, nums_head is the number of attention heads.

The causal self-attention differs from standard self-attention by introducing a lower triangular mask after computing attention scores, allowing each time step to attend only to itself and previous steps, thus preventing future information leakage.

$$\text{attention\_scores}^{(t \times t)} = \frac{QK^T}{\sqrt{d}} \tag{12}$$

$$M_{\text{mask}(ij)} = \begin{cases} 1 & i <= j \\ 0 & i > j \end{cases} \tag{13}$$

For the positions where the values in the M_{mask} are 1, the corresponding attention scores are retained. For the positions where the values in M_{mask} are 0, the attention scores are replaced with negative infinity. These positions, after being passed through the Softmax normalization, will result in attention weights close to 0, effectively masking the influence of future time steps. The masking process for each attention head is illustrated in Fig. 5.

Combining the above process, the process of each attention head through the causal attention mechanism can be summarized as the following equation:

$$\text{head}_i^{(t \times d)} = \text{Softmax}\left(Mask\left(\frac{QK^T}{\sqrt{d}}\right)\right)V \tag{14}$$

Finally, the outputs of all attention heads are concatenated together and passed through a linear layer W_O to restore the original dimension as the following equation. The final result is the new hidden layer sequence H', which is used for subsequent data reconstruction.

$$H'^{(t \times h)} = [Concat(head_1, head_2, ..., head_{nums\_head})]^{(t \times h)} W_O^{(h \times h)} = [h'_1, h'_2, ..., h'_t]^T \tag{15}$$

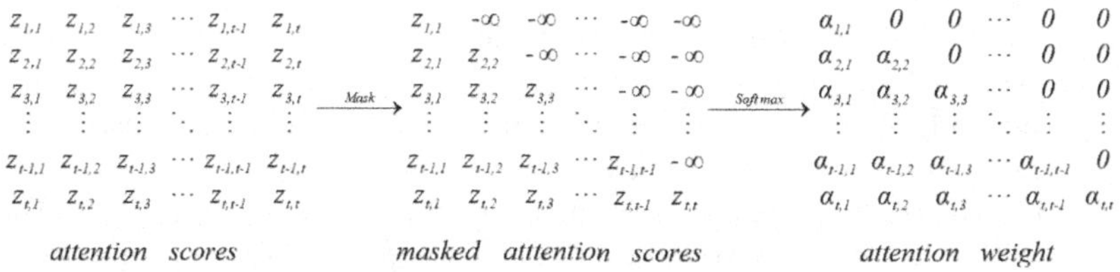

Fig. 5. Masking process of each attention head.

In summary, the encoder module formed by combining TCN, LSTM, and CMA effectively leverages the strengths of all three, enhancing the model's ability to process time series data. TCN is responsible for extracting sequence features and enhancing feature representation through dimensional expansion, LSTM models long-term dependencies, and CMA strengthens the model's learning ability through the masking mechanism. This combination enables the model to perform more efficiently in feature learning and data reconstruction when handling complex time series data.

Decoding. In the model design of this paper, the decoder is composed of LSTM and TCN, which are used to gradually reconstruct the original sequence data from the features extracted by the encoder. Figure 3 shows the specific structure of the decoder.

First, we use the output H' from the causal attention mechanism as the input to the LSTM decoder to gradually generate the decoded sequence. Specifically, at time step t, the LSTM decoder uses the hidden state h'_t, which has been processed by the causal self-attention mechanism at the current time step t, as input, and combines it with the hidden state and cell state from the previous time step $t-1$ to generate the hidden state for the current time step t, while updating the cell state. This process is repeated at each time step until the entire decoded sequence is fully generated.

Next, the TCN is responsible for the final data reconstruction. The output of the LSTM decoder is high-dimensional hidden features, which are mapped back to the original input dimensions through the one-dimensional convolution operation of the TCN.

Through the above process, the decoder can gradually convert the encoder's out-put into a result that is consistent with the original data format, ultimately completing the reconstruction of the input sequence.

Backpropagation. In this model, Mean Squared Error (MSE) is used as the loss function to measure the difference between the input data and the reconstructed data:

$$L = \frac{1}{2N} \sum_{t=1}^{N} (x_t' - x_t)^2 \tag{16}$$

where N represents the total number of samples, x_t is the input data, and x_t' is the reconstructed output of the model.

For each layer of the neural network, its output is expressed as:

$$h^{(l)} = f^{(l)}\left(z^{(l)}\right) = f^{(l)}\left(W^{(l)} h^{(l-1)} + b^{(l)}\right) \tag{17}$$

where $z^{(l)}$ is the result of the linear transformation for this layer, $f^{(l)}$ is the activation function, $W^{(l)}$ is the weight matrix and $b^{(l)}$ is bias vector.

For the output layer, the layer generating the reconstructed data, the result is directly the reconstructed output. Starting from the output layer, the gradient computation for the loss function is:

$$\frac{\partial L}{\partial h^{(l)}} = \begin{cases} \frac{\partial L}{\partial x_t} = x_t' - x_t & \text{output layer} \\ \left(W^{(l+1)}\right)^T \cdot \frac{\partial L}{\partial z^{(l+1)}} & \text{other layers} \end{cases} \tag{18}$$

Based on the chain rule, compute the gradient of the loss function with respect to each parameter and update the model parameters using gradient descent to optimize model performance.

$$W^{(l)} = W^{(l)} - \eta \cdot \frac{\partial L}{\partial W^{(l)}} = W^{(l)} - \eta \cdot \frac{\partial L}{\partial z^{(l)}} \cdot \left(h^{(l-1)}\right)^T = W^{(l)} - \eta \cdot \frac{\partial L}{\partial h^{(l)}} \cdot f'^{(l)}\left(z^{(l)}\right) \cdot \left(h^{(l-1)}\right)^T \tag{19}$$

$$b^{(l)} = b^{(l)} - \eta \cdot \frac{\partial L}{\partial b^{(l)}} = b^{(l)} - \eta \cdot \frac{\partial L}{\partial z^{(l)}} = b^{(l)} - \eta \cdot \frac{\partial L}{\partial h^{(l)}} \cdot f'^{(l)}\left(z^{(l)}\right) \tag{20}$$

where η is the learning rate.

2.3 Anomaly Detection

We utilize the Isolation Forest algorithm to classify the reconstruction error, identifying data points with excessively high errors as anomalies.

The reconstruction error refers to the discrepancy between the original sequence x_t and the reconstructed sequence x_t', which is generated by the autoencoder model. This reconstruction error is subsequently used in conjunction with the Isolation Forest for anomaly detection tasks.

Common anomaly detection algorithms include One-Class SVM (OCSVM), clustering algorithms, and others. OCSVM separates normal samples from anomalous samples by constructing boundaries. However, OCSVM faces the challenge of high computational complexity when handling multidimensional data. Clustering methods, such as K-means, divide the data points into different clusters and classify points that are far from the cluster center as anomalies. However, clustering methods can be affected by uneven data density.

In contrast, Isolation Forest has significant advantages. It does not rely on data distribution, is suitable for handling various forms of data, and has lower time complexity. Isolation Forest constructs a tree structure through random partitioning, enabling it to process large amounts of data in a short time with fast computation, making it suitable for practical applications. Therefore, in this study, we choose Isolation Forest as the method to analyze reconstruction errors, thereby enhancing the accuracy and efficiency of anomaly detection.

The core of the Isolation Forest algorithm is the construction of multiple isolation trees. Anomalous points are more easily isolated under random partitioning. During the construction of each isolation tree, a subset of the sample is randomly selected based on the subsample size. During tree construction, a feature q is randomly chosen at each split, and a split point p is selected from the range of values of that feature, assigning data points to the left or right subtree. This process continues until the data points are isolated or the preset maximum tree depth is reached. All the isolation trees together form the Isolation Forest. The path length of a sample point can be used to determine whether the point is easily isolated. Points with shorter paths are more likely to be anomalous. Figure 6 illustrates the concept of Isolation Forest.

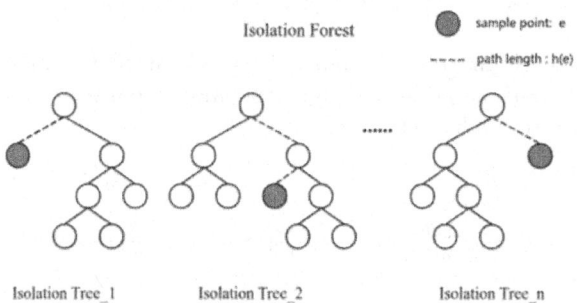

Fig. 6. Isolation Forest.

The anomaly score s(e,n) for each sample point is calculated based on the expected path length, and Isolation Forest uses the following formula to compute the anomaly score for a sample point:

$$s(e,n) = 2^{-\frac{E(h(e))}{c(n)}} \tag{21}$$

where, h(e) is the path length of the data point e in one isolation tree, from the root node to the leaf node. E(h(e)) is the expected path length of data point e across all isolation trees. c(n) is a normalization coefficient that depends on the size of the dataset n, and it is computed as:

$$c(n) = 2H(n-1) - \frac{2(n-1)}{n} \tag{22}$$

$$H(i) = \sum_{k=1}^{i} \frac{1}{k} \approx \ln(i) + \gamma \tag{23}$$

where, H(n) is the n-th harmonic number, γ is the Euler's constant, about 0.57721.

The anomaly score s(e,n) is between 0 and 1. A score closer to 1 indicates a higher likelihood of anomaly, while a score closer to 0 suggests the sample is normal.

3 Experiment

To verify the effectiveness of the proposed anomaly detection method, this study selected two datasets for experimentation. The selected datasets contain anomaly labels, which are used in the model performance analysis phase to evaluate detection effectiveness. However, during the training process, these labeled data are not used to ensure that the model is trained in an unsupervised environment. Table 1 presents detailed information about the dataset.

The Yahoo Webscope S5 dataset is a publicly available benchmark dataset widely used in anomaly detection research. It consists of four subsets: A1, A2, A3, and A4, containing a total of 367 real and synthetic time series data, with all data explicitly labeled with anomalies. In this experiment, we selected the A1-benchmark subset as the experimental object. The data in the A1 subset are all univariate time series. Figure 7 is a display of the A1-benchmark-No.34 time series, where the anomaly points are marked with a circle (○).

The CN24 dataset is provided by China Telecom and contains time series data with multiple dimensions, recording the monitoring metrics of network devices over a two-month period during their operation. The data collection frequency is every 20 min, and data for each port is collected by one or two IP addresses. All anomalies are individually labeled.

This study was conducted on Windows 11 with Python version 3.8, and deep learning was implemented using the PyTorch framework. The hardware environment includes an Intel(R) Core(TM) i5-13400F processor and an NVIDIA GeForce RTX 3060 Ti graphics card.

In this study, Precision, Recall, and F1 score are selected as the evaluation metrics for the proposed algorithm. The metrics are calculated as follows:

$$\operatorname{Re}call = \frac{TP}{TP + FN} \tag{24}$$

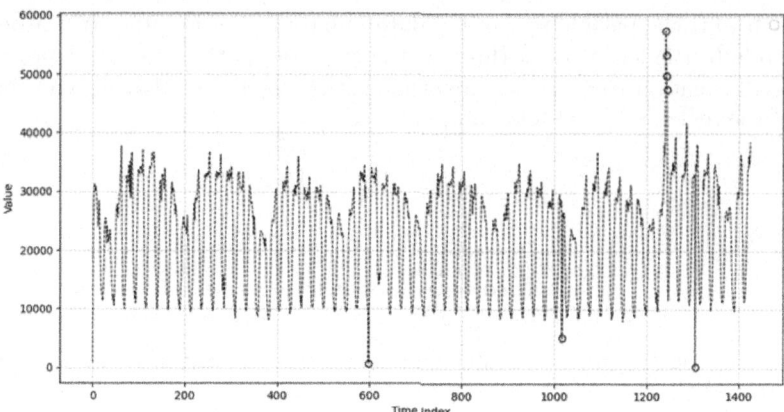

Fig. 7. A1-benchmark-No.34 time series data.

Table 1. Datasets used.

Dataset	Source	Dimensions	Number of data points	Number of anomalies
D1	A1_benchmark_No.3	1	1461	15
D2	A1_benchmark_No.6	1	1439	8
D3	A1_benchmark_No.9	1	1461	8
D4	A1_benchmark_No.24	1	1461	16
D5	A1_benchmark_No.30	1	1461	9
D6	A1_benchmark_No.34	1	1427	7
D7	CN_1**.1**.**0.75	1	4314	22
D8	CN_1**.1**.**1.98	1	4314	21
D9	CN_1**.1**.**6.8/85	2	4314	5
D10	CN_1**.1**.**3.101/106	2	4314	7

$$\text{Precision} = \frac{TP}{TP+FP} \quad (25)$$

$$F1\_Score = 2 * \frac{(\text{Precision} * \text{Recall})}{(\text{Precision} + \text{Recall})} \quad (26)$$

where TP refers to true positives, TN refers to true negatives, FP refers to false positives, and FN refers to false negatives.

All datasets were first normalized, then the sliding window method was applied for data processing. The datasets were then divided into training, validation, and test sets with a ratio of 0.6, 0.2, and 0.2. During the training process, we set the number of epochs to 200 and the learning rate to 0.0001.

Our experiments were compared with two methods: one is a hybrid model combining CNN and LSTM networks, referred to as the C-LSTM model [19], and the other is based on the ratio of loglikelihoods estimated by the dynamic linear model(DLM) [20].

Table 2 shows the F1 scores for three methods on different Yahoo S5 datasets. Both the C-LSTM and DLM methods perform well on most datasets, but C-LSTM performs poorly on D6 dataset, while DLM performs poorly on D3 dataset. In contrast, our proposed method excels on several datasets, with scores exceeding both DLM and C-LSTM on D1, D4, D5, and D6. Specifically, the F1 scores for D1 and D4 datasets reached 1. On D2 and D3 datasets, our method's scores were slightly lower than those of DLM and C-LSTM, but the difference is minor and does not impact the overall performance. Overall, our method performs exceptionally well across most datasets. Figure 8 shows the experimental results of our method on the D6 dataset.

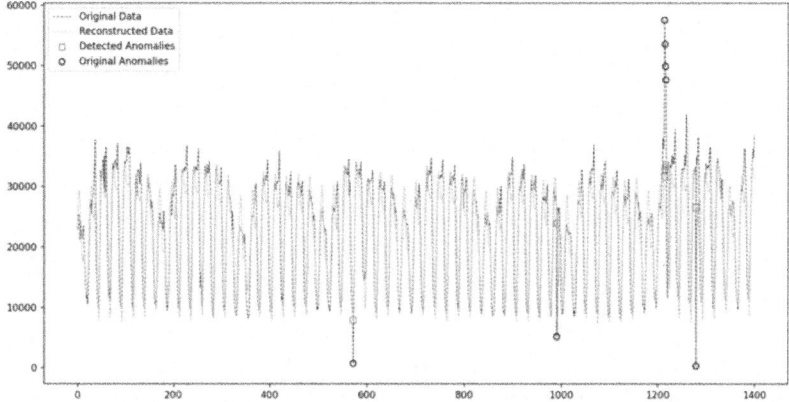

Fig. 8. D6 anomaly detection using proposed method, reconstruction data comparison.

Table 2. F1 scores for each dataset of A1_benchmark.

Dataset	C-LSTM	DLM	proposed
D1	0.96	0.93	1
D2	0.75	0.86	0.82
D3	0.87	0.5	0.82
D4	0.93	0.97	1
D5	0.89	0.71	0.89
D6	0.57	0.71	0.92

Since the anomaly detection for multiple features was not addressed in reference [20], we only used C-LSTM as the baseline method on the China Telecom dataset. The experimental results are shown in Table 3. On the four datasets provided by China Telecom, our proposed method outperforms C-LSTM on all datasets. For the D7 and D8 datasets, our model achieved higher scores on all three metrics, with a significant performance improvement. On the D9 dataset, the performance of both models was relatively average, but our model showed better detection capability. On the D10 dataset, both models performed well, but our model achieved a 100% recall rate.

In addition to the metrics mentioned above, we also conducted a detailed comparison of training times, with the number of epochs set to 200 for all experiments. Ac-cording to the results in Table 4, it can be observed that as the time step increases, our proposed method shows good stability during training, with the training time consistently staying around 30 s to complete the training. In

Table 3. Metric scores of the China Telecom dataset.

Dataset		precision	recall	f1_score
D7	C-LSTM	0.53	0.82	0.64
	proposed	0.68	0.95	0.79
D8	C-LSTM	0.68	0.65	0.67
	proposed	0.86	0.86	0.86
D9	C-LSTM	0.38	0.71	0.50
	proposed	0.46	0.86	0.60
D10	C-LSTM	0.80	0.80	0.80
	proposed	0.83	1	0.91

Table 4. Training time using different method.

Timestep	C-LSTM	proposed
10	63	32
20	95	32
40	165	34

contrast, the training time of the C-LSTM model increases significantly with the time step, and the growth trend is positively correlated with the time step. In conclusion, our method has a significant advantage in terms of time cost, allowing for faster completion of anomaly detection tasks.

Through comparisons with the C-LSTM and DLM methods, our approach demonstrates significant advantages not only in anomaly detection performance but also in training efficiency. Our future studies will explore direct comparisons and potential complementarities between our proposed approach and the latest Transformer-based anomaly detection models, aiming to comprehensively understand and further enhance its applicability in broader scenarios.

4 Conclusion

This paper proposes a novel unsupervised deep learning method for time series anomaly detection. The method introduces a new way to construct an autoencoder, aiming to improve detection accuracy in large-scale data scenarios while maintaining computational complexity. The method integrates TCN, LSTM, and CMA as the encoder of the autoencoder for feature learning, and uses the corresponding LSTM and TCN as the decoder for data reconstruction, restoring the data to its original for-mat. Then, the Isolation Forest method is used to perform isolation analysis on the reconstruction error, completing the anomaly detection task. Experimental results show that the proposed algorithm achieves better performance compared to existing algorithms.

References

1. Chandola, V., Banerjee, A., Kumar, V.: Anomaly detection: a survey. ACM Comput. Surv. **41**(3), 1–58 (2009)
2. Rosner, B.: Percentage points for a generalized ESD many-outlier procedure **25**(2), 165–172 (1983)
3. Kozitsin, V., Kaiser, I., Lakontsev, D.: Online forecasting and anomaly detection based on the ARIMA model. Appl. Sci. **11**(7), 1–13 (2021)
4. Noiboar, A., Cohen, I.: Two-dimensional garch model with application to anomaly detection, In: 2013 IEEE International Conference on Acoustics, Speech and Signal Processing, pp. 6029-6033. IEEE Press (2013)
5. Lansky, J., Ali, S., Mohammadi, M.: Deep learning-based intrusion detection systems: a systematic review. IEEE Access **9**, 101574–101599 (2021)
6. Chalapathy, R., Chawla, S.: Deep learning for anomaly detection: a survey. arXiv preprint: arXiv:1901.03407 (2019)
7. Munir, M., Siddiqui, S.A., Dengel, A.: DeepAnT: a deep learning approach for unsuper-vised anomaly detection in time series. IEEE Access **7**, 1991–2005 (2018)
8. Wang, Z., Cha, Y.: Unsupervised deep learning approach using a deep auto-encoder with a one-class support vector machine to detect damage **20**(1), 1–24 (2016)
9. Jia, Y., Cheng, Y., Shi, J.: Semi-supervised variational temporal convolutional network for IoT communication multi-anomaly detection. In: Proceedings of the 2022 3rd International Conference on Control, Robotics and Intelligent System, pp. 67–73 (2022)
10. Hnamte, V., Nhung-Nguyen, H., Hussain, J., Hwa-Kim, Y.: A novel two-stage deep learning model for network intrusion detection: LSTM-AE. IEEE Access **11**, 37131–37148 (2023)
11. Elsayed, M.S., Le-Khac, N.A., Dev, S., Jurcut, A.D.: Network anomaly detection using LSTM based autoencoder. In: Proceedings of the 16th ACM Symposium on QoS and Security for Wireless and Mobile Networks, pp. 37–45 (2020)
12. Nam, H.S., Jeong, Y.K., Park, J.W.: An anomaly detection scheme based on LSTM autoencoder for energy management. In: 2020 International Conference on Information and Communication Technology Convergence, pp. 1445–1447. IEEE (2020)
13. Yin, C., Zhang, S., Wang, J., Xiong, N.N.: Anomaly detection based on convolutional recurrent autoencoder for IoT time series. IEEE Trans. Syst., Man Cybern.: Syst. **52**(1), 112–122 (2020)
14. Mo, R., Pei, Y., Venkatarayalu, N.V., Joseph, P.N., Premkumar, A.B.: Unsupervised TCN-AE-based outlier detection for time series with seasonality and trend for cellular net-works. IEEE Trans. Wireless Commun. **22**(5), 3114–3127 (2022)
15. Yang, K., Kpotufe, S., Feamster, N.: An efficient one-class SVM for anomaly detection in the Internet of Things. arXiv preprint: arXiv:2104.11146 (2021)
16. Liu, F.T., Ting, K M., Zhou, Z.: Isolation forest. In: Proceedings of the 2008 Eighth IEEE International Conference on Data Mining, pp. 413–422. IEEE Press (2008)
17. Ding, Z., Fei, M.: An anomaly detection approach based on isolation forest algorithm for streaming data using sliding window. In: Proceedings of the 3rd IFAC International Conference on Intelligent Control and Automation Science, pp. 12–17 (2013)
18. Michelucci, U.: An introduction to autoencoders. arXiv preprint: arXiv:2201.03898 (2022)
19. Lu, Y.X., Jin, X.B., Liu, D.J., Zhang, X.C., Geng, G.: Anomaly detection using multiscale C-LSTM for univariate time series. Secur. Commun. Netw., 1–12 (2023)
20. Yoshihara, K., Kei, T.: A simple method for unsupervised anomaly detection: an application to Web time series data. Plos One, 1–25 (2022)

Behaviour Modelling and Wayfinding Error Detection in Low Mountain Hiking

Masaharu Inoue[1]($\boxtimes$), Hidekazu Kasahara[2], and Qiang Ma[1]

[1] Kyoto Institute of Technology, Matsugasaki, Sakyo-ku, Kyoto 6068585, Japan
masaharu645511@gmail.com, qiang@kit.ac.jp
[2] Osaka Seikei University, Aikawa, Higashiyodogawa-ku, Osaka 5330007, Japan
kasahara.hidekazu.k13@kyoto-u.jp

Abstract. Many hikers enjoy hiking in the low mountains near cities in Japan. However, some hikers are careless and do not prepare or plan their hikes well enough due to the low altitude and complexity of the trails, which can lead to lost routes and accidents. In this study, GPS trajectory data of past hiking are converted into series data in grids and used as training data to construct an LSTM behaviour model that predicts the next grid to which the hiker will move. The next grid series is predicted by a beam search method using the grid movement probability estimated by the constructed model. The predicted grid sequences are compared with the actual movement grids to detect wayfinding errors. Furthermore, the computational efficiency is improved by pruning the grids. The method is validated using actual mountaineering data.

Keywords: behaviour model · low mountain hiking · wayfinding error detection

1 Background

Many hikers, mainly middle-aged and older people, enjoy hiking in the low mountains near cities in Japan. In high mountains, climbers are aware of the dangers, so they make detailed mountaineering plans and take on the challenge. Compared to high mountains, low mountains require less preparation, making it easier for even ill-equipped beginners to take on the challenge. But even though they are low in altitude, low mountains are still mountains, and there are dangers unique to low mountains. There is a risk that hikers may be careless and go out with inadequate equipment or take dangerous actions, thinking that they are safe because they are in low mountains. Because of this carelessness, some hikers do not carry the necessary equipment, and some do not make a mountaineering plan themselves. In addition, there are many work paths and old roads in the low mountains, and it is difficult to distinguish them from regular trails. For these reasons, wayfinding errors occur frequently in low mountains, leading to accidents. The Japanese National Police Agency's data [1] on mountain distresses shows that the number of mountain distresses is basically on the increase.

By type of distress, the most common type of distress is wayfinding errors (33.7%), followed by slips (17.3%) and falls (16.9%), which often occur as a result of wayfinding errors and neglecting to pay attention to the path. Therefore, preventing people from wayfinding errors is considered effective in reducing the number of accidents.

Watabe et al. [2,3] proposed a method for detecting deviant behaviour using the difference between the estimated shortest routes to tourists' destinations and their current locations. However, in low mountains, the road network of trails is far more complex than in tourist areas, and the destination itself may change during the hiking. Therefore, the methods, which assumes that the shortest routes to the destinations can be calculated, is difficult to be used for determining wayfinding errors in low mountains.

Hirano et al. [4–6] proposed a method for estimating the hiker's skill level and the difficulty level of each part of mountain area in order to detect wayfinding errors in low mountain hiking. However, the method has not yet been able to detect wayfinding errors using actual hiking data.

Therefore, this study attempts to develop a method for detecting wayfinding errors with the aim of reducing the number of accidents caused by wayfinding errors in low-mountain hiking. The target mountain area is the Daimonji mountain area in Kyoto, Japan, where the trails are complex and wayfinding errors occur frequently. GPS trajectory data from past hikers is converted into a series of grid units as training data, and an LSTM-based behaviour model is constructed to predict which grid the hiker will move to next. The next grid series is predicted by a beam search method using the grid movement probability estimated by the constructed model. The predicted moving grids and the actual moving grids are compared several times in succession, and a consecutive mismatch is identified as a wayfinding error.

However, applying this method directly to hiking trajectory data is computationally inefficient. Therefore, we propose a method to prune the target grids of the beam search by obtaining the distribution of grids and time spent on grids that are prone to get lost in advance. Specifically, the method performs a wayfinding error detection on the trajectory data of a target mountain area in advance, and extracts grids with a high number of wayfinding error detections to prune the target grids for beam search. At the same time, the distribution of time spent on the grid is obtained, and the presence or absence of anomalies in the time spent on the grid is used to prune the beam search target.

2 Related Works

2.1 Wayfinding Error Detection

In their research on wayfinding error, Narumoto et al. [7,8] and Takafuji et al. [9] used the concept of cognitive maps [10] to detect the wayfinding error that occurs when humans become aware that they are lost and notice errors in their cognitive maps. They propose a method to detect wayfinding errors based on a user's realization of cognitive map inconsistencies. However, these methods are based on the assumption that the behaviour changes as a result of the subject's perception of being lost, and thus cannot cope with situations where the subject is unaware that he or she is lost. In low-mountain hiking, there are many work paths and old trails, and it is difficult to distinguish them from regular trails.

2.2 Destination Prediction

Xue [11,12]and Krumm et al. [13] proposed a method to predict inter-grid transitions by dividing the region to be predicted by a grid. These methods have been used to predict transitions between grids. The size of the cell is an issue in these methods, with Xue et al. finding the highest accuracy in predicting destinations when the cell has a side of 2 km, and Krumm et al. finding the highest accuracy in predicting destinations when the cell has a side of 1 km. However, when destinations are more densely populated than this, the cell size that achieves high accuracy with these methods does not limit the number of destinations to be predicted to one. In fact, the mountain area covered in this study is about 4 km on a side, so the cell sizes suggested by these methods cannot be applied.

Ashbrook et al. [14] proposed a method to extract Points Of Interest (POIs) from trajectories using stay points and predict the transitions between them with a Markov model; Besse et al. [15] propose a method that clusters taxi trajectories and predicts a single point on the map as the drop-off location using the likelihood that a new trajectory belongs to each cluster and the centre of the drop-off location of the trajectories in the cluster. Although these methods focus on the characteristic points of the subject's movement, they are considered to be difficult to apply to inter-grid transitions, where a mountain hiker moves along a trail in detail.

A number of studies on NextPOI recommendation [16–18] have been proposed, but they target user behaviour in tourism and daily life. It is difficult to apply them directly to hiking.

2.3 Detection of Deviating Behaviour from the Movement Trajectory Based on Destination Prediction

Watabe et al. [2,3], who conducted destination prediction to detect wayfinding error behaviour, defined detouring, including getting lost, as deviant behaviour and proposed a method to detect deviant behaviour even when the accuracy of destination prediction is low by simultaneously estimating the destination and the degree of deviation using a mixed Gaussian model. They propose a method that can detect deviant behaviour even when the accuracy of destination prediction is low. Watabe et al. induced deviant behaviour by gathering people who easily get lost and having them walk to a specific destination, collected their trajectories, and manually labelled the deviant parts of the trajectories to create a correct dataset. Using this dataset, the accuracy of the proposed method is compared and evaluated by combining destination prediction methods and deviant behaviour detection methods of different difficulty levels to demonstrate the superiority of the proposed method. At first glance, the method of Watabe et al. is consistent with the detection of wayfinding errors in low mountain hiking, where many hikers do not submit a hiking plan and the destination is not given as a given. However, the road network of mountain trails in low mountains is far more complex than in tourist areas, and the destination itself may change during the course of the activity. Therefore, the method of Watabe et al., which assumes that the shortest distance to the destination can be calculated, is difficult to be used as it is in low mountain areas.

2.4 Detection of Low-Mountain Hikers' Wayfinding Error

Hirano et al. [4–6] partitioned the target area into grids by performing two-way clustering of hikers and grids as a bipartite graph, in order to detect wayfinding errors in low mountain hiking. By classifying each grid and hiker based on their skill level, a method is proposed to estimate the level of hikers and the difficulty of hiking each of the low mountains when divided into grids. The applicability of the method for detecting wayfinding errors is mentioned by comparing it with some hiking records, but its validation and evaluation remain a challenge.

2.5 Our Study

Studies on wayfinding errors have been conducted in the past, but there are limited studies focusing specifically on hiking-related wayfinding errors. In this study, we propose a method for detecting wayfinding errors by dividing the target area into grids, modelling mountaineer behaviour by time series prediction using LSTM, performing beam search using the constructed model, predicting the grid in which the hiker moves several times in succession, and comparing it with the actual movement. In order to improve the efficiency of the wayfinding errors detection, the target grids for wayfinding errors detection are pruned, considering both perception and recall.

3 Proposed Method

3.1 Dataset

In this study, we use the GPS movement trajectories of hikers of Mount Daimonji in Kyoto, Japan, provided by Yamap Corporation, as training data. As a dataset, we use the trajectories of 40,537 hikers who climbed Mount Daimonji between 1 January and 31 December 2021. As verification data, we use the GPS trajectories collected by the authors and the GPS trajectories of hikers of Mount Daimonji downloaded from Yamap and YamaReco, Japanese top-2 hiking social networking services. The data set used in this study consists of the wayfinding error data collected by the authors after actually hiking Mount Daimonji, and 10 trajectory data sets from Yamap and Yamareco that can be downloaded as paid members on SNS. The data used were latitude, longitude, altitude and time.

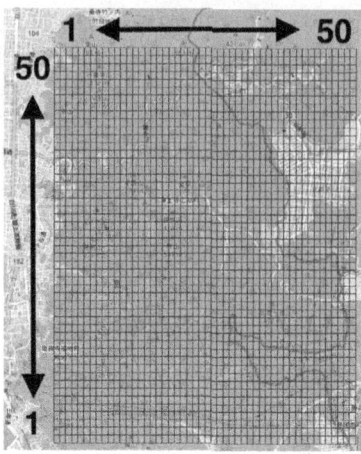

Fig. 1. Target mountain areas divided into grids

3.2 Preprocessing

As a data pre-processing step, the GPS trajectory is first converted into a grid-to-grid transition sequence with four features: latitude ID, longitude ID, time spent label and altitude label.

The target mountain area is divided into a total of 2500 grids of 50 × 50, and each latitude and longitude is given an ID corresponding to a grid. Grid IDs are pairs of (latitude ID, longitude ID). Trajectories outside the target area are deleted. Some trajectories go out of the target area and come back, and simply deleting the trajectory in that part of the area will cause the transition between grids to be skipped. Therefore, when out-of-range trajectories are included, in addition to deleting the out-of-range trajectories, the data is divided into different data before and after the trajectories go out of the target area and after the trajectories return. The generated 2500 grid is shown in Fig. 1.

When changing from a trajectory to a grid sequence, if the grid IDs given at two consecutive points are the same, they are combined into a single grid. The altitude of the merged grid uses the average of the merged grid and the time of day uses the time of the first grid in the merged grid.

The time spent is calculated from the time. The formula for calculating the time spent is as follows.

$$T_i = \frac{1}{2}(t_{i+1} - t_i) + \frac{1}{2}(t_i - t_{i-1}), \quad 2 \leq i \leq n-1 \tag{1}$$

$$T_1 = \frac{1}{2}(t_2 - t_1), T_n = \frac{1}{2}(t_n - t_{n-1}) \tag{2}$$

The T represents the time spent and t the time of day. The time spent includes long stays for breaks and outliers, so grids with stays of more than 3600 s are split in that part of the data.

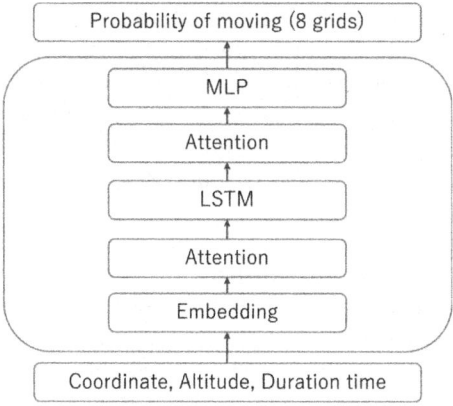

Fig. 2. LSTM behavioural model for low mountain hikers.

The slope is calculated from the altitude. The formula for calculating the slope is as follows.

$$S_i = \frac{1}{2}(h_{i+1} - h_i) + \frac{1}{2}(h_i - h_{i-1}), \quad 2 \leq i \leq n-1 \tag{3}$$

$$S_1 = \frac{1}{2}(h_2 - h_1), S_n = \frac{1}{2}(h_n - h_{n-1}) \tag{4}$$

Note that S represents the slope and h the altitude. This data is used later to detect anomalies in the time spent using the Mahalanobis distance and is not used for training.

The altitude and time spent are also labeled by discretising them, rather than using the values. For the altitude, the average altitude for each of the 2500 grids is calculated and the difference in altitude is divided into 50 equal parts to give labels from 1 to 50. As hikers tend to take longer breaks on certain grids, such as mountain summits, a simple division into equal parts of the time spent would result in large differences in the amount of data contained in each label. Therefore, the average time spent on each of the 2500 grids is determined and divided into 50 equal parts so that the amount of data per label is the same.

For the training data, the grid rows are padded to a length of 10 and all the data are used for training. The 1102496 inter-grid transition sequences of length 10 resulting from the above pre-processing are used for training.

3.3 LSTM Model

The pre-processed training data is trained on the LSTM behavioural model, which is shown in Fig. 2.

Of the 10 grid rows of length used, the first 9 grids are taken as input and the difference between the 9th and 10th grid is calculated to label which of the surrounding 8 grids it moves to from that position, which is the correct label.

Each of the four features is first embedded into a 16-dimensional vector space. After embedding, the four features are concatenated and passed through an attention layer, which allows the model to focus on relevant features or specific grid locations. A batch size of 128 is used, and the hidden layer has 64 dimensions. The attention mechanism is also applied to the output of the LSTM layer, and the final output is reduced to an 8-dimensional representation through the subsequent coupling layers.

CrossEntropyLoss is employed as the loss function to compute the probability of transitioning from the 8-dimensional output to one of the surrounding 8 grids. The loss is calculated by taking the negative logarithm of the predicted probability corresponding to the correct grid, as indicated by the ground truth labels. The accuracy is computed as the percentage of predictions that match the true grid. The formulas for computing the loss and accuracy are given below.

$$\text{Loss} = -\frac{1}{N} \sum_{i=1}^{N} \sum_{j=1}^{8} y_{i,j} \log \hat{y}_{i,j} \tag{5}$$

$$\text{Accuracy} = \frac{1}{N} \sum_{i=1}^{N} \mathbf{1}\left(\arg\max_{j} \hat{y}_{i,j} = y_i\right) \tag{6}$$

N represents the batch size, $y_{i,j}$ the component of the one-hot encoding corresponding to the actual label of sample i (1 for the correct class, 0 otherwise). $\hat{y}_{i,j}$ is the output of the model and represents the probability distribution obtained by Softmax, while $\log \hat{y}_{i,j}$ represents the log of the predicted probability. $\arg\max_j \hat{y}_{i,j}$ represents the predicted labels and y_i the correct answer labels. The $\mathbf{1}$ is an indicator function, which returns 1 if the prediction is correct and 0 otherwise.

3.4 Beam Search and Wayfinding Error Detection

This method compares the predicted movement grid, generated by the LSTM-based behavior model, with the actual movement grid. A hiker is considered to be lost if the actual trajectory deviates from the predicted path for a specified number of consecutive steps. The comparison starts from the first movement grid. Beam search is employed to perform this prediction, with a beam width of 3 and a search depth of 3 steps. Beam search is an efficient heuristic search algorithm for traversing decision trees. Instead of exploring all possible nodes–which would be computationally expensive–it maintains only the top-k most promising candidates (defined by the beam width), thereby significantly reducing computational complexity while still identifying a near-optimal path.

Data 9 to 12 of a sequence data of length 12, used as an example of a beam search, is shown below.

$$\text{sequence\_data} = \begin{bmatrix} 9 & 28 & 10 & 26 \\ 10 & 28 & 5 & 27 \\ 9 & 27 & 2 & 28 \\ 10 & 27 & 12 & 30 \end{bmatrix} \tag{7}$$

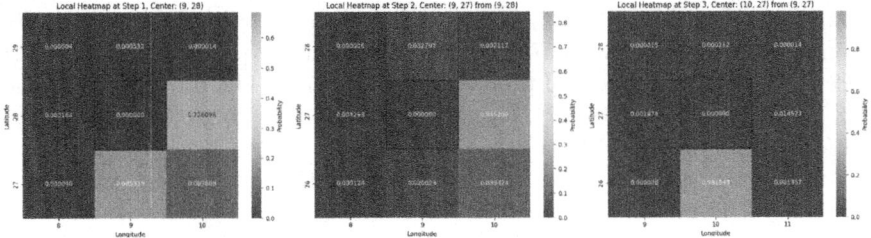

Fig. 3. Probability of surrounding 8 grids in each step

From left to right, each data point consists of a latitude ID, longitude ID, time spent label, and altitude label. In the beam search-based mobile grid prediction task, the first nine data points are used to predict the tenth grid. The probability distribution over the eight surrounding grids at each prediction step is shown in Fig. 3. Example of beam search is shown in Fig. 4.

In Step 1, the model predicts the next grid location based on the input ending at coordinate (9, 28). According to the results in Fig. 3, the top three candidate grids, ranked by predicted probability, are (9, 27), (10, 28), and (10, 27).

In Step 2, the model generates new predictions for each of the three candidate grids selected in Step 1. As time progresses, the oldest data point in the sequence is removed, and the new predicted grid from Step 1 is appended. While four features are required for prediction, only latitude and longitude can be directly predicted; therefore, the average time spent and average altitude associated with each predicted grid are used to complete the feature vector.

For the top candidate from Step 1, (9, 27), the conditional probability of reaching (10, 27) in Step 2 is calculated by multiplying the Step 1 probability (0.685339) by the transition probability to (10, 27) (0.845206), resulting in a conditional probability of 0.579253. This process yields three candidate routes for Step 2: ((9, 27), (10, 27)), ((10, 28), (10, 27)), and ((10, 27), (10, 26)).

In Step 3, predictions are again made for each of the three candidate routes produced in Step 2, using the same data update process. For the route ((9, 27), (10, 27)), which had the highest conditional probability in Step 2, the conditional

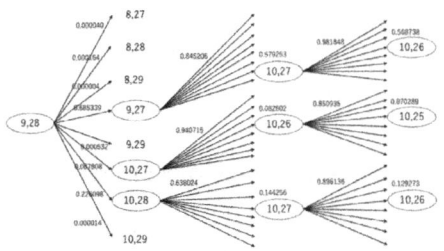

Fig. 4. Illustration of Beam Search Process

probability of reaching (10, 26) in Step 3 is calculated by multiplying the Step 2 conditional probability (0.579253) by the transition probability to (10, 26) (0.981848), resulting in a final probability of 0.568738. The top three candidate routes after Step 3 are: ((9, 27), (10, 27), (10, 26)), ((10, 28), (10, 27), (10, 26)), and ((10, 27), (10, 26), (10, 25)).

Among the candidate routes, the one with the highest conditional probability–((9, 27), (10, 27), (10, 26))–is selected as the predicted route. The grid corresponding to the first step of this route, (9, 27), is treated as the predicted grid.

3.5 Wayfinding Error Detection

The predicted grid is compared against the actual movement grid to detect potential wayfinding errors. In the example, the model predicts the 10th grid to be (9, 27), while the actual movement grid is (10, 28), indicating a mismatch. Consequently, the model proceeds to predict the 11th movement grid using updated input data comprising grids 2 through 10. Although the details of the beam search procedure are omitted here, the predicted 11th grid is (9, 29), which matches the actual 11th grid. Once a match is found between predicted and actual movement, the hiker is no longer considered lost, and the detection process is terminated.

3.6 Improving the Efficiency of Wayfinding Error Detection

To improve computational efficiency, grids prone to wayfinding errors are extracted as prior knowledge through pre-analysis on the training dataset. To enable beam search in this context, the training data is padded to a sequence length of 12. The system then records both the grid IDs where wayfinding errors are detected and the frequency of such occurrences.

Grids with low detection counts are excluded under the assumption that they are unlikely to be associated with wayfinding errors. However, some grids–such as those at mountain summits–may have inflated error counts simply due to high traffic. To address this, a relative score is computed by dividing the number of wayfinding errors by the total number of hikers passing through each grid. Grids are then ranked by this relative score, and those with the lowest scores are eliminated until the remaining set matches the size determined by absolute detection count. This refinement improves detection efficiency by focusing only on grids with relatively high likelihoods of causing disorientation.

3.7 Outlier Detection Based on Time Spent

Grids with unusually long time spent are also treated as potential indicators of wayfinding errors. For this, a kernel density estimation (KDE) approach is employed to model the distribution of time spent per grid. Compared to simple histogram-based approaches, KDE provides a smoother estimation and better captures subtle variations in the data. Grids with significantly higher-than-expected time values are treated as outliers and flagged as high-risk zones for potential confusion.

First, the time spent per grid is aggregated for all training and validation data; for the KDE, the density of each data point is calculated; the formula for the probability density function of the KDE is as follows.

$$\hat{f}(x) = \frac{1}{nh} \sum_{i=1}^{n} K\left(\frac{x - x_i}{h}\right) \quad (8)$$

The n denotes the number of data, h the bandwidth, K the kernel function, x the point for which the probability density is calculated and x_i each data point. The bandwidth is 2 and the kernel function is calculated using a Gaussian function; for the KDE, the outliers are detected by setting the threshold value to the lower 5%.

For the Mahalanobis distance, not only the time spent labels but also the gradient are tabulated for each grid. The Mahalanobis distance is calculated by calculating the mean vector and covariance matrix from the time spent labels and gradients. The formula for calculating the Mahalanobis distance is as follows.

$$D_M = \sqrt{(x - \mu)^T \Sigma^{-1} (x - \mu)} \quad (9)$$

The x represents each data point, μ is the mean vector and Σ is the covariance matrix. For the Mahalanobis distance, anomalies are detected by setting the threshold value to the top 5/

In this study, the computational efficiency is improved by restricting the wayfinding to grids for which anomalies are detected by these two methods.

4 Evaluation

4.1 Objectives and Method of Experiments

To check the accuracy of the wayfinding detection, we evaluate the performance of the LSTM behaviour model and the performance of the wayfinding detection: for the LSTM behaviour model, we evaluate Loss and Accuracy. The Precision, Recall and F1 values are defined as follows.

$$\text{Precision} = \frac{\text{Number of correctly detected wayfinding errors}}{\text{Number of times wayfinding error detection}} \quad (10)$$

$$\text{Recall} = \frac{\text{Number of correctly detected wayfinding errors}}{\text{Actual number of times wayfinding errors}} \quad (11)$$

$$\text{F1 value} = \frac{2 \times \text{Precision} \times \text{Recall}}{\text{Precision} + \text{Recall}} \quad (12)$$

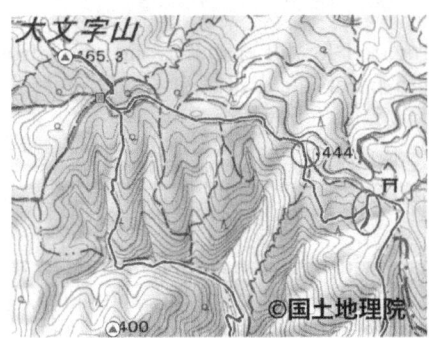 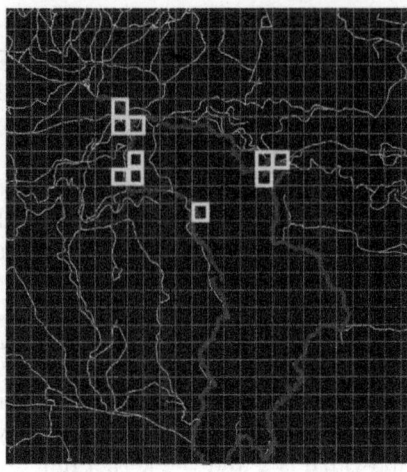

(a) Hiking tracks and lost points (b) Wayfinding error detection results

Fig. 5. An example of wayfinding error detection

As this is the tenth data set to be predicted, no wayfinding errors were detected for the first nine and last two grids of each set of data. First, for each of the validation data, we read the mountain trip records described in the Yamap and Yamareco SNSs, labeled the lost points, and recorded the number of times the path was lost. One example of the validation data is shown in Fig. 5a. There are two wayfinding error points, which are indicated by red circles.

Next, the validation data is padded by 12, the wayfinding error detection is performed and the number of wayfinding error detections is recorded. If a wayfinding error is detected in consecutive grids, it is counted as one time collectively. The results of the wayfinding error detection using the proposed method for the data in Fig. 5a are shown in Fig. 5b. The grids where wayfinding errors were detected are indicated by yellow squares. In this case, the number of wayfinding error detection times is four.

The number of times a wayfinding error can actually be detected is recorded by comparing the grid on which the wayfinding error was detected with the labelled wayfinding errors. In the example, of the two lost points shown in Fig. 5a, only the one on the right was detected correctly. From the correct and incorrect records of the detection results obtained in this way, the Precision, Recall and F1 values are calculated to evaluate the performance of the wayfinding detection. However, performing lost detection on all grids in the target mountain area would be computationally very expensive. Therefore, the lost detection is limited to grids that are prone to get lost and grids where the time spent is anomalous, to evaluate how the detection accuracy changes with the reduction in computational complexity.

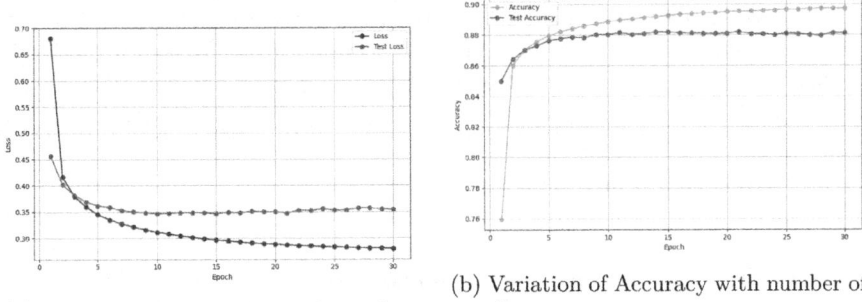

(a) Variation of Loss with epoch number

(b) Variation of Accuracy with number of epochs

Fig. 6. Training performance over epochs

4.2 Performance of LSTM Behaviour Model

The embedding dimension of each of the four features is 16 dimensions, the hidden layer of LSTM is 64 dimensions, the learning rate is 0.001, the batch size is 128 and the number of epochs is 30, and the training and test data are divided 9:1 for training. The loss and accuracy of the training results are shown in Figs. 6a and 6b, respectively.

4.3 Comparison Method

In order to compare the accuracy of the wayfinding detection, the wayfinding detection is carried out in such a way that a person is considered to be lost if he/she returns to the same grid during four moves. For this method, we also evaluate how the accuracy changes while reducing the computational complexity by restricting the lost detection to grids with anomalous time spent.

4.4 Performance of Wayfinding Error Detection

The following seven patterns are tested as variations of the proposed method.
1. The method that searches all grids (LBA)
2. The method that searches the grids that are most likely to getting lost, determined using the absolute value of the number of wayfinding error detection.(LBG)
3. The method that searches grids that are prone to getting lost, determined using the relative value of the number of wayfinding error detections.(LBRG)
4. The method searches grids where the time spent detected by KDE is out of the ordinary.(LBK)
5. The method searches grids with outliers in the time spent detected by the Mahalanobis distance.(LBM)
6. The method searches grids that are prone to getting lost, determined using the absolute value of the number of wayfinding error detection, or grids where the time spent detected by the KDE is out of the ordinary.(LBGoK)

7. The method searches grids that are prone to getting lost, determined using the relevant value of the number of wayfinding error detection, and grids where the time spent detected by the KDE is an anomaly.(LBGaK)

The comparison method performs wayfinding error detection in three cases.

1. The method that searches all grids (NA).
2. The method that searches grids with anomalous time spent detected by KDE (NK).
3. The method that searches grids with anomalous residence times detected by the Mahalanobis distance.(NM)

The accuracy of all these wayfinding error detections is shown in Table 1.

Table 1. Comparison of wayfinding error detection methods

Method	Precision	Recall	F1 value
LBA	0.28125	0.47368	0.35294
LBG	0.29166	0.36842	0.32558
LBRg	0.30769	0.21052	0.25
LBK	0.375	0.15789	0.22222
LBM	0.33333	0.10526	0.16
LBGok	0.29629	0.42105	0.34782
LBGaK	0.4	0.10526	0.16666
NA	0.19444	0.73684	0.30769
NK	0.36363	0.21052	0.26666
NM	0.66666	0.21052	0.32

Table 2. Comparison of computational complexity of wayfinding error detection

Methods	F1 Value	Computational Complexity
LBA	0.35294	827
LBG	0.32558	501
LBRG	0.25	84
LBK	0.22222	81
LBM	0.16	43
LBGoK	0.34782	536
LBGaK	0.16666	46

4.5 Comparison of the Computational Complexity

The number of target grids for wayfinding is used as the computational complexity of the wayfinding because beam search is performed on the number of target grids for wayfinding. The computational complexity of the proposed method for wayfinding is shown in Table 2.

4.6 Results of Comparison

The LBA method had a higher F1 value than all three comparison methods. Comparing the computational complexity, the LBGoK method lowered the F1 value by about 0.005 compared to the LBA method, while it reduced the computational complexity by about 40%. These results show that the LSTM behavioural model and beam search improved the performance of the detection of lost in hiking compared to the compared methods. They also show that detection can be made more efficient by prior knowledge.

4.7 Issue

When comparing the actual trajectory and the wayfinding error detection points, there were cases where the wayfinding error was noticed within the grid and the effect of the wayfinding error did not appear in the grid transitions. This problem could be solved by dividing the grid finer, but this would increase the computational complexity and reduce the conformance rate. There were also problems such as the tendency to be detected as lost by moving back and forth between grids while on a road near a grid boundary, and the tendency to be detected as lost when there are many grid candidates to move to at the top of a mountain. It is necessary to further limit the grids for detecting wayfinding errors.

5 Conclusion

The contributions of this study are summarised as follows;
1) We have constructed a model that learns the behaviour patterns of hikers from hiking trajectory data, and proposed a method for estimating the next movement trajectory in conjunction with beam search.
2) The method attempts to detect wayfinding errors based on the differences from the actual trajectory. The model is used to estimate the next moving trajectory in combination with beam search.
3) The method improves the efficiency of detecting wayfinding errors. It reduces the computational complexity of wayfinding error detection in hiking by detecting grids that are prone to getting lost and grids where the time spent is abnormal, and by pruning the computational targets.
4) The method is verified using a dataset constructed from actual hiking data. The target mountain area for the evaluation experiment is the Daimonji mountain area in Kyoto City, where mountain trails are complex and wayfinding errors occur frequently. The evaluation experiments are to evaluate the performance of the LSTM behavioural model and the performance of the wayfinding error detection.

References

1. Japan National Police Agency. Overview of mountain accidents in 2023 (2024). https://www.npa.go.jp/publications/statistics/safetylife/r05_sangakusounan_gaikyou.pdf
2. Watabe, T., Kasahara, H., Iiyama, M.: Detection of deviating behavior from the movement trajectory based on destination prediction. In: DEIM2020 (2020)
3. Kasahara, H., Watabe, T., Iiyama, M.: Tourist transition model among tourist attractions based on GPS trajectory. J. Smart Tourism 1(2), 19–25 (2021)
4. Rui, H.: A study on the detection of lost trails of low mountain climbers. Master Thesis (2024)
5. Hidekazu, K., Rui, H., Ma, Q.: Way lost in low mountains. In: Proceedings of the 18th National Conference of the Tourism Information Society (2022)

6. Hidekazu, K., Rui, H., Ma, Q.: A study of methods for detecting lost trails of low mountain climbers. In: Proceedings of the 24th Annual Conference of the Tourism Information Society (2023)
7. Ryosuke, N., Sougo, K., Hirozumi, Y., Teruo, H.: A smartphone-based detection method for lost behaviour due to cognitive mapping errors. In: Proceedings of the Multimedia, Distributed Cooperation and Mobile Symposium 2017, vol. 2017, no. 1, pp. 1386–1395 (2017)
8. Ryosuke, N., Sougo, K., Hirozumi, Y., Teruo, H.: Sensing interest behaviour using smartphones. In: Research Report Ubiquitous Computing Systems (UBI), vol. 2018-MBL-86, no. 49, pp. 1–7 (2018)
9. Takumi, T., Yudai, H., Hirozumi, Y., Teruo, H.: A study on the detection of lost behaviour of smartphone users. In: Research Report Ubiquitous Computing Systems (UBI), vol. 2016-UBI-4, no. 1, pp. 1–6 (2016)
10. Golledge, R.G. (ed.): Wayfinding Behavior: Cognitive Mapping and Other Spatial Processes. JHU Press, Baltimore, MD (1999)
11. Andy, Y.X., Rui, Z., Yu, Z., Xing, X., Jin, H., Zhenghua, X.: Destination prediction by sub-trajectory synthesis and privacy protection against such prediction. In: 2013 IEEE 29th International Conference on Data Engineering (ICDE), pp. 254–265 (2013)
12. Andy, Y.X., Jianzhong, Q., Xing, X., Rui, Z., Jin, H., Yuan, L.: Solving the data sparsity problem in destination prediction. VLDB J. **24**(2), 219–243 (2015)
13. Krumm, J., Horvitz, E.: Predestination: inferring destinations from partial trajectories. In: Dourish, P., Friday, A. (eds.) UbiComp 2006. LNCS, vol. 4206, pp. 243–260. Springer, Heidelberg (2006). https://doi.org/10.1007/11853565_15
14. Ashbrook, D., Starner, T.: Using GPS to learn significant locations and predict movement across multiple users. Pers. Ubiquit. Comput. **7**(5), 275–286 (2003)
15. Besse, P.C., Guillouet, B., Loubes, J.-M., Royer, F.: Destination prediction by trajectory distribution-based model. IEEE Trans. Intell. Transp. Syst. **19**(8), 2470–2481 (2018)
16. Li, P., de Rijke, M., Xue, H., Ao, S., Song, Y., Salim, F.D.: Large language models for next point-of-interest recommendation. In: Proceedings of the 47th International ACM SIGIR Conference on Research and Development in Information Retrieval, pp. 1463–1472 (2024)
17. Wang, B., Zhang, Y., Ma, Y., Jin, Y., Xu, Y.: Sequence-aware long- and short-term preference learning for next POI recommendation. In: SA-LSPL (2024)
18. Feng, S., Lyu, H., Chen, C., Ong, Y.-S.: Zero-shot generalization of LLMs for next POI recommendation, Where to Move Next (2024)

Explainable Time Series Anomaly Detection by Dynamic Mode Decomposition

Shun Kawakami[1], Toshiyuki Amagasa[2], and Savong Bou[2]

[1] Graduate School of Science and Technology, University of Tsukuba, Tsukuba, Japan
[2] Center for Computational Sciences, University of Tsukuba, Tsukuba, Japan
savong-hashimoto@cs.tsukuba.ac.jp

Abstract. With the rapid spread of time series data, real-time anomaly detection techniques have become increasingly important to ensure reliability. Existing anomaly detection methods are excellent at detecting anomalies, but are limited in their ability to explain the underlying causes of the anomaly. In this study, we propose a new approach that uses dynamic mode decomposition to explain the dynamic characteristics behind the data points identified as anomalies by their singular values. This method computes the singular values that explain the anomaly by comparing the predictions computed by dynamic mode decomposition with the anomalous observed values. While conventional methods explain the anomalies using dimensional subspaces, the proposed method uses singular values to explain the anomalies, allowing for explanations that take into account periodicity and trends. Experiments proved the explanatory power of the proposed method by identifying dimensions with strong influence of singular values as explanations and comparing their error rates with those of dimensions with actual anomalies.

Keywords: Sliding window · Explainable anomaly detection · Streams

1 Introduction

Time-series data [1–8,12,14,16] is used in a wide range of industries, including manufacturing, the financial sector, and network security, for real-time analysis. The detection of anomalies in time series [9,20] has attracted particular attention and has a wide range of important applications, including predictive maintenance of equipment in the manufacturing industry, detection of fraudulent transactions in the financial sector and detection of intrusions in network security.

Explainability in anomaly detection systems has become an important issue for decision making. In recent years, the accuracy of anomaly detection has improved dramatically with the development of machine learning technologies [22] such as deep learning. However, while these methods are highly accurate, their decision-making process has become a black box, making it difficult for humans to understand the basis of the detection results. Especially in areas where false positives can have serious consequences, such as criticalbreak infrastructure and medical equipment, it is essential for experts to understand and properly evaluate the system's judgment. Therefore, there is a strong need for an accountability methodology that can be applied to advanced anomaly detection methods.

Various methods have been proposed to transform the black box of machine learning into a more digestible form and to increase client confidence in the output. Outlying Aspect Mining (OAM) [19,21] is for finding subspaces that distinguish the outliers from the normal data. The subspaces are ranked using distance- or density-based outlier scores. A method [15] gives feature subspaces a form that is useful for human analysts to make decisions. Provides an easy-to-understand indication of which features contribute to the anomaly by calculating sequential explanations using an outlier scoring scale. Other methods, such as LIME and SHAP [10,17], approximate the original complex model with a simple linear model around specific data points and explain how each feature affects the predicted results based on the coefficients of this linear model; SHAP is based on the Shapley value of cooperative game theory. It considers each feature quantity as a player and the predicted outcome as a reward, and calculates how much each player contributed to the reward. However, since these methods focus on static data, they do not sufficiently take into account various dynamic characteristics such as periodicity and trends of time series data.

As a solution, this paper proposes a new method of explaining the behavior of complex time-series by identifying Dynamic Mode Decomposition (DMD) [13], which is developed in the field of fluid dynamics [11] to extract dynamic modes that characterize temporal behavior. DMD is suitable for real-time analysis due to its efficient algorithm based on linear algebra. Furthermore, each mode obtained by singular value decomposition has the advantage of representing time-series characteristics such as periodicity and trend in an interpretable form.

The proposed method utilizes each mode obtained by the singular value decomposition of DMD to quantitatively explain the dynamic structure behind the anomaly. The importance of the singular value representing each mode is calculated based on its contribution to the prediction error, and the singular value with the highest contribution is output as an explanation. The singular values correspond to specific periodicity or increase/decrease patterns in the data, and can directly indicate the dynamical characteristics of the time series that cause the anomaly. While conventional methods such as LIME and SHAP provide local explanations in the feature space, the proposed method provides an explanation that takes into account the global behavior of the time series. This allows for a more intrinsic understanding of the underlying causes of anomalies and is expected to improve client confidence in the decisions of the system.

2 Dynamic Mode Decomposition

Dynamic Mode Decomposition (DMD) [13] is a data-driven analysis method to extract dynamic properties from time series data. If the time series is represented as a state vector $x_k \in \mathbb{R}^d$, the time evolution in discrete time is represented by:

$$x_{k+\Delta k} = A x_k \qquad (1)$$

where A is the operator governing the time evolution and can be regarded as an infinite-dimensional Koopman operator finite-dimensional approximation.

To compute the operator from actual data, n time series data are constructed as the following two snapshot matrices:

$$X = [x_1\ x_2\ \cdots\ x_{n-1}] \in \mathbb{R}^{d \times (n-1)}, \quad Y = [x_2\ x_3\ \cdots\ x_n] \in \mathbb{R}^{d \times (n-1)} \quad (2)$$

Using these matrices, the time evolution of the system is expressed by:

$$Y = AX \quad (3)$$

The optimal operator A such that equation (3) is satisfied can be calculated by using the Moore-Penrose pseudo-inverse matrix as follows:

$$A = YX^\dagger \quad (4)$$

However, directly computing pseudo-inverse matrices for high-dimensional data is computationally expensive and impractical. Therefore, the singular value decomposition of X is used for efficient computation.

The singular value decomposition of matrix X is expressed as follows:

$$X = U\Sigma V^T \quad (5)$$

where $U \in \mathbb{R}^{d \times r}$ and $V \in \mathbb{R}^{(n-1) \times r}$ are normal orthogonal matrices and $\Sigma \in \mathbb{R}^{r \times r}$ is a matrix with singular values as diagonal components. r is the rank number, usually chosen according to the characteristics of the data.

The low-rank approximated operator $\tilde{A}$ can be efficiently computed as follow:

$$\tilde{A} = YX^\dagger = YV\Sigma^{-1}U^T \quad (6)$$

The obtained operator $\tilde{A}$ can then be used to predict future states as follows:

$$x_{t+1} = \tilde{A}x_t \quad (7)$$

3 Proposed Method

This chapter explains the proposed explainability method using DMD.

3.1 Overview

A new explanatory method for anomaly detection in time series is proposed. The method has two main abilities: (1) capture the time-series-specific features that cause anomaly, and (2) analyze the contribution of each singular value to the anomaly, which is obtained through DMD's singular value decomposition.

The architecture of the proposed method is shown in Fig. 1. First, an anomaly detection system is used to determine anomalous values (1). If an abnormality is determined, the past data points corresponding to n steps from that data point are used to decompose the time series data using DMD (2). Next, for each

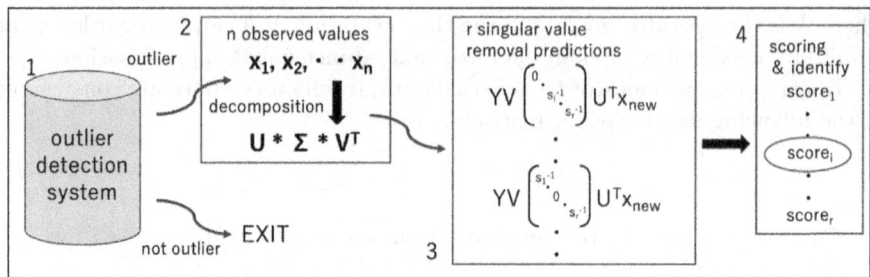

Fig. 1. Architecture of the proposed method

singular value obtained from the DMD singular value decomposition, predictions are calculated that eliminate the influence of that singular value (3). These predictions are the predictions with certain dynamic characteristics removed, taking advantage of the fact that the predictions by DMD are linearly coupled. Then, each singular value removal prediction is scored using the prediction error between the predicted value and the actual anomaly, and the singular value with the lowest score is identified (4). With this score, the influence of the dynamic characteristics on the singular value can be compared, and the one that contributed the most to the observed value being determined as abnormal can be calculated. The explanatory singular value can therefore be interpreted as representing the dynamic characteristic that caused the anomaly value to deviate most significantly from its true value.

This method allows for explanations based on dynamic characteristics unique to time series (e.g., periodicity and trends) that cannot be captured by conventional methods such as OAM and LIME. The proposed method also has the advantage to be used in combination with any anomaly detection system.

3.2 Data Structure and Preprocessing

The proposed method manages time series using a sliding window X_C of length $n+1$ to calculate operator $\tilde{A}$ in DMD when the observed value is determined to be anomalous. The following process is done each time a new data X_{new} arrives:

1. Sliding window X_C update
2. Determination of X_{new} using anomaly detection system
3. If judged as anomaly, process to find explainability

In Process 1, X_C is updated to hold the observed values and new data points for the n steps needed to calculate the operator $\tilde{A}$. Unnecessary n+1 steps previous data points are discarded and X_{new} is added. Thus, X_C continues to hold n+1 data points of past data points and new data points for n steps. In Process 2, an optional abnormality detection system judges the new data point for abnormality. This determines whether to perform Process 3 or wait until the new data point arrives. Then, in Process 3, the explainability process, the data in X_C is first preprocessed. In order to allow comparison between features with different scales, the following normalization process is performed in the preprocessing:

$$X_{norm}[i] = \frac{X_C[i] - min(X_C[i])}{max(X_C[i]) - min(X_C[i])} \qquad (8)$$

where $min(X_C[i])$ and $max(X_C[i])$ are the minimum and maximum values in $X_C[i]$. This process is performed in all dimensions in X_C.

From this normalized data X_{norm}, two snapshot matrices X and Y are constructed based on the formula (2), which forms the underlying matrix for capturing the dynamic characteristics of the system:

$$X = X_{norm}[\,0\,:\,n-2\,], \quad Y = X_{norm}[\,1\,:\,n-1\,] \qquad (9)$$

3.3 Calculation of Predicted Values

Proposed method computes the singular value that serves as the explanation by comparing the predictions without the influence of each singular value. This is called SVRP. To compute SVRP, singular value decomposition is applied to the matrix X formed by Eq. 9, which is needed to compute the operator $\tilde{A}$:

$$U_r, \Sigma_r, V_r^T = svd(X) \qquad (10)$$

Here, the singular value decomposition is computed up to an arbitrary rank r. This yields basis vectors characterizing the dynamic properties and singular values representing their importance.

Next, the matrix with the absolute value of singular value i set to 0 is computed from the obtained Σ_r:

$$\Sigma_{\bar{i}} = \begin{pmatrix} s_1 & & 0 \\ & \ddots & \\ & s_i & \\ 0 & & \ddots \\ & & & s_r \end{pmatrix} - \begin{pmatrix} 0 & & 0 \\ & \ddots & \\ & s_i & \\ 0 & & \ddots \\ & & & 0 \end{pmatrix} = \begin{pmatrix} s_1 & & 0 \\ & \ddots & \\ & 0 & \\ 0 & & \ddots \\ & & & s_r \end{pmatrix} \qquad (11)$$

Finally, $\Sigma_{\bar{i}}$ is used to compute the SVRP of singular value i:

$$\bar{x}_i = Y V_r \Sigma_{\bar{i}}^{-1} U_r^T x_{last} \qquad (12)$$

where x_{last} is one element before x_{new}, i.e., the last element of Y.

The processing for all singular values i in Eqs. 11 and 12 is performed, and the results are stored in an array in descending order of absolute value.

3.4 Score Calculation

The score is computed using SVRP to quantitatively evaluate the singular values that explain the outliers, which is based on following theoretical considerations.

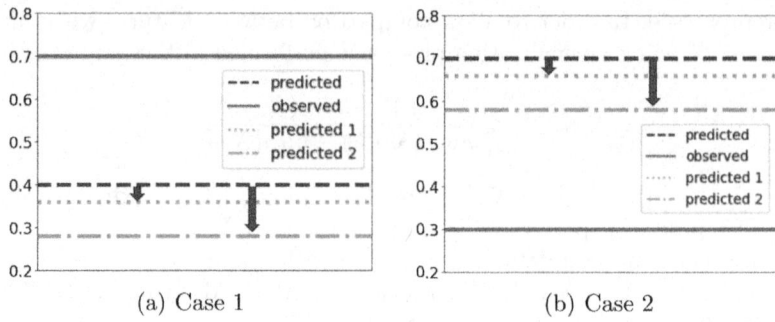

Fig. 2. Example of the relationship between observed and predicted values and singular value removal predictions.

Explaining singular value is a singular value with dynamic characteristics that are not included in the anomaly. In other words, it is the singular value that contributes the most to being determined as an anomaly. The score for each singular value is calculated using the SVRP and the prediction error of the anomalous observation. We now consider when this prediction error is the singular value contributing to the anomaly using the following two cases:

Case 1. The observed value exceeds the predicted value:
In Fig. 2(a), the observed value at which the system detected an anomaly was 0.70. In this situation, the DMD predicted value (predicted) was 0.40, the SVRP for singular value 1 (predicted 1) was 0.36, and the SVRP for singular value 2 (predicted 2) was 0.28. In the case of large observed values, the larger change due to a particular singular value (arrows in the figure) contributed to moving closer to the observed value, while the smaller change contributed to moving away from the observed value. Therefore, the larger the prediction error with respect to the SVRP, the closer to the observed value, and the smaller the prediction error, the further away from the observed value. The singular value 1 has a smaller prediction error and thus contributes to the anomaly in the observed value.

Case 2. The observed value is less than the predicted value:
In Fig. 2(b), the observed value at which the system detected as anomaly was 0.30. In this situation, the DMD predictions (predicted) were 0.70, the SVRP for singular value 1 (predicted 1) was 0.66, and the SVRP for singular value 2 (predicted 2) was 0.58. In the case of small observed values, the smaller change (arrows in the figure) due to a singular value contributes to moving closer to the observed value, while the larger change contributes to moving away from the observed value. It can be said that the larger the prediction error for the SVRP, the closer to the observed value, and the smaller the prediction error, the further away from the observed value. In this example, the singular value 2 has a smaller prediction error and contributes to the anomaly in the observed value.

Thus, regardless of the relationship between observed and predicted values, singular values with small prediction errors with respect to SVRP are more

$$scores = \begin{pmatrix} sc_1 \\ sc_2 \\ sc_3 \\ \ldots \\ sc_r \end{pmatrix} \begin{array}{l} \text{Return the index of} \\ \text{the smallest scores} \\ \Rightarrow i_{exp} = \\ argmin_i(scores[i]) \\ \text{where } i \in r \end{array} \qquad pred\_list = \begin{pmatrix} \bar{x}_{11}, \bar{x}_{12}, \ldots, \bar{x}_{1d} \\ \bar{x}_{21}, \bar{x}_{22}, \ldots, \bar{x}_{2d} \\ \ldots \\ \bar{x}_{r1}, \bar{x}_{r2}, \ldots, \bar{x}_{rd} \end{pmatrix}$$

$\text{pred}_{exp} = $
$\text{pred\_list}[i_{exp}]$ $\Downarrow$ Get all values at index i_{exp}
$pred_{exp} = (\bar{x}_{i_{exp}1} \; \bar{x}_{i_{exp}2} \; \cdots \; \bar{x}_{i_{exp}d})$

(a) i_{exp} is obtained from $scores \in \mathbb{R}^{r \times 1}$ (b) $pred_{exp}$ is from $pred\_list \in \mathbb{R}^{r \times d}$

Fig. 3. How $i_{exp} \in \mathbb{R}^{1 \times 1}$ and how $pred_{exp} \in \mathbb{R}^{1 \times d}$ are obtained

$$pred\_list = \begin{pmatrix} \bar{x}_{11}, & \bar{x}_{12}, & \ldots, & \bar{x}_{1d} \\ \bar{x}_{21}, & \bar{x}_{22}, & \ldots, & \bar{x}_{2d} \\ \ldots \\ \bar{x}_{i_{exp}1}, & \bar{x}_{i_{exp}2}, & \ldots, & \bar{x}_{i_{exp}d} \\ \ldots \\ \bar{x}_{r1}, & \bar{x}_{r2}, & \ldots, & \bar{x}_{rd} \end{pmatrix}$$

$\Downarrow \sum \frac{1}{r-1}$ Average of all values except the one at index i_{exp}
$pred_{avg} = (p_1, p_2, \ldots, p_d)$

(a) $pred_{avg}$ is from $pred\_list \in \mathbb{R}^{r \times d}$

$pred_{avg} = (p_1, p_2, \ldots, p_d)$
$pred_{exp} = (\bar{x}_{i_{exp}1}, \bar{x}_{i_{exp}2}, \ldots, \bar{x}_{i_{exp}d})$
$rate\_list = (r_1, r_2, \ldots, r_d)$
Return the indices of the k biggest values in rate_list
argsort(rate_list)[d-k:] $\Downarrow$ $i_k \in d$ and $i_{k-1} \leq i_k$ and $i_{k-1} \neq i_k$
$evaluation\_indices = (i_1, i_2, \ldots, i_{k-1}, i_k)$

(b) How $evaluation\_indices$ is obtained

Fig. 4. How $pred_{avg} \in \mathbb{R}^{1 \times d}$ and how $evaluation\_indices \in \mathbb{R}^{1 \times k}$ are obtained.

suitable for explanation. Based on this theoretical consideration, the error obtained as follows is used to calculate the score of singular values:

$$e_{x_i} = x_{new} - \bar{x}_i \in \mathbb{R}^d \tag{13}$$

where e_{x_i} is the prediction error vector with respect to SVRP. The score for each singular value is calculated by applying the get_score function to these vectorized errors. The specific methods (variance, norm) for this scoring function are analyzed and discussed in Sect. 5.3.

Algorithm 1 summarizes a series of processing steps related to the explainability of the outliers described so far. How $i_{exp} \in \mathbb{R}^{1 \times 1}$ is obtained from the $scores \in \mathbb{R}^{r \times 1}$ is elaborated in Fig. 3(a). The algorithm systematically shows the overall workflow of the proposed method, from data preprocessing to the generation of scores for each singular value.

4 Framework for Assessing Accountability

Various methods have been proposed for explainability in anomaly detection, but methods for quantitatively evaluating their explainability are not well established. In particular, it is an important issue to objectively evaluate the quality of the explanations obtained in methods that capture dynamic characteristics, such as the proposed method. This paper proposes a new framework to quantitatively evaluate the explainability by singular values.

Algorithm 1. ODE

INPUT: X_C has n records and time-series data TD, **OUTPUT:** i_{exp}, $pred\_list$

1: **for** x_{new} from TD **do**
2:   $X_C.pop$, $X_C.push(x_{new})$
3:   **if** x_{new} is outlier **then**
4:     $X_{min} = min(X_C)$, $X_{max} = max(X_C)$
5:     $X_{norm} = (X_C - X_{min}) / (X_{max} - X_{min})$
6:     $Y = X_{norm}[0 \sim n-2]$
7:     $X = X_{norm}[1 \sim n-1]$
8:     $U, \Sigma, V^T \leftarrow svd(X, r)$
9:     $scores = []$, $pred\_list = []$
10:    **for** $i \in \{1, \ldots r\}$ **do**
11:      $\Sigma_{\bar{i}} = \Sigma - \Sigma[i][i]$
12:      $\bar{x}_i = Y \times V \times \Sigma_{\bar{i}} \times U^T \times x_{new}$
13:      $scores \leftarrow get\_score(x_{new} - \bar{x}_i)$
14:      $pred\_list \leftarrow \bar{x}_i$
15:    **end for**
16:    $i_{exp} \leftarrow argmin_i(scores[i])$ where $i \in r$
17:    print "explainable of x_{new} is $explainable$"
18: **end if**
19: **end for**

Each singular value obtained by the equation (10) captures a particular pattern of variation in the data, each of which is associated with a particular dimension of the time series. Therefore, we propose to quantify the validity of the explanations by measuring the agreement between the k dimensions in which the observed values are highly anomalous and the k dimensions in which the singular values that explain them are strongly associated.

Algorithm 2 shows the processing procedure for identifying the dimensions to which the explanatory singular values for a single anomaly are strongly related. $pred\_list \in \mathbb{R}^{r \times d}$ is an array containing r singular value removal predictions, and i_{exp} represents an index of singular values that are explanatory in $pred\_list$ (Fig. 3(b)). These two were obtained in a series of processes in Algorithm 1. The singular value removal prediction for i_{exp} is stored in $pred_{exp} \in \mathbb{R}^{1 \times d}$ (line 1) and the average of singular value removal predictions for indexes other than i_{exp} is stored in $pred_{avg} \in \mathbb{R}^{1 \times d}$ (line 2) (Fig. 4(a)). Next, the difference between the two sequences is analysed to identify the dimension associated with the singular value that is the explanation. The relevant dimension is the dimension that contributed to the selection of the explanatory singular value. In other words, when the explanatory singular value is compared to other singular value removal predictions for each dimension, the dimension with the smallest prediction error is the dimension that contributed to the selection. Therefore, the differences between $pred_{exp}$ and $pred_{avg}$ are taken for each dimension (lines 4–10). The method of calculating these differences can be classified into the following two cases, depending on the relationship between the observed and predicted values:

Algorithm 2. Evaluation

INPUT: i_{exp} and $pred\_list$ has r $\bar{x}_i$ k, **OUTPUT:** score

1: $pred_{exp} = pred\_list[i_{exp}]$
2: $pred_{avg} = (\sum_{i=1}^{r} pred\_list[i] - pred_{exp}) / (r - 1)$
3: $rate\_list = []$
4: **for** $i \in \{1, \ldots, d\}$ **do**
5:     **if** $X_{true}[i] >= X_{pred}[i]$ **then**
6:         $rate\_list \leftarrow pred_{exp}[i] - pred_{avg}[i]$
7:     **else**
8:         $rate\_list \leftarrow pred_{avg}[i] - pred_{exp}[i]$
9:     **end if**
10: **end for**
11: $evaluation\_indices = \text{argsort}(rate\_list)[d\text{-}k\text{:}]$

Case 1. (The observed value exceeds the predicted value): The larger the predicted singular value, the smaller the prediction error will be and the closer it will be to the observed value. Therefore, the difference is calculated as follows:

$$pred_{exp} - pred_{avg} \tag{14}$$

Case 2. (The observed value is below the predicted value): The smaller the predicted singular value, the smaller the prediction error will be and the closer it will be to the observed value. Therefore, the difference is calculated as follows:

$$pred_{avg} - pred_{exp} \tag{15}$$

Finally, the dimensions that are strongly related to the singular value are identified by comparing the differences (the k biggest ones) (line 11) (Fig. 4(b)).

5 Experiments

Experiments are conducted on synthetic and real-world datasets to validate the effectiveness of the proposed method. The following items are verified.
1. Experiments on the maximum value of evaluation per experimental data set
2. Experiments on the accuracy per scoring function
3. Experiments with varying parameters
4. Running time to find the singular value that explains
An Apple M1 Pro, 16 GB machine was also used in the experiment and the results of 1000 executions are shown in box-and-whisker diagrams.

5.1 Datasets

Synthetic dataset: sd1 is defined as follows:

$$S(t) = \sin 0.5 * t + \sin 5 * t + 0.01 \cdot t + \lambda \cdot \varepsilon$$

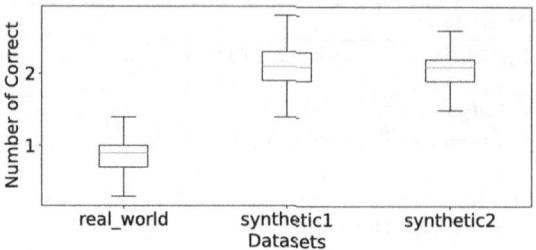

Fig. 5. Maximum value of evaluation per dataset

Here, the first term expresses the long-term periodicity and the second term the short-term periodicity. The third term simulates the long-term upward trend of the time series by adding a linear trend. The fourth term, random Gaussian noise $\varepsilon \sim N(0, 1)$ scaled by a coefficient $\lambda = 0.3$, simulates data noise and various shapes. This synthetic dataset represents time series data with multiple periodicity and trend combinations, simulating complex real-world time series.

Synthetic dataset: sd2 is defined as follows:

$$S(t) = \begin{cases} \sin\left(\frac{t-t_0}{\omega}\right) + \lambda \cdot \varepsilon & S_{rand} = 0 \\ \cos\left(\frac{t-t_0}{\omega}\right) + \lambda \cdot \varepsilon & S_{rand} = 1 \end{cases}$$

where S_{rand} is a random sheet of 0 or 1. The time delay $t_0 \in [50, 100]$ and frequency $\omega \in [40, 50]$ simulate various periodic cycles, while the random Gaussian noise scaled by a factor λ simulates noise and variability in the data as in sd1. The frequencies of two sine waves are highly correlated if they are similar and almost in phase. By randomly selecting the frequency and phase of each time series, one would expect some pairs to be highly correlated and others to be less so. sd1 and sd2 both allow the number of data points and the number of dimensions to be set freely.

Real-world dataset: rd1 is EDENISS2020 [18]. This dataset contains time-series data recorded from 97 sensors installed in a closed-loop research greenhouse built at the initiative of the German Aerospace Centre for a future long-term space mission called EDEN ISS. This data set records a variety of values, including temperature, humidity and CO_2.

The three data sets are without anomaly labels. Therefore, the data points that are anomalies are randomly generated at intervals greater than a certain number of consecutive data points. Using the given N_a and N_k, the N_k dimension of N_a data points out of the total data points is multiplied by 1.5. In this way, anomalies are generated for the three datasets to evaluate the proposed method.

5.2 Maximum Value Experimental per Dataset

To assess the basic performance of the proposed method, we tested how accurately the singular values identify the anomalous dimensions in each dataset.

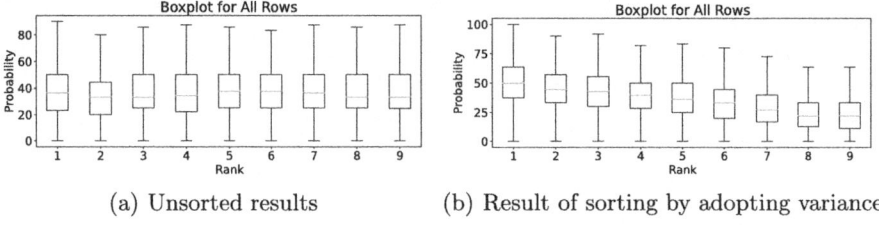

(a) Unsorted results (b) Result of sorting by adopting variance

Fig. 6. Evaluation results for each sort order using dataset:rd1

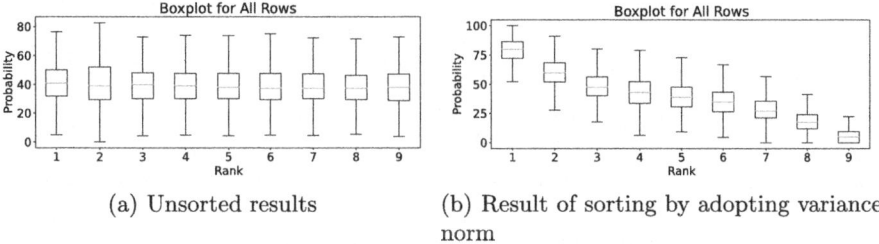

(a) Unsorted results (b) Result of sorting by adopting variance norm

Fig. 7. Evaluation results for each sort order using dataset:sd1

The maximum number of matches was selected from all calculated singular values and the accuracy was compared for each dataset. Parameters were set as d = 30 (synthetic datasets only), n = 30, r = 10, k = 5, $N_a = 10$ and $N_k = 5$.

The Fig. 5 shows the results for each dataset. The experimental results show that the average number of dimension matches in each dataset was rd1: 0.87, sd1: 2.11 and sd2: 2.08. The synthetic dataset, where the periodicity and trend are not too complex, correctly identifies on average more than two of the five dimensions in which anomalies occur, showing a satisfactory performance as explained by a single singular value. On the other hand, the number of matches in the real-world dataset is about half that of the synthetic dataset. This may be due to the more complex periodicity and trend of the real-world data and the fact that the number of dimensions is more than three times larger than that of the synthetic data. Considering these conditions, the fact that an average of 0.87 dimensions can be identified in the real-world data suggests that the proposed method has practical explanatory power.

5.3 Comparison of Sorts

In the proposed method, the identification of singular values, which is the explanation based on the prediction error between the singular value removal predictions and the observed values, is highly dependent on the scoring function for its accuracy. In this experiment, norm and variance were employed as scoring functions and their performance was compared. Parameters were set to d=30 (synthetic datasets only), n = 30, r = 10 and k = 5. Relative accuracy (%), which is the number of dimensional matches per sort order normalised by the

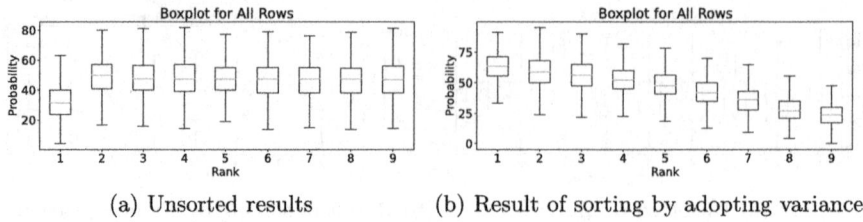

(a) Unsorted results (b) Result of sorting by adopting variance

Fig. 8. Evaluation results for each sort order using dataset:sd2

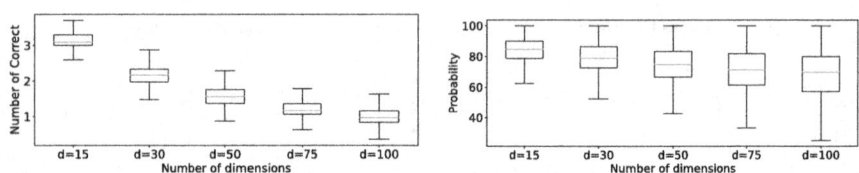

(a) Maximum value of the evaluation result (b) Sorting by adopting variance norm

Fig. 9. Experiments with varying number of dimensions

maximum number of matches, is used as the evaluation metric, with three types of sorting: no sorting, sorting by norm and sorting by variance.

The results are in Figs. 6, 7, 8. First, it is observed that without sorting, the performance is almost equal for all ranks. This suggests that the singular values suitable for the explanation are uniformly distributed. On the other hand, from Fig. 7(b), sorting by norm shows superior performance on dataset:sd1, which only has non-negative values, achieving a relative accuracy of about 80% with respect to the singular values that are explanatory. It can also be observed that the relative accuracy decreases significantly as rank decreases. In contrast, sorting by variance performed below norm overall. The difference was particularly pronounced for dataset:sd1, where the relative accuracy of the singular values that explain the data was about 60%. However, for the other datasets (rd1, sd2), the performance differences due to the scoring function were relatively small. This means that the performance is excellent for positive values of the norm, but not so good for both positive and negative values. On the other hand, it can be confirmed that the variance is stable for both positive and negative values.

5.4 Experiments with Varying Parameters

To assess the scalability of the proposed method, experiments with varying parameters were conducted. Specifically, the number of dimensions and the number of ranks in the singular value decomposition were varied using dataset:sd1.

Experiments with Varying Number of Dimensions. Figure 9 shows the results of varying the number of dimensions of the data points. In this exper-

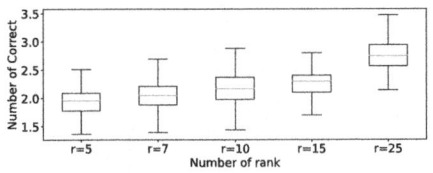 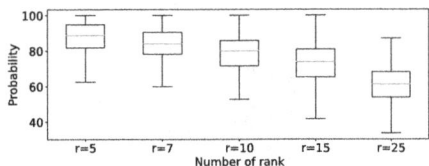

(a) Maximum value of the evaluation result (b) Sorting by adopting variance norm

Fig. 10. Experiments with varying number of ranks

iment, the number of dimensions was varied from 15, 30, 50, 75 and 100. The Fig. 9(a) shows the maximum number of singular value matches and the Fig. 9(b) shows the results of the relative accuracy (%) normalised by the maximum number of matches, which explains the norm adopted. The Fig. 9(a) shows that the correctness rate of the proposed method decreases as the number of dimensions increases. However, when looking at the error rate per number of dimensions, it can be seen that even though the number of dimensions is 6.7 times higher, the error rate is still about one-third, indicating that the error rate has not decreased that much. The Fig. 9(b) also shows that the average relative accuracy does not change that much from the number of dimensions. This means that the sorting is well done. However, the worst case relative accuracy has gradually approached zero because the number of matches is close to 1 when the number of dimensions is 100, even for the best singular values, so it is thought that the relative precision of the singular values that explain the results is zero at times.

Experiments with Varying Number of Ranks. Figure 10 shows the results of varying the number of ranks in the singular value decomposition. The number of ranks was varied from 5, 7, 10, 15 and 25. The Fig. 10(a) shows the maximum number of singular value matches, and the Fig. 10(b) shows the results of the relative accuracy (%) of the number of singular value matches normalised by the singular value with the highest number of matches, which explains the norm adopted. The Fig. 10(a) shows that increasing the number of ranks also increases the number of matches of the proposed method. This is because increasing the number of ranks increases the range of characteristics expressed by the singular values and makes it easier for singular values to appear that are more suitable for the explanation. On the other hand, the Fig. 10(b) shows that the quality of sorting decreases as the number of ranks is increased. This is because the number of singular values increases, making it more difficult to sort and select the best singular value. However, it can be seen that the rate of decrease in probability is less than the rate of increase in the number of ranks, so it can be said that the quality of sorting by norm itself is not that bad.

5.5 Execution Time of the Proposed Method

To assess the efficiency of the proposed method, the execution time required to find the explanatory singular value for a single anomaly was measured. In

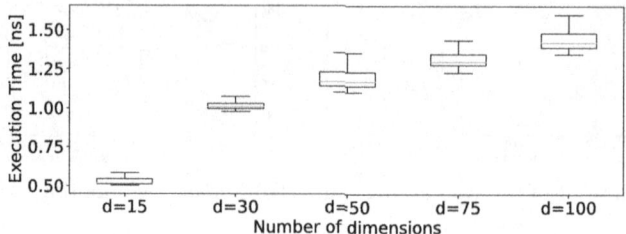

Fig. 11. Execution time for varying number of dimensions

the experiments, the execution time was measured by varying the number of dimensions from 15, 30, 50, 75 and 100 using the dataset:sd1.

Figure 11 shows the execution time results. The median execution times for each number of dimensions were 0.52, 1.00, 1.17, 1.30 and 1.42 [ns]. Overall, the execution times are very small, indicating that real-time processing is possible. It can also be confirmed that the rate of increase in execution time is relatively slow when the dimension is increased, which means that the scalability of the proposed method is not problematic.

6 Conclusions

This paper proposes a singular value-based explainability method for anomaly. The proposed method used dynamic mode decomposition to calculate the predictions excluding each singular value, and then calculated the singular value that explains the prediction error between each of them and the actual observed values by calculating and comparing scores using norms and variances.

Experiments confirmed the effectiveness of the method on both synthetic and real-world datasets, and showed that using the norm to score prediction errors produced better results than using variance.

Future work exists on developing better scoring functions and methods to better specify the latent dynamics represented by the singular values.

Acknowledgements. This paper is based on results obtained from the project, "Research and Development Project of the Enhanced infrastructures for Post-5G Information and Communication Systems" (JPNP20017), commissioned by the New Energy and Industrial Technology Development Organization (NEDO), JST CREST Grant Number JPMJCR22M2, and JSPS KAKENHI Grant Numbers JP23K24949 and JP23K16888.

References

1. Bou, S., Amagasa, T., Kitagawa, H.: Scalable keyword search over relational data streams by aggressive candidate network consolidation. Inf. Syst. **81**, 117–135 (2019)

2. Bou, S., Amagasa, T., Kitagawa, H.: Intrans: fast incremental transformer for time series data prediction. In: DEXA, pp. 47–61 (2022)
3. Bou, S., Amagasa, T., Kitagawa, H.: Finformer: fast incremental and general time series data prediction. IEICE Trans. Inf. Syst. **E107.D**(5), 625–637 (2024)
4. Bou, S., Amagasa, T., Kitagawa, H.: O(1)-time complexity for fixed sliding-window aggregation over out-of-order data streams. IEEE TKDE **36**(11), 6745–6757 (2024)
5. Bou, S., Amagasa, T., Kitagawa, H., et al.: PR-MVI: efficient missing value imputation over data streams by distance likelihood. In: iiWAS, pp. 338–351 (2022)
6. Bou, S., Amagasa, T., Kitagawa, H., et al.: Efficient missing value imputation by maximum distance likelihood . In: IEEE BigData, pp. 331–338 (2023)
7. Bou, S., Kitagawa, H., Amagasa, T.: L-bix: incremental sliding-window aggregation over data streams using linear bidirectional aggregating indexes. Knowl. Inf. Syst. **62**(8), 3107–3131 (2020)
8. Bou, S., Kitagawa, H., Amagasa, T.: CPiX: real-time analytics over out-of-order data streams by incremental sliding-window aggregation. IEEE Trans. Knowl. Data Eng. **34**(11), 5239–5250 (2022)
9. Breunig, M., Kriegel, H., Ng, R., Sander, J.: LOF: identifying density-based local outliers. In: SIGMOD (2000)
10. Cheong, M., Wu, M.C., Huang, S.H.: Interpretable stock anomaly detection based on spatiotemporal relation networks with genetic algorithm. IEEE Access **9** (2021)
11. Demo, N., Ortali, G., Gustin, G., Rozza, G.: An efficient computational framework for naval shape design and optimization problems by means of data-driven reduced order modeling techniques. BUMI **14**, 211–230 (2021)
12. Kawakami, S., Bou, S., Amagasa, T.: Lsix: a scheme for efficient multiple continuous window aggregation over streams. In: Big Data Analytics and Knowledge Discovery: 26th International Conference, Naples, Italy, pp. 322–328 (2024)
13. Kutz, J.N., Brunton, S.L., Brunton, B.W., Proctor, J.L.: Dynamic mode decomposition: data-driven modeling of complex systems. SIAM **121** (2016)
14. Mahdavinejad, M.S., Rezvan, M., Barekatain, M., Adibi, P., Barnaghi, P., Sheth, A.P.: Machine learning for internet of things data analysis: a survey. Digital Commun. Netw. (2018)
15. Mokoena, T., Marivate, V., Celik, T.: Why is this an anomaly? Explaining anomalies using sequential explanations. Pattern Recogn. **121**, 108227 (2022)
16. Morales, G., Bifet, A., et al.: IoT big data stream mining. In: ACM SIGKDD (2016)
17. Park, S., Moon, J., Hwang, E., Lecouat, B., Chandrasekhar, V.: Explainable anomaly detection for district heating based on Shapley additive explanations. In: The 2020 International Conference on Data Mining Workshops (2020)
18. Rewicki, F., Norman, R.V.: The EDEN ISS 2020 telemetry dataset
19. Samariya, D., Aryal, S., Ting, K.M., Ma, J.: A new effective and efficient measure for outlying aspect mining. In: International Conference on Web Information Systems Engineering (2020)
20. Wang, H., Jiang, W., Deng, X., Geng, J.: A new method for fault detection of aero-engine based on isolation forest. Measurement **185**, 110064 (2021)
21. Zhang, J., Lou, M., Ling, T.W., Wang, H.: HOS-miner: a system for detecting outlyting subspaces of high-dimensional data. In: Proceedings of the 30th International Conference on Very Large Data Bases (VLDB'04) (2004)
22. Zhou, C., Paffenroth, R.C.: Anomaly detection with robust deep autoencoders. In: ACM International Conference on Knowledge Discovery and Data Mining (2017)

Exploring Quantum Bootstrap Sampling for AQP Error Assessment: A Pilot Study

Feng Yu[✉] and Raya Jahan

Department of Computer Science and Information Systems, Youngstown State University, Youngstown, OH, USA
fyu@ysu.edu, rjahan01@student.ysu.edu

Abstract. Error assessment for Approximate Query Processing (AQP) is a challenging problem. Bootstrap sampling can produce error assessment even when the population data distribution is unknown. However, bootstrap sampling needs to produce a large number of resamples with replacement, which is a computationally intensive procedure. In this paper, we introduce quantum bootstrap sampling (QBS) framework to generate bootstrap samples on a quantum computer and produce an error assessment for AQP query estimations. The quantum circuit design is included in this framework.

Keywords: Quantum Computing · Bootstrap Sampling · Approximate Query Processing · Qubit · Superposition

1 Introduction

In the era of big data, computing complex queries requires a tremendous amount of time, posing emerging challenges in many computing fields. Approximate Query Processing (AQP) [9–11] aims to reduce the original big data into small data by employing sampling methods and statistical estimators to provide prompt and accurate approximated query answers.

A well-known problem for AQP research is to estimate the error of a query estimation. The existing methods in this research usually require strict assumptions on the original dataset. Bootstrap sampling [4,23] is a nonparametric statistical method that can produce estimations of the error for a given AQP answer even when the statistics of the original dataset are unknown. However, bootstrap sampling requires producing a large number of bootstrap resamples, which are random samples with replacement, based on the sample data, or sample synopses, used by AQP methods. After that, estimations on the bootstrap resamples shall be produced, named bootstrap replications. This procedure can be computationally intensive when a large number of bootstrap samples is needed, and even worse when the sample dataset is also large.

Different from classical computers, which use classical bits, quantum computers are developed to compute based on quantum bits, or qubits. Unlike classical bits, which can only be either 0 or 1 at a time, a qubit can be both $|0\rangle$ and $|1\rangle$ at the same time, named a superposition. The superpositions are employed by a quantum computer to achieve quantum parallelism to tackle computing-intensive problems. For example, the large number factoring problem, which forms the security backbone of modern cryptographic methods, such as RSA, can be promptly solved by Shor's algorithm [19] on a quantum computer.

The contribution of this research is threefold. First, we explore using a quantum computer to implement bootstrap resampling with replacement on a given sample of data. Second, we introduce quantum bootstrap sampling (QBS), a bootstrap sampling framework used for AQP error assessment implemented using quantum computing. Finally, we design the quantum circuit to implement the introduced quantum bootstrap sampling framework. Experimental results show that the proposed quantum circuit can correctly produce desired output.

The paper is structured as follows. Section 2 includes the background. Section 3 includes the problem statement. Section 4 introduces the framework of quantum bootstrap sampling. We present experimental results in Sect. 5. The related work is included in Sect. 6. The conclusion and future work are included in Sect. 7.

2 Background

Bootstrap sampling is a non-parametric statistical method often employed when error assessment is required for an estimator based on a statistical sample. A unique procedure in bootstrap sampling is *resampling*, which will generate many new random samples with replacement, named bootstrap samples, from the given sampled dataset. Usually, the larger the number of bootstrap samples generated, the better the accuracy of the error assessment. However, this process can be computationally intensive if many bootstrap samples are needed.

Given a sample $\vec{y} = (y_1, y_2, ..., y_n)$ from an unknown distribution F, a bootstrap sample $\vec{y}^* = (y_1^*, y_2^*, ..., y_n^*)$ is a resampled collection obtained by randomly sampling n times with replacement from $\vec{y}$. For example, if $n = 5$, we might obtain various bootstrap samples with different combinations such as $\vec{y}_1^* = (y_5, y_3, y_1, y_2, y_1)$, $\vec{y}_2^* = (y_2, y_5, y_4, y_1, y_2)$, $\vec{y}_3^* = (y_3, y_3, y_2, y_3, y_4)$ etc. After summarizing the frequency of each element sampled, we obtain the distribution of a bootstrap sample, $\hat{F} = (\hat{f}_1, \hat{f}_2, ...)$, where $\hat{f}_k = \#\{y_i^* = y_k\}/n$.

Figure 1 depicts the examples of bootstrap samples and bootstrap distribution.

A typical application of bootstrap sampling is to estimate the standard error of a sample estimator even when the population distribution is unknown. Suppose that we are interested in a parameter $\theta = t(F)$ calculated based on $\vec{y}$. Usually, we do not have all the information of F, but can only calculate an estimation of θ from the given sample $\vec{y}$ denoted by $\hat{\theta} = s(\vec{y})$. For each bootstrap sample $\vec{y}^*$ we can generate a *bootstrap replication* of $\hat{\theta}$, denoted by

$$\hat{\theta}^* = s(\vec{y}^*)$$

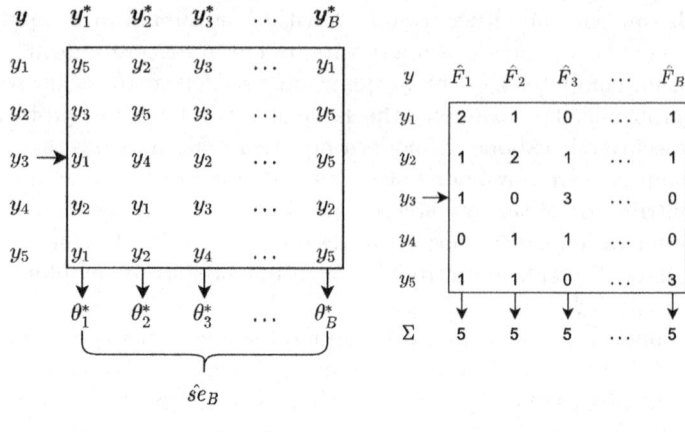

Fig. 1. Example: Bootstrap Sampling

For example, when $\hat{\theta}$ is the sample mean $\bar{y}$, a bootstrap replication $\hat{\theta}^*$ is the sample mean on a bootstrap sample $\bar{y}^*$.

After generating several B independent bootstrap samples, we can obtain the standard error of all $\hat{\theta}^*$, i.e. $\widehat{se}_B(\hat{\theta}^*)$, called the *bootstrap estimation of the standard error*. When $B \to \infty$, $\widehat{se}_B(\hat{\theta}^*) \to se_{\hat{F}}(\hat{\theta}^*)$.

$se_{\hat{F}}(\hat{\theta}^*)$ is called the *ideal bootstrap estimation* of the ground truth standard error of $\hat{\theta}$, $se_F(\hat{\theta})$. We say that $se_{\hat{F}}(\hat{\theta}^*)$ is a *plug-in* estimate of $se_F(\hat{\theta})$ that uses the empirical distribution $\hat{F}$ in replacement of the unknown distribution F.

$\widehat{se}_B(\hat{\theta}^*)$ can be calculated as,

$$\widehat{se}_B(\hat{\theta}^*) = \left[\frac{1}{B-1} \sum_{i=1}^{B} \left(\hat{\theta}^*(i) - \bar{\theta}^* \right)^2 \right]^{\frac{1}{2}} \quad (1)$$

where $\bar{\theta}^* = \sum_{i=1}^{B} \hat{\theta}^*(i) / B$.

3 Problem Statement

3.1 Selection Query Estimation

In this research, the considered query formulation, Q, is as follows:

```
SELECT Agg(attributes) FROM table WHERE conditions;
```

The Agg is an analytical aggregation function. In this pilot study, we focus on the COUNT function in the analytic query. A COUNT query can be considered as a simpler case of SUM and AVG queries in AQP. [24] introduced a comprehensive AQP framework for common aggregate functions such as SUM and AVG.

3.2 Query Size Estimation

To estimate the query Q using approximate query processing, it will first be executed on the sample S of the original table. Each sample tuple $u_i \in S$ will produce a sample query result y_i based on the aggregation function. If the aggregation function is COUNT and the primary key is included in the attribute collection, then y_i will either be 1 if u_i satisfies the selection condition, or 0 otherwise. The query result Y_s on the sample S can be calculated as $Y_s = \sum_{i=1}^{n} y_i$, where $n = |S|$ or the sample size.

Given the size of the original table R as N, the sample fraction $f = n/N$ and the ground truth query result of Q on the original table R can be estimated as

$$\widehat{Y} = \frac{Y_s}{f} \tag{2}$$

3.3 Error Assessment for Query Estimation

Equation 2 is an unbiased estimation of the query result of Q. The next step is to assess the error produced by the estimation. Bootstrap sampling is a useful tool that can be employed to assess the query estimation based on the given sample query results.

When the aggregation is COUNT, the sample query result $S_Q = \{y_i\}_{i=1}^{n}$ includes the tuple contribution, y_i either 1 or 0, depending u_i satisfying the query condition or not. We can perform bootstrap sampling on S_Q and generate bootstrap samples $\{\vec{y}_j\}_{j=1}^{B}$ where B is the total number of bootstrap samples. Each $\vec{y}_j = \{y_{j,i}\}_{i=1}^{n}$ is a bootstrap resample of S_Q.

Using Eq. 1 on each $\vec{y}_j$, $j = 1, ..., B$, we obtain the bootstrap replications,

$$\widehat{Y}_j = \frac{Y_{\vec{y}_j}}{f} \tag{3}$$

For example, if the aggregation is COUNT, then the estimator is

$$\widehat{Y}_j = \frac{1}{f} \sum_{i=1}^{n} y_{j,i} \tag{4}$$

The bootstrap sampling is repeated for B times, and a collection of bootstrap replications is obtained, denoted by $\widehat{Y}_B = \{\widehat{Y}_j\}_{j=1}^{B}$. The bootstrap standard deviation is computed as

$$\widehat{se}_B = \left[\frac{1}{B-1} \sum_{j=1}^{B} (\widehat{Y}_j - \overline{\widehat{Y}}_B)^2 \right]^{\frac{1}{2}} \tag{5}$$

where $\overline{\widehat{Y}}_B$ is the sample mean of all bootstrap replications $\widehat{Y}_B$.

Suppose the significance level is denoted as α, where α is a probabilistic value. Common choices of α include 5% for a 90% level of significance and 2.5%

for a 95% level of significance, respectively. The standard method for bootstrap CI (confidence intervals) is calculated as

$$\left(\widehat{Y} - z^{(1-\alpha)} \cdot \widehat{se}_B, \widehat{Y} + z^{(1-\alpha)} \cdot \widehat{se}_B\right) \tag{6}$$

where $\widehat{Y}$ is the query estimation and $z^{(1-\alpha)}$ is the $100(1 - \alpha)$th percentile of a standard normal distribution. For example, for 90% level of significance, $z^{(.95)}$=1.645, and for 95% level of significance, $z^{(.975)}$=1.960.

3.4 Quantum-Based Bootstrap Sampling for Error Assessment

In this research, we would like to investigate how to use quantum computing to accelerate bootstrap sampling for error assessment of query estimations.

The most computationally intensive procedure in bootstrap sampling is to generate a large number of bootstrap samples. Each bootstrap sample consists of random samples with replacements from a given sample of data. For the problem of approximate query processing, the sample data shall be the tuple sample results in $S_Q = \{y_i\}_{i=1}^n$.

The technical questions include the following:

1. *Quantum bootstrap resampling*: How to employ the quantum method to generate random samples with replacement given a set of tuple sample results. This will involve generating random sampling IDs with replacement using quantum computing.
2. *Quantum bootstrap replication*: How to compute bootstrap replications based on the quantum bootstrap samples generated. We need to translate the randomly sampled tuple IDs into the query tuple results.

4 Quantum Bootstrap Sampling

This research introduces a *hybrid* framework, including both classical and quantum computing, to accelerate bootstrap sampling for AQP error assessment. Two of the internal processes will employ quantum computing techniques, including the bootstrap random index generation for bootstrap resampling and the query tuple result computation for bootstrap replication computation. After obtaining many bootstrap replications, the total bootstrap standard deviation can be efficiently computed on a classical computer by Eq. 1.

4.1 Quantum Resampler

We employ quantum superpositions to produce resampled tuple IDs with replacement from a given set of sample tuple results S_Q. Given an n-qubit quantum system and a pure state of $|0\rangle^{\otimes n}$, a superposition can be produced by using the Hadamard gates $H^{\otimes n}$, which will be $\psi_1 = (H|0\rangle)^{\otimes n}$. Each qubit of ψ_1 is

$H|0\rangle = \frac{1}{\sqrt{2}}(|0\rangle + |1\rangle)$, or a one-qubit superposition, that can be measured either $|0\rangle$ or $|1\rangle$ each with a 50% probability, respectively.

The n-qubit superposition bit string converted to decimal will represent random numbers for resampled tuple IDs with the same probability. In addition, the quantum superpositions will not inhibit repetitions of random index numbers. Observed in different quantum measurements, the same index number generated from the superpositions can be repeated. This enables *sampling with replacement* of the tuples in the given sample dataset. Therefore, the quantum superposition can generate uniform random samples with replacement from the given sample data, which can be used for bootstrap sampling.

With the bootstrap sample index number $|i\rangle$ generated, we employ QRAM [2] to convert $|i\rangle$ to the tuple sample results, y_i, in S_Q. QRAM can store S_Q into qubit format and encode an index number $|i\rangle$ to the tuple data $|y_i\rangle$ in S_Q. A common QRAM implementation is using the Bucket-Bridge approach [5]. Our focus in this work is to use QRAM for tuple retrieval.

After passing the $|i\rangle$ through the QRAM we get $\psi_2 = 2^{-\frac{n}{2}} \sum_{i=1}^{n} |i\rangle|y_i\rangle$ which represents the tuple qubits in a quantum bootstrap sample B_i. Assuming we measure after QRAM, the measurement of ψ_2 will be an equal chance of $|y_i\rangle$ in S_Q, which is encoded in quantum. The module of quantum resampling can be considered as one circuit named *Quantum reSAmpler (QSA)*.

4.2 Quantum Bootstrap Replication

In order to obtain a bootstrap resample $\vec{y}_j = \{y_{j,i}\}_{i=1}^{n}$, we need to perform n times of resampling the same size as the sample S_Q. We can either repeatedly run the QSA for n times or create n QSA in parallel. For a simpler design of quantum circuits, we use the parallel QSA design, which comprises n QSA that can simultaneously create n of $|y_{j,i}\rangle$'s.

To sum the tuple result values in a quantum bootstrap sample, we employ a quantum counter [6,8], denoted by QC, which can perform counting on qubit inputs. In this research, the aggregation function is COUNT and the tuple result $y_{j,i}$ is either 1 if the tuple satisfies the filtering condition or 0 otherwise. The quantum counter QC will count the sum of the corresponding value $y_{j,i}$ in a quantum bootstrap sample encoded in $|y_{j,i}\rangle$, which can be either $|0\rangle$ or $|1\rangle$. The quantum counter comprises control registers and counter registers. Each time there is a $|1\rangle$ input to the control registers, the counter registers will increment by $|1\rangle$, which represents a binary counting number in Least Significant Binary (LSB) form.

After passing ψ_2 through QC, we will obtain the assembled state of $\psi_3 = \sum_{i=1}^{n} \alpha_i |i\rangle |y_{j,i}\rangle |\#\{y_{j,i} \in \vec{y}_j | y_{j,i} = 1\}|$, where α_i is the amplitude of the quantum bootstrap sample $\vec{y}_j$ generated by the quantum resampler, and $\sum_{i=1}^{n} \alpha_i^2 = 1$. After measuring ψ_3 at the end of the quantum circuit and translating the quantum value to binary using LSB, we obtain the classical sum of $y_{j,i}$ in $\vec{y}_j$, which is $Y_{\vec{y}_j}$. Using a classical computer, we can convert $Y_{\vec{y}_j}$ to $\hat{Y}_j$ by Eq. 3, $\hat{Y}_j = Y_{\vec{y}_j}/f$, where $f = n/N$ is the sampling ratio.

Algorithm 1. Quantum Bootstrap Sampling

Input: $S_Q = \{y_i\}_{i=1}^{n}$: Sample tuple results
Output: $\widehat{Y}_{B_i}$: A bootstrap replication of query estimation
1: $\psi_1 \leftarrow (H|0\rangle)^{\otimes n}$ ▷ Generate a superposition using Hadamard gates
2: $\psi_2 \leftarrow \text{QRAM}(\psi_1)$ ▷ Pass ψ_1 through the QRAM
3: $\psi_3 \leftarrow \text{QUANTUM\_COUNTER}(\psi_2, n, \log_2(n+1))$ ▷ Pass ψ_2 through the quantum counter
4: $Y_{B_i} \leftarrow$ Measuring ψ_3 ▷ Measure a quantum output
5: **return** $\widehat{Y}_{B_i} \leftarrow \frac{Y_{B_i}}{f}$ ▷ Compute on a classical computer

---

Algorithm 2. QUANTUM_COUNTER$(\cdot, p, q)$

Input: p: number control qubits; q: a factor number of count qubits
Output: QC: a quantum circuit of the quantum counter
1: QC ← a quantum circuit with p control qubits $s_0, \ldots, s_{p-1}$ and q counter qubits $u_0, \ldots, u_{q-1}$
2: **for** $i \leftarrow p-1$ **to** 0 **do**
3:    **for** $j \leftarrow q-1$ **to** 0 **do**
4:       Add an MCX gate with s_i, $u_0,\ldots,u_{j-1}$ as control and u_j as the target qubit
5:    **end for**
6: **end for**
7: **return** QC ▷ generate quantum circuit

Algorithm 1 demonstrates the quantum bootstrap sampling procedure. We can repeat the procedure many times as required and accumulate a collection of classical values of bootstrap replications $Y_{\bar{y}_j}$ or Y_{B_i}. We can employ a classical computer to calculate the bootstrap standard deviation and the confidence interval.

Algorithm 2 demonstrates the construction of a quantum counter. The arguments p and q determine the number of control qubits and counter qubits. The control qubits can only be either $|0\rangle$ or $|1\rangle$ state. The counter qubits are initialized as $|0\rangle$ state. The designed quantum counter accumulates how many control qubits are in the state $|1\rangle$. Given p control qubits, $q = \log_2(p+1)$ counter qubits will be sufficient to count all possible inputs.

4.3 Quantum Circuit Design

Figure 2 includes the circuit design of the quantum sampling framework. Figure 2a depicts the design of the quantum resampler (QSA). The input states are initialized in $|0\rangle^{\otimes n}$. A superposition is produced after the Hadamard gates to simulate the randomly sampled indices with replacement. After that, the QRAM will translate the random sampling indices to the sample tuple results.

Figure 2c illustrates the overall quantum bootstrap sampling framework. The sampled tuple results from multiple QSA in parallel will be fed into a quantum counter QC. The quantum counter will count the total of non-zero tuple results in each bootstrap resample, which can be measured and converted to a classical bit value in LSB format to produce a bootstrap replication.

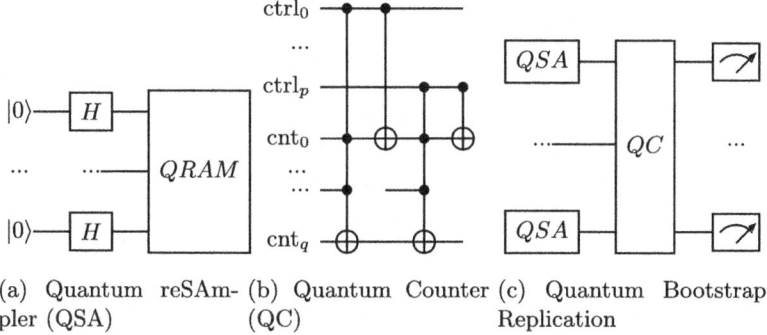

(a) Quantum reSAmpler (QSA)   (b) Quantum Counter (QC)   (c) Quantum Bootstrap Replication

Fig. 2. Quantum Circuit Design

4.4 Complexity

Algorithm 1 can instantly compute a bootstrap replication after each measurement. The complexity of each measurement is $O(1)$. To generate M bootstrap replications, the total complexity will be $O(M)$.

4.5 SUM and AVG

To process SUM and AVG queries, actual numeric values $y_i \in \mathbb{R}$, instead of binary values (0 or 1), can be employed for input sample $S = \{u_1, u_2, ..., u_n\}$. These values will be encoded in QRAM using a fixed-width binary format, for instance, using 5 qubits for values up to 31. The QRAM can map an address state $|i\rangle$ to a data value $|y_i\rangle$ using Toffoli and X gates. For example, address $|001\rangle$ can be mapped to data $|10100\rangle$ for $y_1 = 20$. To accumulate QRAM outputs, we can employ the quantum ripple-carry adder circuit [3] that can add QRAM outputs to an accumulated sum.

5 Experiment

5.1 Experiment Design

The quantum bootstrap sampling includes two procedures, namely quantum bootstrap resampling and quantum counting. We designed two experiments to simulate the two procedures. We used the IBM Qiskit [7], a Python-based high-level quantum design software, to implement the proposed quantum circuit and conduct simulated experiments. Because of the current limitation in quantum computing, the experiments involved a maximum of three qubits, which can process bootstrap sampling of $n = 2^3 = 8$ different tuples. The source code of the experiments is available on GitHub[1].

5.2 Quantum Resampling Test

Figure 3 shows a circuit implementation for quantum sampling with replacement. We initialize our experiment with three qubits in the $|0\rangle$ state. In this circuit, each $|0\rangle$ goes through the Hadamard gate, resulting in the superposition of $\frac{1}{\sqrt{2}}(|0\rangle + |1\rangle)$ that generates all possible paths simultaneously. The superposition undergoes a QRAM, which encodes the superposition into a quantum memory address and retrieves the data state from that address.

For demonstration, the QRAM in this experiment is initialized using a static array where each odd address has a value of 1, and each even address has a value of 0, namely Array $= [0, 1, 0, 1, 0, 1, 0, 1, 0]$. In the AQP application for a COUNT query, value 0 means the tuple result is 0 or not selected, and 1 means the tuple result is 1 or selected. The values of 0 and 1 will be encoded as $|0\rangle$ and $|1\rangle$, respectively.

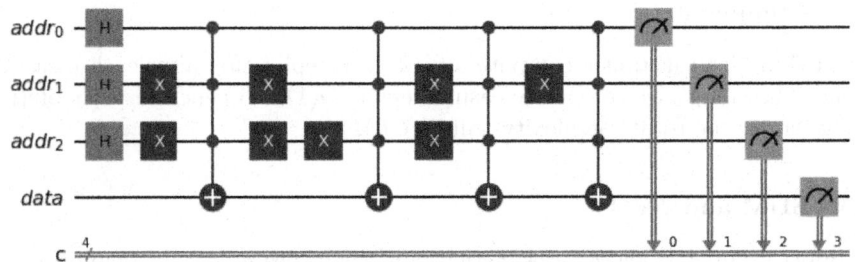

Fig. 3. An Implemented QRAM Circuit for Bootstrap Resampling

The QRAM implemented in this experiment leverages insights from the Bucket-Brigade scheme [5], where we employ three qubits to demonstrate how classical data associated with a register's address is encoded and retrieved through the properties of superposition and entanglement.

The procedure for finding the data qubit values (sample tuple results y_i) corresponding to the superposition address register states is as follows:

1. To set up the register, we employ three address qubits (addr$_i$) representing 2^3=8 memory locations. We have one data qubit (data) that stores the output value corresponding to the selected address. A qubits are initialized as $|0\rangle$ states. A classical register (c) of size four is used to store the measurement results from both the address and data qubits.

---

[1] Experiment code: https://github.com/YSU-Data-Lab/quantum_bootstrap.git

2. Using three Hadamard gates, the address qubits are converted into the superposition states, generating all possible memory addresses in parallel.
3. We employ Toffoli gates (also known as Multi-Controlled NOT gate or MCX gate) [13] to flip the data qubit according to the states of address qubits. X (NOT) gates are employed to control the address qubits to match the bitstring pattern in the data array for the QRAM. For example, when the address qubits after Hadamard gates are $|000\rangle$, the data qubit shall be measured as $|0\rangle$; whereas for address qubits $|001\rangle$, the data qubit shall be $|1\rangle$.

In this experiment, we have executed 1024 shots for generating samples with replacement using Qiskit's AerSimulator [7]. Measurements on all addresses and data qubits have been performed accordingly. Due to Qiskit's little-endian output generation format, we reverse the bit ordering of address qubits to make it human-readable.

Table 1 shows all possible QRAM inputs and outputs. Each row correlates to a possible state from sampling measurement results. When the address is odd, the data qubit returns 1 which is consistent with the encoding mechanism.

Figure 4 depicts the frequency of each quantum state being sampled. After 1024 experiment shots, the data qubit value having 1 (odd addresses) has similar frequencies to even addresses. It verifies that the functionalities of superposition are intact.

Table 1. Measured QRAM Output States

Address (Binary)	Address (Decimal)	Data Qubit	Count
000	0	0	128
010	2	0	135
100	4	0	125
110	6	0	131
001	1	1	129
011	3	1	116
101	5	1	126
111	7	1	134

5.3 Quantum Counter Test

Figure 5 depicts a simplified quantum counter circuit to count the number of qubits in $|1\rangle$ state from the output of a quantum resampler. Our implementation is inspired by Jiang et al. [8]. The output count result is stored in binary format, representing a ripple-carry adder-like binary counter. The output number represents the total of tuple results of 1, or selected, in the COUNT query, which can be used to compute the bootstrap replication.

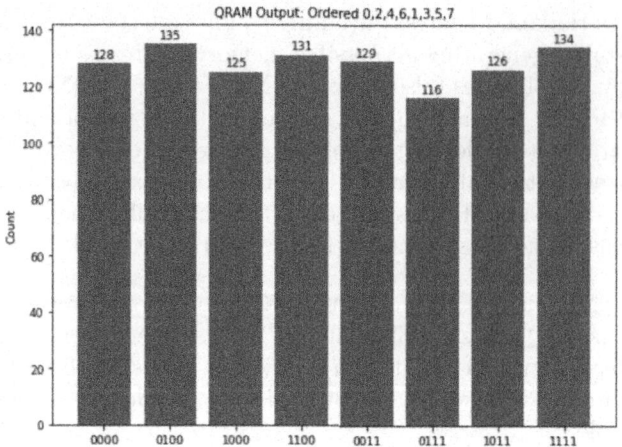

Fig. 4. A Distribution of QRAM Output States

The setup of the quantum counter circuit is as follows. In the previous test, we employed three qubits representing the sample data size of $n = 8$. In the implemented quantum counter, the number of control qubits is eight and the number of counter qubits is $4 = \log_2(8+1)$. The function of the quantum counter is to count the total of $|1\rangle$ states in the control qubits. The control qubits are also initialized as $|0\rangle$ and undergo the Hadamard gates to generate a superposition to simulate all possible states of the control qubits.

Initially, all counter qubits are initialized as $|0\rangle$. The counter qubits are connected to each control qubit using Toffoli and control-NOT gates. The numbers in the counter qubits are represented in the LSB format. For each $|1\rangle$ state in a control bit, the value in the counter bits will be incremented by 1. To simulate the quantum counter for each quantum bootstrap sample, we run the quantum counter circuit using the AerSimulator with one shot each time.

Table 2 depicts an example quantum counter output during one simulation. The input control qubits are $|00011111\rangle$, namely the control qubits 0 to 4 are in the $|1\rangle$ state. The output of the quantum counter, or the counter qubits, is $0101_2 = 5$. Therefore, the quantum counter correctly encoded the total $|1\rangle$'s in the control qubits into the counter qubits.

Table 2. One Result of Quantum Counter Simulation

Description	Value
Full bitstring measured	$\|010100011111\rangle$ (last bit to first bit)
Control bits measured	$\|00011111\rangle$ (5 ones)
Counter bits measured	$\|0101\rangle$ (binary of value 5)

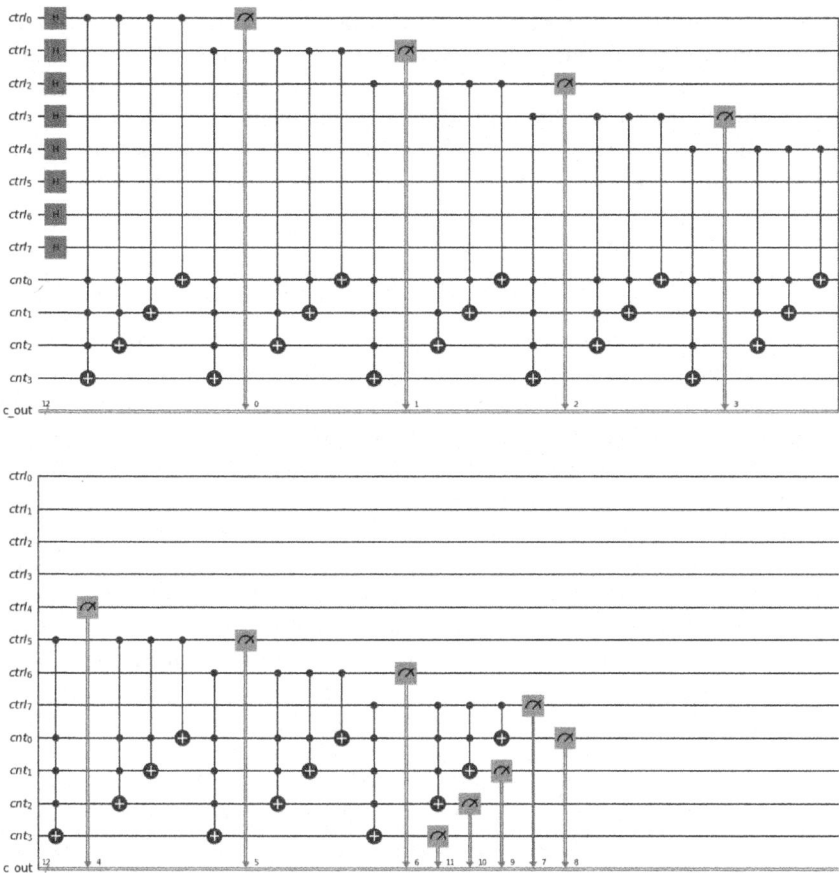

Fig. 5. A Simplified Quantum Counter to Count Qubits ($n \leq 8$) in $|1\rangle$ State

6 Related Work

Wu et al. [22] proposed a hybrid classical-quantum sampling framework known as an amplifier for sampling-based approximate query processing (AQP) to identify normal and rare groups. It amplifies the signals of the highly selective and skewed distributed rare groups iteratively by implementing Grover's diffusion operator and QRAM. Employing a diffusion operator, the scheme demonstrated the power of quadratic acceleration $O(\sqrt{N})$ to achieve balanced sampling across various data categories. The classical sampling approach categorizes a set of normal and rare groups, whereas the quantum sampling method was used to amplify the rare groups in [22]. The classical sampling-based AQP is considered unsuitable for Group-By queries with high selectivity. Quantum computing can enhance a specific desired quantum state through amplitude amplification without the necessity of data pre-processing. This research has conducted several experiments where it is observed that, with the increased number of amplification iterations, the probability of obtaining desired tuples also increased remarkably. Oracle was used to mark the index 0 of the desired tuple, and the diffusion operator was used to amplify its probability and spread the quantum state around the mean. Finally, using QRAM, they encoded the quantum state $|i\rangle$ to $|t_i\rangle$ according to the address mapping. This work was mainly on addressing the single-table sampling problem for integer values instead of strings.

[12, 18, 20, 21] studied optimization of sampling for AQP. However, how a quantum algorithm helps optimize bootstrap sampling for faster AQP of large-scale databases is still understudied.

Phalak et al. [16] compared various types of QRAM (Bucket-Brigade, Fan-Out, Flip-Flop, EQGAN, Approximate PQC) with classical RAM, focusing on its explanatory perspective, advantages, and significance in exponential speedup for algorithms such as Fourier Transform, discrete logarithm, and pattern recognition. When it comes to database searching, discovering differences in elements, faster data retrieval, and storing of classical and quantum data, QRAM can access multiple memory locations at once due to superposition. QRAM uses quantum swap operations and qubits ($|0\rangle$ and $|1\rangle$) for read and write operations, whereas classical RAM uses bits (0 or 1) and read/write signals. For n address registers, bucket-brigade QRAM [5] is efficient due to the $O(n)$ gate activation mechanism in comparison to RAM having $O(2^n)$. The output of QRAM is the data state of a specific address state and can be retrieved by performing entanglement operations using SWAP or CNOT gates. The quantum switches in this type of QRAM have three states $|.\rangle$ or waiting state, $|0\rangle$, and $|1\rangle$. For Bucket-Brigade QRAM, we can sequentially access the data states from the most significant bit to the least significant bit. Additionally, Fanout QRAM [15] uses qubits that control exponential $O(2^n)$ switches for single and superposition of all the addresses. Furthermore, Flip-Flop QRAM uses a multi-controlled rotation gate to store data in the address register qubit. EQGAN [14] is an entanglement-based model that optimizes training the data on a Quantum Neural Network (QNN) [1] having a classification accuracy of around 65%. Approximate PQC-based QRAM [17] can store complex image data and binary data in sequential order. Loading images from QRAM and sending them to QNN is way faster than without QRAM. Qudit-based memory architectures eliminate the need for ancillary qubits, which helps in reducing hardware overhead.

Jiang et al. in [8] introduced a simplified quantum counter to solve NP-hard exact cover problem, which reduces the number of quantum gates by $(4mb - 4m)[\pi/4\sqrt{N/M}]$ and circuit depth by $2mb[\pi/4\sqrt{N/M}]$, where b is counting qubits, m is quantum counters, N is the total number of possible input instances and M is the number of target inputs (or solutions). One question is to find the number of solutions M. To address this challenge, this study utilizes the quantum counting algorithm to estimate the number of solutions to the given problem. The quantum counting algorithm depends on quantum phase estimation to derive θ of the Grover iterator to find M. Quantum Fourier Transform (QFT) was utilized to transform states to the computational basis of $|0\rangle$ and $|1\rangle$ for measuring counting qubits. θ can be found by these measurement results. According to this work, having θ and N, the number of solutions M can be found by $M = N\sin^2(\theta/2)$. In the proposed quantum circuit, one quantum counter and one inverse quantum counter for resetting were employed. There are two types of qubit registers, namely controller and counter, in this design. The counter acts as an accumulator, and its value increases by 1 when the control qubit is $|1\rangle$ to determine how many times each element in the universal set is covered by sets in the collection. The oracle of the algorithm uses the MCX gate to control counter qubits and X gates were used to flip the ancilla qubit if every counter gives a value of 1.

7 Conclusion and Future Work

In this work, we introduced quantum bootstrap sampling, which can be used to accelerate error assessment for AQP query estimations. The framework begins by producing indices for random sampling with replacement. After that, we employed a QRAM and a quantum counter to produce the bootstrap replication. We implemented the prototypes of quantum circuits, including both the quantum resampler and the quantum counter, using IBM Qiskit. The experiments showed that bootstrap resamples can be instantly generated based on testing data. The implemented quantum counter was observed functioning correctly given inputs from the quantum resamplers. In the future, we will further investigate using quantum bootstrap sampling for more types of aggregate functions, such as SUM and AVG. We plan to design experiments using more complicated real datasets.

Acknowledgement. This research is partially supported by the Research Advancement Grant, Research Professorship Award, and Student Small Grant at Youngstown State University.

References

1. Abbas, A., Sutter, D., Zoufal, C., Lucchi, A., Figalli, A., Woerner, S.: The power of quantum neural networks. Nature Computational Science **1**(6), 403–409 (2021)
2. Arunachalam, S., Gheorghiu, V., Jochym-O'Connor, T., Mosca, M., Srinivasan, P.V.: On the robustness of bucket brigade quantum ram. New J. Phys. **17**(12), 123010 (2015)
3. Cuccaro, S.A., Draper, T.G., Kutin, S.A., Moulton, D.P.: A new quantum ripple-carry addition circuit (2004), https://arxiv.org/abs/quant-ph/0410184
4. Efron, B., Tibshirani, R.J.: An introduction to the bootstrap. CRC press (1994)
5. Giovannetti, V., Lloyd, S., Maccone, L.: Quantum random access memory. Phys. Rev. Lett. **100**(16), 160501 (2008)
6. Heidari, S., Farzadnia, E.: A novel quantum lsb-based steganography method using the gray code for colored quantum images. Quantum Inf. Process. **16**(10), 242 (2017)

7. Javadi-Abhari, A., Treinish, M., Krsulich, K., Wood, C.J., Lishman, J., Gacon, J., Martiel, S., Nation, P.D., Bishop, L.S., Cross, A.W., et al.: Quantum computing with qiskit. arXiv preprint arXiv:2405.08810 (2024)
8. Jiang, J.R., Wang, Y.J.: Using a simplified quantum counter to implement quantum circuits based on grover's algorithm to tackle the exact cover problem. Mathematics **13**(1), 90 (2024)
9. Li, F., Wu, B., Yi, K., Zhao, Z.: Wander join and xdb: Online aggregation via random walks. ACM Trans. Database Syst. **44**, 2:1–2:41 (1 2019)
10. Ling, Y., Sun, W., Rishe, N.D., Xiang, X.: A hybrid estimator for selectivity estimation. IEEE Trans. Knowl. Data Eng. **11**, 338–354 (1999)
11. Liu, Q.: Approximate query processing. In: LIU, L., ÖZSU, M.T. (eds.) Encyclopedia of Database Systems, pp. 113–119. Springer US (2009)
12. Lund, A.P., Bremner, M.J., Ralph, T.C.: Quantum sampling problems, bosonsampling and quantum supremacy. npj Quantum Information **3**(1), 15 (2017)
13. Nielsen, M.A., Chuang, I.L.: Quantum computation and quantum information. Cambridge university press (2010)
14. Niu, M.Y., Zlokapa, A., Broughton, M., Boixo, S., Mohseni, M., Smelyanskyi, V., Neven, H.: Entangling quantum generative adversarial networks. Phys. Rev. Lett. **128**(22), 220505 (2022)
15. Park, D.K., Petruccione, F., Rhee, J.K.K.: Circuit-based quantum random access memory for classical data. Sci. Rep. **9**(1), 3949 (2019)
16. Phalak, K., Chatterjee, A., Ghosh, S.: Quantum random access memory for dummies (2023), https://arxiv.org/abs/2305.01178
17. Phalak, K., Li, J., Ghosh, S.: Approximate quantum random access memory architectures. arXiv preprint arXiv:2210.14804 (2022)
18. Singh, D., Muraleedharan, G., Fu, B., Cheng, C.M., Newton, N.R., Rohde, P., Brennen, G.K.: Proof-of-work consensus by quantum sampling. Quantum Science and Technology (2023)
19. Wang, Y.: Quantum computation and quantum information (2012)
20. Wild, D.S., Sels, D., Pichler, H., Zanoci, C., Lukin, M.D.: Quantum sampling algorithms, phase transitions, and computational complexity. Phys. Rev. A **104**(3), 032602 (2021)
21. Wocjan, P., Abeyesinghe, A.: Speedup via quantum sampling. Physical Review A-Atomic, Molecular, and Optical Physics **78**(4), 042336 (2008)
22. Wu, S., Shi, M., Zhang, D., Zhao, J., Yuan, G., Chen, G.: When quantum computing meets database: A hybrid sampling framework for approximate query processing. IEEE Transactions on Knowledge and Data Engineering (2024)
23. Yu, F., Cal, S., Cheng, E., Kerns, L., Xiong, W.: Non-parametric error estimation for σ-aqp using optimized bootstrap sampling. International Journal for Computers & Their Applications **29**(1) (2022)
24. Yu, F., Hou, W.C.: CS*: Approximate query processing on big data using scalable join correlated sample synopsis. In: 2019 IEEE International Conference on Big Data (Big Data). pp. 583–592. IEEE (2019)

AI-Driven Semantic Data Quality Assessment and Scoring for Relational Databases

Antony Seabra[1,2](✉)[iD], Claudio Cavalcante[1,2][iD], Nicolaas Ruberg[1][iD], and Sergio Lifschitz[2][iD]

[1] BNDES - Área de Tecnologia da Informação, Rio de Janeiro, Brazil
{amede,cfrag,nic}@bndes.gov.br
[2] PUC-Rio - Departamento de Informática, Rio de Janeiro, Rio de Janeiro, Brazil
{amedeiros,cfraga,sergio}@inf.puc-rio.br

Abstract. This paper introduces an AI-driven approach for automated data quality assessment, scoring, and correction in relational databases, leveraging the power of Large Language Models (LLMs). Traditional data quality checks often rely on syntactic rules, overlooking the semantic context for accurate evaluation. Our method employs LLMs to perform semantic analysis of data, enabling a deeper understanding of data integrity and consistency beyond structural compliance. By analyzing data content and metadata, the LLM generates a comprehensive data quality score, reflecting the semantic accuracy and completeness of the database. Furthermore, the system provides actionable correction suggestions to address identified data quality issues. This scoring and correction system empowers data governance professionals to efficiently identify, prioritize, and remediate data quality issues. We demonstrate the efficacy of our approach through experiments on real-world relational databases, showcasing the ability of LLMs to detect subtle semantic anomalies and provide practical solutions for improved data quality management.

Keywords: Data Quality · Large Language Models · Scoring System · Relational Databases · Reliability · Integrity

1 Introduction

The integrity and reliability of data within relational databases are paramount for informed decision-making and the smooth operation of modern applications. However, ensuring data quality remains a significant challenge, particularly as databases grow in size and complexity. Traditional data quality assessment methods often rely on predefined rules and constraints, focusing primarily on syntactic correctness. While effective for detecting structural anomalies, these methods frequently overlook semantic inconsistencies that can severely impact data usability and trustworthiness. The increasing availability of powerful Large Language Models (LLMs) presents a unique opportunity to address these limitations by leveraging their ability to understand and interpret natural language and semantic relationships.

This paper proposes an AI-driven approach to automate semantic data quality assessment and scoring for relational databases, utilizing the advanced capabilities of LLMs. By incorporating semantic analysis, our method transcends the limitations of traditional syntactic checks, enabling a more comprehensive and nuanced evaluation of data quality. Additionally, we introduce a novel scoring system that quantifies the semantic accuracy and completeness of data, providing data governance professionals with actionable insights to prioritize and address data quality issues effectively. The primary contributions of this work are as follows:

Semantic Data Quality Assessment with LLMs: We present a methodology that leverages LLMs to perform semantic analysis of relational database content, enabling the detection of subtle inconsistencies and anomalies that are often missed by traditional methods.

Automated Data Quality Scoring System: We develop a quantitative customizable scoring system that reflects the semantic accuracy and completeness of the data, providing a clear and actionable metric for data governance professionals.

LLM-Driven Data Correction and Remediation: Our approach not only identifies data quality issues but also generates actionable correction suggestions, facilitating automated data remediation.

Practical Application for Data Governance: We demonstrate the practical application of our approach in 255 real-world relational database scenarios, showcasing its ability to streamline data quality management and improve data trustworthiness.

This paper is structured as follows: Sect. 2 details our proposed methodology for semantic data quality assessment and scoring, and Sect. 3 presents the experimental evaluation and results. Finally, Sect. 4 concludes the paper and outlines future directions.

2 Methodology

We categorize data quality issues into five key dimensions, each representing a key aspect of data integrity. These categories are assigned weights (w_i) based on their relative importance, with higher weights indicating more critical issues (Table 1).

2.1 LLM-Generated Code for Assessment

For each data quality category, we employ a customizable prompt, tailored to each data quality category, to guide the LLM in creating dynamic assessment scripts. This approach leverages the LLM's ability to interpret database schemas and translate natural language descriptions of data quality criteria into executable Python code. The process begins with the retrieval of the database

Table 1. Data Quality Assessment Categories

Category	Description
Completeness (w_1)	Measures the presence of all required data values. For example, a customer record missing a phone number would indicate incompleteness if it is required.
Accuracy (w_2)	Assesses the correctness of data values, ensuring they reflect real-world entities. For instance, a product price listed as "−10" would be an accuracy error.
Consistency (w_3)	Evaluates uniformity of data across tables/columns. Example: foreign key "CustomerID" in "Orders" table not found in "Customers" table.
Validity (w_4)	Checks whether data values conform to predefined formats, constraints, or business rules. A phone number stored in an incorrect format is an example of a validity issue.
Semantic Coherence (w_5)	Evaluates the contextual meaning and relationships between data elements, ensuring they make sense. A product description that contradicts its category illustrates a semantic coherence problem.

schema, which serves as the foundation for code generation. Thus, the LLM operates on metadata and schema information, not the raw data itself, during code generation. For instance, when evaluating completeness, the LLM generates Python code that will, upon execution, query the database for null values in designated columns. The generated Python code performs the following steps:

1. Connects to the relational database using appropriate database connectors.
2. Executes SQL queries, generated within the Python code via Text-to-SQL, to retrieve data relevant to the assessment category.
3. Applies data validation and analysis functions to identify data quality issues.
4. Calculates a score (s_i) for each category, representing the percentage of data that meets the quality criteria.
5. Generates suggestions for corrections for the data quality issues, in the form of SQL queries.

In addition to these core assessment functions, the generated Python code also interfaces with the LLM's natural language processing capabilities. For instance, when analyzing Semantic Coherence, the code sends retrieved data segments to the LLM for contextual analysis, allowing the system to evaluate the semantic consistency of the data against its inherent context and external knowledge sources. This enables the identification of logical inconsistencies and semantic anomalies that traditional rule-based methods would miss.

Furthermore, the LLM is used to generate Python code that will, when executed, suggest data correction operations. These operations are based on the identified data quality issues and the database schema. This code will contain SQL queries to fix the found issues. The LLM is used to generate the best correction possible, based on the context and the database schema.

2.2 Main Workflow

As shown in Fig. 1, The figure illustrates the architecture of our LLM-powered data quality assessment and correction system, outlining the flow of information and processes from audit request to audit report. The system leverages Large Language Models (LLMs) to automate and enhance data quality evaluation and correction within relational databases.

The process begins with an Audit Request (1), initiating the data quality assessment. This request triggers the establishment of a Database Connection (2), which facilitates interaction with the target database. The system then proceeds to retrieve the necessary database schema information using the Metadata Extractor (3). This extracted metadata is key for the subsequent code generation and data quality evaluation steps.

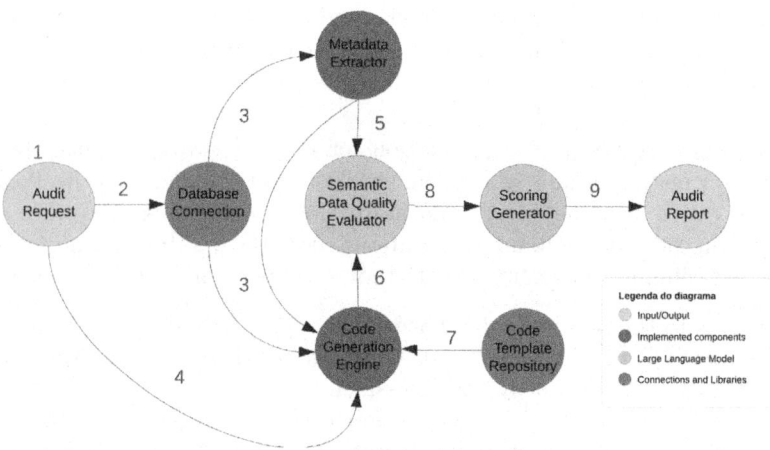

Fig. 1. Data Quality Assessment and Scoring Methodology.

The Metadata Extractor (3) forwards the schema information to the Semantic Data Quality Evaluator (5) and the Code Generation Engine. The Code Generation Engine, which is a core component utilizing an LLM, generates Python code (6) based on the retrieved schema and predefined data quality criteria. This code is stored in the Code Template Repository (7), acting as a library of assessment scripts.

The generated Python code, upon execution, performs the actual data quality assessment. It connects to the Database Connection (3), executes SQL queries to retrieve relevant data, applies validation and analysis functions, and identifies data quality issues. The results of this evaluation are then sent to the Semantic Data Quality Evaluator (8).

The Semantic Data Quality Evaluator (8) not only analyzes the results but also leverages the LLM's natural language processing capabilities to assess semantic coherence. It sends retrieved data segments to the LLM for contextual analysis, identifying logical inconsistencies and semantic anomalies beyond traditional rule-based methods.

Based on the evaluation, the Scoring Generator (8) calculates a comprehensive data quality score, reflecting the semantic accuracy and completeness of the database. This score, along with detailed findings and correction suggestions (generated by the LLM), are compiled into the final Audit Report (9).

The system also incorporates a feedback loop where the results from the Semantic Data Quality Evaluator can be used to refine and improve the generated code within the Code Generation Engine. This iterative process allows for continuous improvement and adaptation of the assessment scripts.

3 Evaluation

To evaluate the effectiveness of our LLM-based data quality scoring methodology, we conducted experiments on 255 real-world relational databases. These databases varied in size, schema complexity, and application domains, providing a comprehensive assessment of our approach's generalizability. For each database, we applied our methodology, generating Python code for data quality assessment and calculating the overall data quality score (S). We also tracked the actions taken to address the identified data quality issues and measured the percentage of problems successfully resolved.

3.1 Experimental Setup

Unlike traditional benchmark-driven approaches, our method was applied across 255 real-world databases from a large bank, each with distinct schemas, data types, and business rules, requiring adaptive and context-sensitive assessments. Instead of relying solely on fixed datasets or predefined rules, we leveraged an LLM fine-tuned for Python code generation and semantic analysis to dynamically generate tailored assessment code for each case. While completeness and accuracy evaluations followed structured methodologies–such as null value checks and outlier detection using Isolation Forest–our semantic coherence evaluation was uniquely powered by OpenAI's API, allowing the model to interpret and score data in context, capturing nuanced inconsistencies that rule-based methods cannot detect. This highlights a key difference from established approaches: our framework adapts to heterogeneous environments and includes semantic evaluation, for which no standard benchmarks currently exist.

3.2 Results

The following table presents the median values of each data quality category score and the percentage of problems solved after applying our methodology across the 255 databases (Table 2):

The results demonstrate the effectiveness of our LLM-based methodology in assessing and improving data quality in real-world databases. The median scores indicate that semantic coherence was the most challenging category, highlighting the complexity of capturing and assessing semantic relationships. However, the problem resolution rates show that our methodology facilitated significant improvements in all categories.

The variability in database schemas and data characteristics underscores the LLM's adaptability and code generation capabilities. The detailed explanations of the assessment processes illustrate the framework's ability to tailor assessment scripts to specific data quality criteria and database structures.

The high problem resolution rates suggest that our methodology provides actionable insights for data governance professionals, enabling them to identify and address data quality issues effectively. The quantitative scores and detailed

Table 2. Median Data Quality Scores and Problem Resolution Rates

Category	Median Score (%)	Problems Solved (%)
Completeness	45	65
Accuracy	38	58
Consistency	52	70
Validity	55	62
Semantic Coherence	20	68

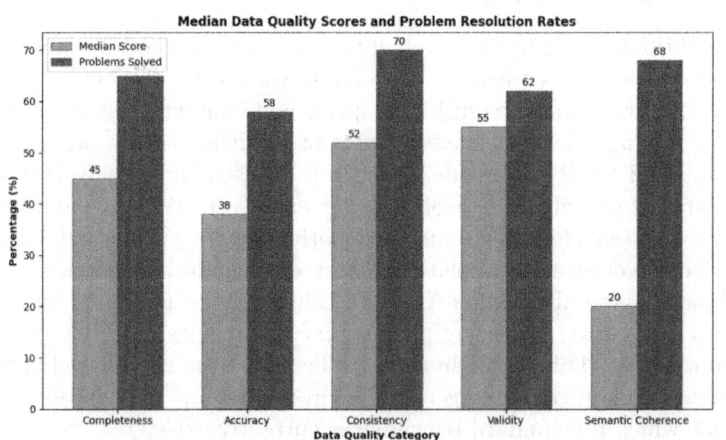

Fig. 2. LLM-Based Data Quality Issues Solved

audit reports facilitate targeted remediation efforts, leading to improved data trustworthiness and usability (Fig. 2).

3.3 Scalability of the Proposed Solution

The proposed LLM-based data quality assessment framework exhibits strong scalability, making it suitable for large and complex enterprise database environments. Experiments conducted with up to 255 real-world databases demonstrated that the solution maintains efficient performance as the workload increases. Specifically, we evaluated the following key metrics:

- **Execution Time:** Grew linearly with the number of databases, indicating a consistent and predictable processing rate.
- **Resource Utilization:** Remained stable across experiments, highlighting efficient management of CPU and memory.
- **Throughput:** Maintained a high and consistent rate of table processing, even with increasing database size and complexity.

These results confirm that the framework can scale effectively across varying volumes and complexities of data. Simulations with even larger datasets reinforced the observed linear scaling behavior, underscoring the framework's potential for enterprise-scale data governance tasks.

3.4 Comparative Analysis with Established Benchmarks

Our approach to data quality assessment differs from established benchmarks by emphasizing semantic coherence–a dimension often overlooked in traditional methods. While conventional techniques typically assess data quality using metrics like precision, recall, and F1-score, focusing on syntactic errors and constraint violations through curated datasets and formal rules (e.g., functional dependencies), they fall short in evaluating the contextual meaning and interrelationships within data. In contrast, our LLM-based framework leverages the language model's capacity to interpret data semantically, enabling the identification of inconsistencies that lie beyond formal constraints. Although the lack of standardized benchmarks for semantic coherence limits direct comparisons, our evaluation highlights the complementary strengths of LLMs in providing a more context-aware and holistic view of data quality.

4 Conclusions and Future Work

Our proposed framework distinguishes itself from traditional data quality assessment methods by leveraging LLMs to generate customized scripts capable of evaluating complex and often overlooked dimensions such as semantic coherence. While existing approaches typically rely on static rules and perform well in detecting syntactic errors or constraint violations, they struggle to capture contextual inconsistencies within the data. In contrast, our LLM-based method adapts to diverse schemas across 255 real-world databases and effectively interprets the meaning and relationships among data elements, significantly enhancing semantic evaluation. This contextual understanding is a key differentiator, especially given the absence of standardized benchmarks for semantic coherence. Looking forward, we aim to address this gap through the development of reproducible evaluation methodologies and further improvements in LLM reasoning and script generation, expanding the framework's applicability across data types and domains.

References

1. Batini, C., Scannapieco, M.: Erratum to: Data and Information Quality: Dimensions, Principles and Techniques. In: Data and Information Quality. DSA, pp. E1–E1. Springer, Cham (2016). https://doi.org/10.1007/978-3-319-24106-7_15
2. Redman, T.C.: Data Quality: The Field Guide. Butterworth-Heinemann (2013)
3. Vaswani, A., et al.: Attention is all you need. In: Advances in Neural Information Processing Systems, pp. 5998–6008 (2017)
4. Devlin, J., Chang, M.W., Lee, K., Toutanova, K.: BERT: pre-training of deep bidirectional transformers for language understanding. In: Proceedings of NAACL-HLT, pp. 4171–4186 (2018)
5. Brown, T., et al.: Language models are few-shot learners. In: Advances in Neural Information Processing Systems, pp. 1877–1901 (2020)
6. Zhong, V., Xiong, C., Socher, R.: Seq2SQL: generating structured queries from natural language using reinforcement learning. arXiv preprint arXiv:1709.00103 (2017)
7. Yu, T., et al.: Spider: a large-scale human-labeled dataset for complex and cross-domain semantic parsing and text-to-SQL task. arXiv preprint arXiv:1809.08887 (2018)
8. Rahm, E., Do, H.H.: Data cleaning: problems and current approaches. IEEE Data Eng. Bull. **23**(4), 3–13 (2000)
9. Papastefanatos, G., Vassiliadis, P., Vassiliou, Y.: Data error detection via conditional functional dependencies. In: Proceedings of the VLDB Endowment, vol. 13, no. 12, pp. 2487–2500 (2020)
10. Li, Y., Tang, B., Li, G., Zou, L.: ShadowGNN: graph neural network for text-to-SQL. In: Proceedings of the VLDB Endowment, vol. 15, no. 11, pp. 2323–2336 (2022)
11. Wang, Y., Li, G., Feng, J.: Data transformation with large language models. In: Proceedings of the ACM Web Conference 2023, pp. 1974–1984 (2023)
12. Krishnan, A., Li, G., Feng, J.: Large language models for data imputation and error correction. In: Proceedings of the ACM SIGMOD International Conference on Management of Data, pp. 1234–1245 (2023)
13. Li, Z., Wang, Y., Li, G.: Automating data transformation pipeline generation with large language models. In: Proceedings of the VLDB Endowment, vol. 16, no. 1, pp. 123–135 (2023)

Network Anomaly Detection Using Gramian Angular Field Transformation and Vision Transformer

Jaroslaw Kobiela(✉)

Institute of Computer Science, University of Opole, Opole, Poland
jaroslaw.kobiela@uni.opole.pl

Abstract. Traditional network anomaly detection methods struggle in dynamic IoT environments. We propose a novel approach for transforming time series into images using the Gramian Angular Field (GAF) for classification with a Vision Transformer (ViT). On a public IoT dataset, our model achieved 97.72% accuracy, 97.71% F1-score, and exceptionally high recall (99.34%), identifying nearly all attacks. However, this strength resulted in a higher false-alarm rate (3.85%) than that of the benchmark models. This approach is promising for prioritizing attack detection and requires further false alert reduction.

Keywords: Network Anomaly Detection · Vision Transformer (ViT) · Gramian Angular Field (GAF) · Cybersecurity · Machine Learning for Security · Network Traffic Analysis

1 The Network Security Threat Landscape

Sophisticated and widespread cybersecurity threats are continually evolving, rendering traditional signature-based protection insufficient for dynamic networks. This study proposes an advanced machine learning framework for network anomaly detection that focuses on identifying unusual attack patterns. A core challenge is balancing high threat identification performance with minimizing false alarms, which is crucial for the effectiveness of anomaly detection systems.

2 Existing Approaches to Network Anomaly Detection

Network anomaly detection has evolved from statistical methods [1] to advanced ML/DL owing to dynamic environments and zero-day threats [2]. Early ML (SVM [3], Random Forests) improved accuracy/real-time performance [4, 5]. DL models, such as LSTMs/GRUs, capture temporal aspects [6, 7], whereas CNNs excel in features. Hybrid architectures (CNNs, transformers, and attention) exhibit high performance [8–10]. Notably, MFVT [11] leveraged Vision Transformers with CNNs, demonstrating the potential of visual models. Persistent challenges include interpretability, scalability, zero-day detection [7, 12], and balancing the detection rates with false alarms [4, 5].

3 Proposed Methodology for Detecting Anomalies in Network Traffic

3.1 Data Pre-processing and Feature Engineering

The analysis utilized publicly available data [13] from the IoT Research Lab (IoTRL) at the Army Cyber Institute (ACI), which replicates a typical IoT home environment. Data collected over one week encompassed normal traffic and various cyberattacks (e.g., DoS, spoofing, brute force), available in NetFlow and PCAP formats. Pre-processing involved loading CSVs files, standardizing formats, validating numeric values, imputing missing data (KNNImputer), and scaling numerical features (Min-MaxScaler).

3.2 Research Methodology

Our anomaly detection methodology (Fig. 1) transforms time-series network data into GAF images [14, 15] for ViT classification. The raw data (normal traffic, DoS/spoofing attacks) were resampled (30s intervals), forming overlapping 150s windows. The key selected features (flow bytes/s, avg. Packet length, SYN/PSH flags, and avg. Flow IAT [16–18]) within these windows were GAF-transformed. This generated three normalized RGB channels: red (GASF [14]–temporal relations), green (GADF [15]–value variability), and blue (quadratic interpolation). A pretrained ViT-Small-Patch16-224, adapted for binary classification (80/20 train/val split), was trained for 50 epochs (early stopping) using augmentations, mixup, soft target cross-entropy loss, and AdamW (LR = 0.0001, batch = 16). The performance was evaluated (accuracy, loss, and confusion matrix) using Grad-CAM for interpretability.

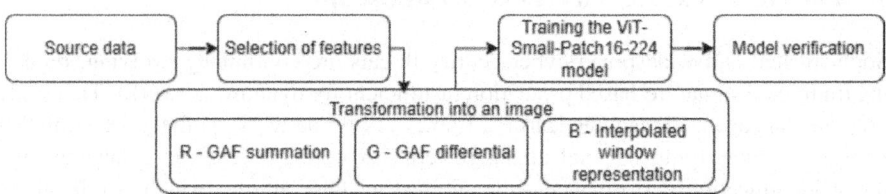

Fig. 1. Data Processing and Modeling Pipeline with GAF and ViT

4 Experiments and Evaluation of Model Performance

The accompanying diagrams (Fig. 2) show the dynamics of the Vision Transformer (ViT) learning process during the classification of network traffic into 'benign' and 'attack' categories. Analysis of the loss function and accuracy (Fig. 2, left and right panels, respectively) monitored on the training and validation sets across epochs demonstrated strong model generalization. The validation accuracy consistently reached 95–97%, with the validation loss remaining lower than the training loss, indicating effective regularization and good generalization without overfitting.

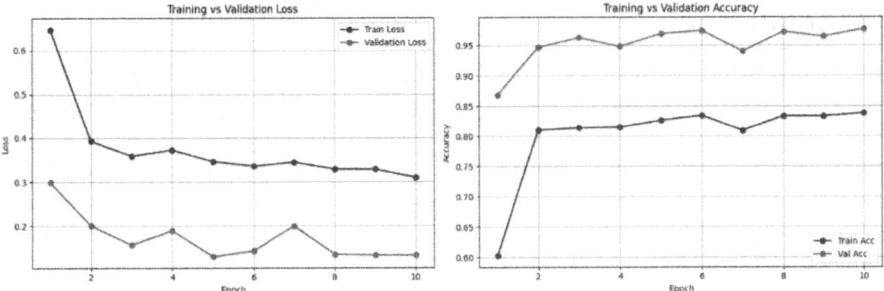

Fig. 2. Training and Validation Loss and Accuracy

4.1 Data Sets and Evaluation Indicators

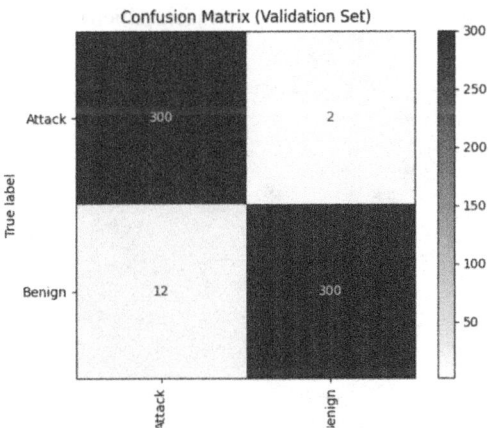

Fig. 3. Confusion Matrix

Model evaluation used a predefined validation dataset of 614 samples (150s time windows), comprising 302 'Attack' and 312 'Benign' class network traffic segments. The confusion matrix (Fig. 3) details the GAF + ViT classification results, yielding an overall accuracy of $(300 + 300)/(300 + 300 + 2 + 12) \approx 97.72\%$. For the 'Attack' class, the key metrics were precision 96.15%, recall 99.34% (FN = 2), and F1-score 97.71%. This exceptionally high recall confirms effective attack identification, albeit with a higher false alarm rate (FP = 12) than missed attacks.

4.2 Analysis of Misclassifications

Analysis of misclassified GAF images corresponding to validation set samples challenged by the GAF + ViT model indicated that false positives (FPs) were predominant (12 cases where normal movement was labelled as an attack) over false negatives (FNs, two cases where an attack was considered benign). Despite the model's high performance, these errors highlight the difficulties in distinguishing certain GAF-transformed patterns. False alarms may stem from GAF images of rare normal motion sequences resembling typical attack patterns. The few false negatives (FNs), despite the model's high sensitivity, suggested that some attacks generated atypical GAF patterns, possibly due to visual feature overlap at classification boundaries or GAF transformation specificity for the given data.

4.3 Activity Maps

The average activation intensity across the ViT layers showed stable value ranges (−0.45 to + 0.35) and activity patterns. The expected feature complexity progression in the later ViT layers was not clearly observed, potentially due to the training techniques (e.g., layer normalization and dropout). Consistent vertical/horizontal activity bands across layers suggest early, stable feature recognition. However, directly linking these GAF visual features to raw network traffic for full interpretability remains a challenge that requires further investigation.

4.4 PCA and t-SNE Analysis

PCA and t-SNE visualized the validation data distribution (Fig. 4), assessing 'Benign'/'Attack' class separability and misclassified sample characteristics (green: benign, blue: attack, red: misclassified).

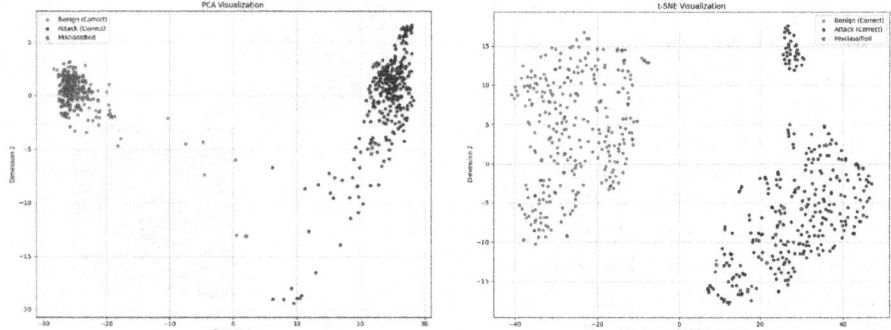

Fig. 4. PCA and t-SNE Visualization of Classification Results

PCA (Fig. 4, left) showed clear group separation; benign samples formed a compact cloud and attacked a more diffuse one. t-SNE (Fig. 4, right), reflecting local/nonlinear structures, confirmed high separability; 'Benign' formed a well-defined cluster, 'Attack' a distinct, possibly subgrouped one. Critically, the misclassified (red) points in both visualizations were almost exclusively at the inter-cluster boundaries, indicating model errors in ambiguous borderline cases near the decision surface. Both methods confirm the high performance of GAF + V iT and the successful distinction of most instances, with errors mainly at the class decision boundaries.

5 Conclusions

We proposed and evaluated a GAF + ViT model that transforms time-series network data into GAF images for the ViT classification. Comparative experiments on a public IoT dataset against RF, XGBoost, and CNN-LSTM ensured comparability through identical data/splits, consistent data representation approaches, class weights, and standard evaluation metrics.

The performance (Table 1) showed that all models achieved > 97% accuracy. Although CNN-LSTM was marginally the best overall (99.19% accuracy/99.18% F1) and XGBoost offered the highest precision (99.67%) and lowest FAR (0.0033), our GAF + ViT achieved high accuracy (97.72%) and exceptional recall (99.34%, FN = 2), excelling in sensitivity. However, this strength resulted in lower precision (96.15%) and notably higher FAR (3.85%).

Table 1. Performance Comparison of Classification Model

Model	Random Forest	XGBoost	CNN-LSTM	GAF + ViT
Accuracy	0.9886	0.9902	0.9919	0.9772
Precision	0.9934	0.9967	0.9935	0.9615
Recall	0.9837	0.9837	0.9902	0.9934
F1-Score	0.9885	0.9902	0.9918	0.9771
False Alarm Rate	0.0065	0.0033	0.0065	0.0385
TP	302	302	304	300
FP	2	1	2	12
FN	5	5	3	2
TN	305	306	305	300

The near-complete attack detection of GAF + ViT is promising, although its higher FAR necessitates further optimization (e.g., refining GAF/features) to improve precision/reduce FAR and match the balanced performance of the leading models.

6 Future Research Directions

Future work should prioritize GAF + ViT precision/FAR reduction via threshold optimization, advanced feature engineering, GAF/ViT tuning, and data augmentation. The key direction is a two-stage model (GAF + ViT for recall and XGBoost as an FP reducer). Crucial steps include assessing generalizability (diverse datasets, zero-day attacks), improving XAI for GAF-raw traffic linkage, and evaluating computational efficiency (runtimes) for real-time deployment.

References

1. Romansky, R.: Statistical analysis of empirical network traffic data from program monitoring **14** (2022)
2. Chatterjee, A., Ahmed, B.S.: IoT anomaly detection methods and applications: a survey. Internet Things **19**, 100568 (2022). https://doi.org/10.1016/j.iot.2022.100568
3. Ma, Q., Sun, C., Cui, B., Jin, X.: A novel model for anomaly detection in network traffic based on kernel support vector machine. Comput. Secur. **104**, 102215 (2021). https://doi.org/10.1016/j.cose.2021.102215

4. Kumar, S., et al.: DDoS detection in SDN using machine learning techniques. Comput. Mater. Contin. **71**(1), 771–789 (2022). https://doi.org/10.32604/cmc.2022.021669
5. Aswathy, M.C., Rajkumar, T.: Real time anomaly detection in network traffic: a comparative analysis of machine learning algorithms. Int. Res. J. Adv. Eng. Hub IRJAEH **2**(07), 1968–1977 (2024). https://doi.org/10.47392/IRJAEH.2024.0269
6. Fotiadou, K., Velivassaki, T.-H., Voulkidis, A., Skias, D., Tsekeridou, S., Zahariadis, T.: Network traffic anomaly detection via deep learning. Information **12**(5), 215 (2021). https://doi.org/10.3390/info12050215
7. ALMahadin, G., et al.: VANET network traffic anomaly detection using GRU-based deep learning model. IEEE Trans. Consum. Electron. **70**(1), 4548–4555 (2024) https://doi.org/10.1109/TCE.2023.3326384
8. Ji, C., Yu, H., Dai, W.: Network traffic anomaly detection based on spatiotemporal feature extraction and channel attention. Processes **12**(7), 1418 (2024). https://doi.org/10.3390/pr12071418
9. Li, R.: Analysis of machine learning application in campus network traffic anomaly detection. Appl. Math. Nonlinear Sci. **9**(1), 20241261 (2024). https://doi.org/10.2478/amns-2024-1261
10. Zhao, X., Huang, C., Wang, L.: MTC-NET: a multi-channel independent anomaly detection method for network traffic. Biomimetics **9**(10), 615 (2024). https://doi.org/10.3390/biomimetics9100615
11. Li, M., Han, D., Li, D., Liu, H., Chang, C.-C.: MFVT: an anomaly traffic detection method merging feature fusion network and vision transformer architecture. EURASIP J. Wirel. Commun. Netw. **2022**(1), 39 (2022). https://doi.org/10.1186/s13638-022-02103-9
12. Manokaran, J., Vairavel, G.: DL-ADS: Improved grey wolf optimization enabled AE-LSTM technique for efficient network anomaly detection in internet of thing edge computing. IEEE Access **12**, 75983–76002 (2024). https://doi.org/10.1109/ACCESS.2024.3405628
13. Bastian, N. B., Bierbrauer, D. B., McKenzie, M. M., Nack, E. N.: ACI IoT network traffic dataset. IEEE DataPort. (2023) https://doi.org/10.21227/QACJ-3X32
14. Yokkampon, U., Mowshowitz, A., Chumkamon, S., Hayashi, E.: Anomaly detection using autoencoder with Gramian Angular summation field in multivariate time series data (2022)
15. Sinanc Terzi, D.: Gramian angular field transformation-based intrusion detection. Comput. Sci. **23**(4) (2022). https://doi.org/10.7494/csci.2022.23.4.4406
16. Early, J. P., Brodley, C. E., Spafford, E. H.: Behavioral feature extraction for network anomaly detection'
17. Krupski, J., Iwanowski, M., Graniszewski, W.: Extraction of minimal set of traffic features using ensemble of classifiers and rank aggregation for network intrusion detection systems. Appl. Sci. **14**(16), 6995 (2024). https://doi.org/10.3390/app14166995
18. A. V, R. S, N. L R, and S. R.: Network intrusion detection system using optimized feature selection. In: 2024 2nd International Conference on Artificial Intelligence and Machine Learning Applications Theme: Healthcare and Internet of Things (AIMLA), Namakkal, India: IEEE, pp. 1–10 (2024). https://doi.org/10.1109/AIMLA59606.2024.10531381

Machine Learning/Artificial Intelligence Applications

Identifying Multimodal Sarcasm Based on Incongruous Knowledge Capturing and Contrastive Learning

Yan Zhu[1]( ), Chang Liu[1], and Yiqiang Peng[2]

[1] School of Computing and AI, Southwest Jiaotong University, Chengdu, China
yzhu@swjtu.edu.cn
[2] Xihua University, Chengdu, China

Abstract. People could use various types of subtle metaphorical language forms to express individual opinions or negative emotion in daily life, such as sarcasm, satire or irony. Sarcasm employs contradictory and incongruous elements to convey the difference between reality and expectation. Detecting precisely whether comments, opinions or conversations have sarcastic intention is crucial for understanding the talkers' feeling and attitude. Since the complexity of sarcasm requires certain contextual information to catch on ironic meaning correctly, how to leverage multimodal information to grasp sarcastic remarks has become a hot research topic. A Multimodal Sarcasm Detection method based on Contrastive Learning (MSDCL) is proposed in this paper to exploit multimodal information and incongruity knowledge for improving performance. MSDCL extracts fine-grained features from text and image and captures contradictory semantics between text and image by using the multi-head self-attention mechanism. A supervised contrastive learning is implemented to better learn intra- and inter-class relationships among sarcastic and non-sarcastic data, where the classification loss and contrastive learning loss are integrated to guide the multimodal data embedding. The experiment results show MSDCL outperforms the compared methods.

Keywords: Multimodal Sarcasm Detection · Incongruous Knowledge Capturing · Multi-Head Self-Attention Mechanism · Contrastive Learning

1 Introduction

As a complex but maybe frequently-used figures of speech, sarcasm demonstrates ambivalent, implied, and obscure features. Analyzing speaker's true emotion and intention from sarcastic expressions will greatly help with tasks such as sentiment detection, public opinion analysis and tracking, intelligent human-computer interaction, and so on.

Many scholars have already conducted researches on text-only sarcasm detection, such as papers [1–4]. Nowadays image as a modality provides important contextual information to the corresponding text for optimizing sarcasm mining. However, the multimodality brings new challenges, which mainly indicate the following three aspects:

- The feature extraction process is complex. Text and image modalities are often independent of each other, which require different modeling methods to capture respective information.
- Learning incongruous features between multimodalities is difficult, because not all inter- and intra-modal incongruities are evident and can convey the sarcasm clearly.
- Understanding semantics in-depth is quite important. Many researches only focus on feature extraction from multimodal image-text pairs of sarcasm samples, but ignore somewhat the inherent commonalities among different sarcasm samples.

Existing researches on multimodal sarcasm detection have considered many key aspects such as multimodality representation and fusion [5, 6], incongruous feature capturing and multimodal information extension [7, 8]. However, the intra- and inter-class relationships inherent in sarcastic and non-sarcastic samples should be exploited in depth. These relationships are concealed in the intrinsic commonalities among similar samples from the same class and also in the subtle differences between similar samples from different classes. To address the issues, MSDCL approach (**M**ulti-modal **S**arcasm **D**etection Based on **C**ontrastive **L**earning) is proposed, which detects sarcasm based on multimodal image-text pairs, inconsistent knowledge and contrastive learning to improve satiric semantics understanding. The main contributions are as follows:

- Extract and integrate fine-grained multimodal information, incl. Text, image, and OCR (Optical Character Recognition) to deal with the 1st challenge above.
- Capture the incongruous information of multimodal features with multi-head self-attention mechanism and conduct finer granular analysis on sarcasm to handle the 2nd challenge.
- Exploit sarcastic semantics by supervised contrastive learning and a positive and negative pair construction mechanism for addressing the 2nd and 3rd issue above.

The rest of the paper is organized as follows. Section 2 introduces the related researches. The details of MSDCL are addressed in Sect. 3. Section 4 analyzes the performance of MSDCL based on the verification experiments. Finally, we sum up the contribution and discuss possible future work in Sect. 5.

2 Related Work

2.1 Multimodal Sarcasm Detection

Researchers have been studying sarcasm in terms of multimodal information, because multimodal data transmits richer information than pure text. Cai et al. [5] released the first multimodal sarcasm dataset and proposed HFM for multimodal sarcasm detection. They introduced a pre-trained image recognition model to obtain image attributes and proposed three ways of fusion, i.e., early fusion, representation fusion, and modality fusion, to integrate different modal data. Xu et al. [9] pointed out the work in [5] mainly focused on the multimodal data fusion and did not address the issues on multimodal contrast and semantic context understanding. They proposed a D&R Net, where the D-Net captured the contrast information from text and image modal, and the R-Net learned the semantic association between image and text via a cross-modality attention method.

Their method was more effective for detecting sarcasm than the related methods. Pan et al. [7] designed a BERT-based model to obtain the representation of text and the hashtags within the text. A co-attention matrix was applied to model the incongruity between text and hashtags as the intra-modality incongruity. A self-attention mechanism was used to capture the inter-modality incongruity. This study only used text for identifying the sarcasm in image but ignored semantic gaps between images and texts. Wang et al.[6] constructed a bridge layer to map image feature from the ResNet space to the BERT space that only accepts text as input, in order to fill the semantic gap between images and texts. Their method belongs to a kind of multimodal transformers [10]. However, they have not paid close attention to the incongruity learning.

2.2 Contrastive Learning

Contrastive learning originated from self-supervised learning and aims to improve training effectiveness based on self-supervised training with a lot of unlabeled data [11]. Currently, the idea of contrastive learning has been applied to the supervised and semi-supervised learning.

A typical contrastive learning model is SimCLR proposed by Chen et al. [12], which consists of four parts, i.e., a data augmentation module, an encoder module, a projection head module, and a contrastive loss module. SimCLR framework performed data augmentation on one same image as positive pairs and treated two images augmented from the different original images as negative pairs. SimCLR maximized the consistency between positive pairs by the contrastive learning on positive and negative pairs to obtain an encoder, which provided excellent feature representation ability for downstream tasks.

Gao et al. [13] proposed SimCSE, which used random dropout to construct positive pairs and focused on text-based contrastive learning. However, the ineffective data augmentation in constructing positive pairs due to the discreteness of text data is an issue.

Khosla et al. [14] introduced supervised contrastive learning, where multiple positives per an anchor sample were considered, unlike self-supervised contrastive learning using only one single positive. The positive/negative relationships between samples were linked by labels, namely, the positive pairs of an anchor sample were no longer produced by the data augmentation methods on anchor samples, but selected from the samples belonging to the same label as that of the anchor sample. Their supervised contrastive loss was more stable to hyper parameter settings.

Mai et al. [15] proposed the HyCon contrastive learning framework for trimodal representation for the task of multimodal sentiment analysis. This framework consists of three parts: intra-modal contrastive learning (IAMCL), inter-modal contrastive learning (IEMCL), and semi-contrastive learning (SCL). Three losses, IAMCL, IEMCL and SCL, were designed to learn inter-/intra- modal dynamics in supervised and unsupervised manners comprehensively to obtain cross-modal embeddings. However, they have not learnt the semantics of the multimodal samples from different classes fully.

3 Methodology of MSDCL

MSDCL is implemented to detect sarcasm tendency based on dissonant knowledge learning and supervised contrastive learning (SCL). It composes of 5 components: embedding of image and text, multimodality integration, incongruity knowledge capturing based on Multi-head Self-Attention Mechanism (MHSA), intra-class and inter-class commonalities and nuances learning via SCL, and sarcastic semantics detection. The framework is demonstrated in Fig. 1.

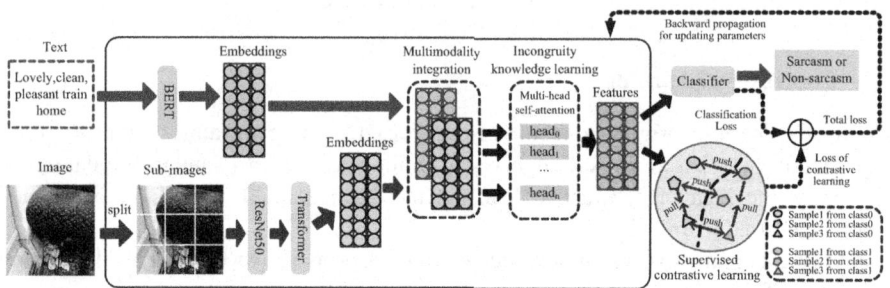

Fig. 1. The framework of MSDCL mechanism.

In Fig. 1, the multimodal image-text dataset D $\{<t_1,img_1>, <t_2,img_2>,..., <t_{N_D}, img_{N_D}>\}$ is input into MSDCL, where N_D is the number of samples in D and one sample $<t_i, img_i>$ consists of a piece of text and a unique associated image. The embeddings of a text and its corresponding image are produced separately. The SCL is applied to pull the distance between similar samples from the same class as closer as possible (e.g., triangle, pentagon, and oval with the same color), and push away the similar samples from different classes as farther as possible (e.g., the triangles with different color), in order to intensively learn multiple semantics from samples and understand the sarcastic tendency.

3.1 Text Embedding

BERT [16] is an excellent pre-trained language model that can generate different embedding representations based on the context of a word. The text modality is embedded by BERT, which input layer adds a special start token "[CLS]" to the text. Token embedding $e_t \in R^{d_{BERT}}$ is the one-hot vector of a word, where R denotes the set of real numbers and d_{BERT} is the hidden vector dimension of BERT. Segment embedding $s_t \in R^{d_{BERT}}$. is used to understand the boundaries and relationships between different segments or sentences in a text. Positional embedding $p_t \in R^{d_{BERT}}$ ensures the order of words and adds positional information to the input sequence.

The three types of embedding vectors are combined (as Eq. (1)) to form input sequence in text form.

$$X_{text} = [e_t^0 + s_t^0 + p_t^0, \ldots, e_t^{N_t} + s_t^{N_t} + p_t^{N_t}] \qquad (1)$$

where N_t represents the length of the text sequence, e_t^0, s_t^0, and p_t^0 represent the token, segment, and positional vectors for "[CLS]" token, and e_t^i, s_t^i, and p_t^i represent the token, segment, and positional vectors for the ith word.

The vector representation obtained through BERT processing is shown in Eq. (2):

$$f_t^i = BERT(e_t^i + s_t^i + p_t^i) \qquad (2)$$

where f_t^i is the representation vector of the ith word containing contextual information.

Reimers et al. [17] found that the average word vectors of BERT output sequence perform better than the embedding vector of the "[CLS]" token. Compared to sentence-level features, BERT can produce better representation at the word level. Therefore, we use the word vectors extracted by BERT rather than the "[CLS]" sentence vector. The final text representation is $f_t \in R^{N_t \times d_{BERT}}$.

3.2 Image Embedding

Inspired by Lu et al. [18], size transformation is performed on the image sample X_{img}, which can be segmented into M sub-images to retain more information and represented by Eq. (3):

$$X_{img} = \{X_{img}^1, X_{img}^2, \cdots, X_{img}^M\} \qquad (3)$$

where $X_{img}^j \in R^{C \times H \times W}$, C is the number of color channels of image, H and W denotes the height and width of sub-images, respectively.

An original image sample is resized to 512 × 512 and split into 64 sub-images, each with a height and width of 64. Each sub-image X_{img}^j is processed by *ResNet50* network for feature extraction to obtain the feature vector $e_{img}^j \in R^{d_{ResNet}}$, where d_{ResNet} is the dimension of output hidden layer of *ResNet50*. This process can be represented by Eq. (4):

$$e_{img}^j = ResNet50(X_{img}^j) \qquad (4)$$

The feature e_{img} of an image is represented by combining the feature sequences of all sub-images, $e_{img} \in R^{M \times d_{ResNet}}$. e_{img} is passed through a fully connected layer to convert the image feature dimension d_{ResNet} to d_{Bert}, which is also the dimension of text features. The Transformer network is used to capture the interaction between the sub-images to obtain the final image representation $f_{img} \in R^{M \times d_{BERT}}$.

$$f_{img} = Transformer(W_{img} e_{img} + b_{img}) \qquad (5)$$

where W_{img} and b_{img} are trainable parameters.

3.3 Multimodal Incongruity Learning

The multi-head self-attention (MHSA) mechanism is used to capture incongruity information from both text and image representations, because the mechanism can conduct association analysis on the representations in terms of many aspects and fully learn the incongruous information from intra- and inter-multimodality.

The text and image representation f_t and f_{img} are firstly concatenated and fused to obtain a single image-text representation f_{fusion} for incongruity learning.

$$f_{fusion} = f_t \oplus f_{img} \tag{6}$$

where $f_{fusion} \in R^{(N_t+M) \times d_{BERT}}$ represents the image-text feature and $\oplus$ denotes the concatenation operation.

The multimodal feature containing incongruity is produced by MHSA mechanism in Eq. (7).

$$f_e = (head_1 \oplus head_2 \oplus \cdots \oplus head_{N_{head}})W_{head} \tag{7}$$

where N_{head} represents the number of attention heads used in MHSA, W_{head} denotes the parameter matrix required for linear transformation. $Head_i$ indicates the calculation result from the ith group of self-attention mechanism. The calculation process of $head_i$ is shown as:

$$head_i = softmax\left(\frac{[W_Q^i f_{fusion}]^T [W_K^i f_{fusion}]}{\sqrt{d_k}}\right)[W_V^i f_{fusion}]^T \tag{8}$$

where $W_Q^i, W_K^i, W_V^i \in R^{d_{BERT} \times d_k}$ indicates the learnable weight matrices used in the ith self-attention mechanism group, and d_k represents the matrix dimension processed by each head.

f_e contains rich information about multimodalities and their incongruities to a certain degree. A local max-pooling strategy is applied to extract the important parts of f_e and a local avg-pooling strategy is used to synthesize the important features of each part. During this processing the vector dimension of f_e is reduced for classification. The multimodal feature vector f is output.

$$f = AvgPool(MaxPool(f_e)) \tag{9}$$

3.4 Contrastive Learning for Optimization

The features extracted above are still not very comprehensive, because sarcasm is an expression with implicit strong meanings that is in contradiction to the literal meaning, where there is some deep semantic knowledge. Context analysis can reveal the nuance in the similar expressions but from sarcastic and non-sarcastic classes respectively, as well as the intrinsic uniformity in the similar expressions from the same class. For example, your friend shows up an hour late to your lunch date, and you ironically greet her/him with 'Wow, you're soo punctual!' But a similar non-ironic expression may be 'Wow, you're very punctual.'

In order to recognize the semantic nuance from context, our MSDCL applies contrastive learning to enhance the training for widening the discrepancy between sarcastic and non-sarcastic features and close the gap between the features from the same class.

Since many nuances and intrinsic similarity may have been dampened during the integration of multimodal features above, the contrastive learning module processes multimodal features independently. Two tasks are accomplished in contrastive learning.

- Learning the intrinsic uniformity in the similar samples from the same class through minimizing the distance between them in the feature space, i.e., similar sarcastic samples have closer distances, so have the similar non-sarcastic samples.
- Capturing the nuance between the similar samples but from sarcastic and non-sarcastic classes respectively by maximizing the distance between them in the feature space, namely, although a sarcastic and a non-sarcastic sample may have very similar expression, the subtle differences between them should be extracted and widened for enhancing the distinction.

The contrastive learning is conducted based on 3 steps:

(1) The **positive and negative pairs** are constructed on a given sample x_s in the multimodal image-text dataset D, which is shown in the **Algorithm** below. If the similarity between x_s and another sample x_i ranks within top N_{pair} and x_i has the same label as x_s, x_s and x_i form a positive pair. If the similarity between x_s and another sample x_j ranks within top N_{pair} and x_j has the opposite label of x_s, x_s and x_j form a negative pair. N_{pair} denotes the finally determined pair number. Line 11 ~ 12 of the **Algorithm** denote the work mentioned above, where $\sim label_s$ means the opposite label of that of x_s.

The semantics of the image modality is textually processed to address the similarity between image and text samples. The method provided by Cai et al. [5] is used to extract image attributes, which describe the image content at word-level granularity.

Furthermore, analyzing the dataset provided by [5], the result shows that 54.73% of the images contain text information, which is very helpful for further understanding images. Tesseract (an open source OCR engine) is used to extract the short text from an image, if it has one. The text, image attributes, and OCR result of a sample x_s are combined into a complete sentence. SpaCy (an open source library for NLP in Python) is used to measure the similarity between samples. Line 1~10 of the **Algorithm** accomplish the task.

Algorithm: Constructing Positive and Negative Pairs.
Input: Dataset D, image attribute set $Attr$, image OCR set $Chara$, N_{pair}.
Output: Positive pair set Pos, Negative pair set Neg.
1   **INITIALIZE** $FPair$, Pos and Neg;
2   **FOR** x_s=<$text_s$, $image_s$> in D **DO**
3        $x_s^{Attr} \leftarrow Attr(image_s), x_s^{Chara} \leftarrow Chara(image_s)$;
4        $t_s \leftarrow text_s + x_s^{Attr} + x_s^{Chara}$;
5   **FOR** x_j=<$text_j$, $image_j$> $(x_j \neq x_s)$ in D **DO**
6   **IF** both (s, j) and (j, s) not in $FPair$ **THEN**
7               $x_j^{Attr} \leftarrow Attr(image_j), x_j^{Chara} \leftarrow Chara(image_j); t_j \leftarrow text_j + x_j^{Attr} + x_j^{Chara}$;
8               $FPair[(s, j)] \leftarrow FPair[(j,s)] \leftarrow spaCySim(t_s, t_j)$;
9   **END IF**
10   **END FOR**
11       $Pos[x_s] \leftarrow topNSim(FPair, Label_s, x_s)$;
12       $Neg[x_s] \leftarrow topNSim(FPair, \sim Label_s, x_s)$;
13 **END FOR**

(2) **Encoder** is a key of contrastive learning and applies MHSA module to transform the visual and textual information to vector representations for embedding samples. It is optimized with the combination of classification loss and contrastive loss.

(3) The contrastive loss is used to optimize the visual and textual features f extracted by the encoder for x_s. The function is shown in Eq. (10). Based on Eq. (10), if the similarity between x_s and a negative sample decreases and the similarity between x_s and a positive sample increases, the contrastive loss $Loss_{CL}$ decreases. Therefore, this approach shortens the distance between the similar samples from the same class and opens the distance between the similar samples but from different classes in the feature space.

$$Loss_{CL} = -\frac{1}{N_{pair}} \left(\sum_{i=1}^{N_{pair}} \log \frac{\exp(\cos(f_{x_s}, Pos_{x_s}^i))}{\sum_{j=1}^{N_{pair}} \exp(\cos(f_{x_s}, Pos_{x_s}^j)) + \exp(\cos(f_{x_s}, Neg_{x_s}^j))} \right) \tag{10}$$

where f_{x_s} represents the visual and textual features of x_s extracted by the Encoder. $\cos(\cdot)$ is the cosine similarity function. Pos_{x_s} or Neg_{x_s} are the feature set of positive or negative pairs for sample x_s, respectively.

3.5 Sarcasm Discovery

The overall loss of MSDCL (Eq. 12) consists of contrastive loss (Eq. 10) and classification loss (Eq. 11) and is applied to improve the comprehensive features iteratively.

$$Loss_{CS} = -(y_i Log\, p_i + (1-y_i)Log\,(1-p_i)) \qquad (11)$$

$$Loss = (1-\theta)Loss_{CS} + \theta Loss_{CL} \qquad (12)$$

where θ is a weight parameter that adjusts the contribution of the two loss components, y_i is the true label of the ith sarcastic sample, and P_i is the prediction probability of sarcasm.

In order to recognize a sarcastic sample, the features optimized above are transformed with the fully connected neural network, which output is transformed by softmax to the corresponding probability of detection result.

4 Experiments and Analysis

4.1 Dataset and Baseline Models

The widely used multimodal image-text sarcasm dataset MMSD (Twitter) [5] is applied in the experiments, where each sample consists of one text and one corresponding image. The numbers of sarcasm and non-sarcasm are 10560 and 14075, respectively.

The following 8 different baseline methods are compared with MSDCL in this paper.

a) **SIARN**(single-dimensional intra-attention recurrent network) and **MIARN**(multi-dimensional intra-attention recurrent network) were developed in paper [19] and MIARN was improved based on SIARN. MIARN models the semantics between each word in the input sequence and extracts the contrast and incongruity between word pairs within sentences with the multi-dimensional intra-attention mechanism. Their key idea is that each word pair is projected down to a low dimensional vector before computing the affinity score, which can capture multiple views.
b) **SMSD**(self-matching sarcasm detection) [20] captures the contradictory features of word pairs using self-matching matrices. The model introduces Bi-LSTM to extract bidirectional sentence-level features and uses low-rank bilinear pooling to remove redundant information between word-level features and sentence-level features.
c) **ViT**(vision transformer) [21] is an application of Transformer model for image feature extraction. The output vector of the token "[CLS]" is used to detect sarcasm.
d) **HFM** [5] uses Bi-LSTM to extract textual features and ResNet to extract visual features. HFM introduces early fusion, representation fusion, and modality fusion to integrate the multimodal features and perform sarcasm detection.
e) **Res-BERT** [7] treats textual statements as a combination of input sequence and hashtag texts. BERT is used to extract features from both parts of text. The extracted features are then concatenated with the image features extracted by ResNet and used for classification.
f) **D&R Net** [9] applies ResNet to extract image features and Bi-LSTM to extract text features. The decomposition network (D-Net) and relation network (R-Net) are introduced to capture the incongruity information between image and text features.
g) **Image-Text Fusion** [6] uses multimodal features for sarcasm classification, where a bridge layer processes image features extracted by ResNet and maps them into BERT space. BERT layer fuses text and image features and jointly trains detection model by vector space transformation and 2D-Intra-Attention method.

4.2 Parameter Settings and Evaluation Metrics

In experiments the PyTorch deep learning framework and NVIDIA Tesla P40 environment are launched. Text is processed by WordPiece in BERT, where if the length is less than the length of the text sequence N_t, "[PAD]" is padded; if it exceeds N_t, truncation is conducted. The training objective is to minimize the total losses in Eq. (12). Adam is used to optimize the parameters (Table 1).

Table 1. Key hyperparameters and their values in experiments

Text sequence length N_t	Head number of MHSA N_{head}	Final pair number determined by CL N_{pair}	Loss adjustment weight θ	Number of sub-images M	Initial learning rate	batch size
256	8	5	0.2	64	3e-5	32

Commonly used metrics such as ACC, Precision, Recall, and F1 are used for assessing the sarcasm detection performance.

4.3 Experimental and Analytical Study

As shown in Table 2, the proposed MSDCL outperforms the related methods in terms of most of the criteria. The approaches based on multimodal integration generally have a better detection performance than those employing single modality, which indicates that multimodal information is very helpful.

HFM tends to focus on the fusion of multimodal features. However, the Bi-LSTM network is not sufficient to capture important information from sentences. Moreover, the incongruity information in image-text sarcasm is ignored to a large degree, so that the sarcasm semantics may not be understood deeply.

Res-BERT can obtain high-quality sentence representation using BERT and thus improves the performance. Nevertheless, it does not consider the incongruity factors in sarcasm detection and lacks interpretability as HFM.

The D-Net in D&R Net method captures the incongruity information from text and image modal, and the R-Net learns the semantic association between image and text features. However, the semantics of images and texts has a gap and their features are in different feature spaces. It is difficult for D-Net to learn incongruity from both modalities. Besides, sub-image features representing the overall image in MSDCL can obtain more detailed information than the D&R Net model.

Different from the Res-BERT, Image-Text Fusion [6] utilizes BERT to combine the raw text and the image features learned by ResNet for capturing multimodal image-text information. In the BERT training process, incongruity information between text and image modalities are extracted by self-attention. The method exploits the idea of multimodal transformers [10] and shows big improvement in sarcasm detection with the second place, while in our MSDCL, contrastive learning is leveraged to understand

Table 2. Experimental result of multimodal Sarcasm Detection.

Modality	Method	ACC	F1	P	R
Text	SIARN	80.57	75.63	75.55	75.70
	MIARN	82.48	77.36	79.67	75.18
	SMSD	80.90	75.82	76.46	75.18
Image	ViT	67.83	63.43	57.93	70.07
Text + Image	HFM	83.44	80.18	76.57	84.15
	Res-BERT	84.80	81.57	78.87	84.46
	D&R Net	84.02	80.60	77.97	83.42
	Image-Text Fusion	85.41	82.14	79.92	84.49
	MSD (this paper)	84.23	81.73	81.45	82.27
	MSDCL (this paper)	**85.85**	**84.20**	**84.18**	84.22

semantic differences between sarcasm and non-sarcasm in terms of image and text modals, so that the sarcasm detection has been optimized.

4.4 Analysis on the Number of Positive and Negative Pairs in Contrastive Learning

This experiment is designed for finding the suitable N_{pair}. The detection criteria will be the worst, when MSDCL does not select positive and negative pairs ($N_{pair} = 0$) and the contrastive learning has not been conducted yet, as shown in Table 3. The overall performance becomes better and better with the increasing of N_{pair} value. This is because the more the samples for contrastive learning are put into the usage, the more the commonality and difference between samples are learned, but the more the memory is required for computation. Therefore N_{pair} value is set as 5 due to tradeoff.

Table 3. The influence of N_{pair} value on MSDCL performance.

N_{pair}	ACC	F1	P	R
0	84.23	81.73	81.45	82.27
1	84.57	82.45	82.76	82.20
3	85.02	82.50	82.22	82.95
5	**85.85**	**84.20**	**84.18**	**84.22**

In addition, the functionality of contrastive learning can be demonstrated clearly with diagrams, where t-SNE (t-distributed Stochastic Neighbor Embedding) is used to create a low-dimensional space that high-dimensional samples are mapped to. Its advantages are to reduce the multimodal feature dimensions for visualization and simultaneously maintain the similar distribution of data in the low dimensional space. The effect of

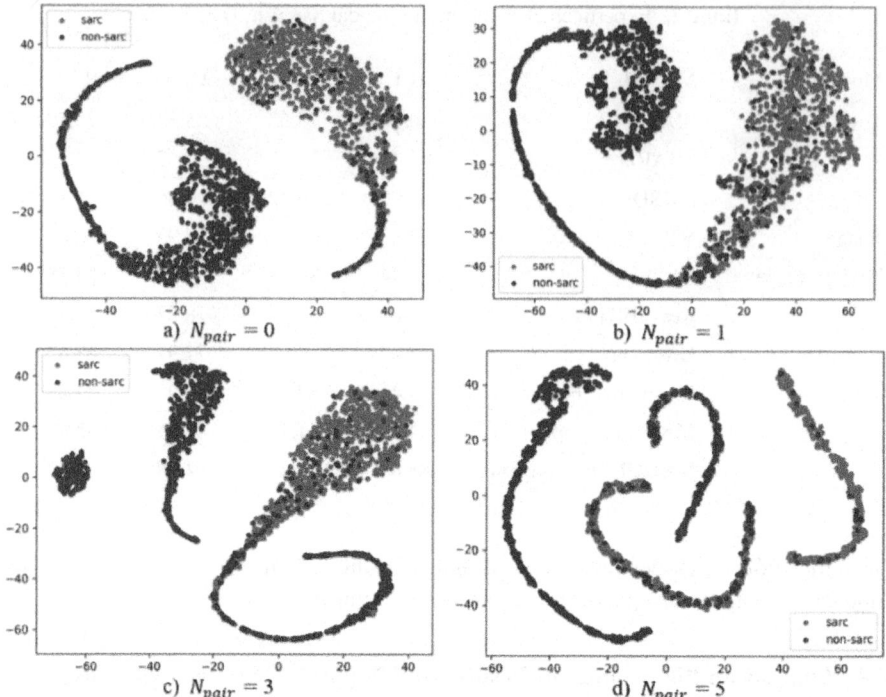

Fig. 2. Visual analysis on contrastive learning

contrastive learning with different N_{pair} values by using testing samples embedded via MSDCL is shown in Fig. 2. The red dot or blue dot denotes a sample with true sarcasm label or true non-sarcasm label. With the increasing of N_{pair}, the distance between samples from the same class is gradually closing to each other to form very compact clusters, which indicates the contrastive learning can improve sarcasm detection greatly.

4.5 Ablation Experiment

An ablation experiment is designed to verify the effectiveness of the components in MSDCL for detecting sarcasm, which results are in Table 4. ***T, V, I, CL*** denotes the textual module, the visual module, the image-text incongruity discovery, and the integrated framework with contrastive learning (***CL***), respectively.

It is shown ***T*** has more powerful articulation and sarcastic intention can be mined more accurately than ***V***. Directly concatenating ***T*** and ***V*** (***T + V***) results in slight performance reduction. The possible reasons are that simple concatenation hinders from extracting incongruity between text and image and dilutes the important information in the textual features, since visual features may contain fuzzy incongruity information but a lot of noise.

The performance becomes better by integrating ***I*** module, because the incongruous knowledge in the textual and visual modalities has been effectively utilized for sarcasm detection.

Table 4. The effectiveness of different MSDCL components for sarcasm detection.

Method	ACC	F1	P	R
V	65.84	63.81	64.12	63.66
T	82.65	81.65	82.19	81.28
T + V	81.24	80.91	80.74	81.96
T + V + I	84.23	81.73	81.45	82.27
T + V + I + CL	**85.85**	**84.20**	**84.18**	**84.22**

The contrastive learning is further integrated ($T + V + I + CL$), which exploits the similar and different semantic knowledge of sarcasm and non-sarcasm data, so that the improvement is achieved.

5 Conclusion

The proposed MSDCL approach has several strengths: (1) Extract fine-grained features from text and image and integrate multimodal features by multi-head self-attention mechanism. (2) Exploit a contrastive learning framework for learning the common and distinct knowledge thoroughly and better understanding the semantics of sarcasm. (3) Optimize the algorithm by combining classification loss and contrastive loss for improving detection performance.

Although our approach achieves good performance based on the limited resources, the enhancement of MSDCL is imperative from many aspects, such as generalization based on multiple datasets, analysis of computational costs, statistical significance testing, and the leveraging of LLMs.

In addition to the utilization of textual modality as well as visual and OCR feature, speaking voice as a new modality could be explored for more accurate identification in the future, because the intonation between sarcastic and non-sarcastic expression is subtly different and implies rich semantics. Besides, some researches provide enlightenment on generating additional visual and textual data [22] or rewriting sarcastic/non-sarcastic samples to augment counterfactual data [23] for improving contrastive learning. Inspired by this direction, multimodal data, including text, image, voice, or comment, could be enhanced and extended to mine incongruity knowledge accurately.

Disclosure of Interests. The authors have no competing interests to declare that are relevant to the content of this article.

References

1. Hazarika, D., Poria, S., Gorantla, S., et al.: Cascade: contextual sarcasm detection in online discussion forums. In: Proc of COLING, pp.1837–1848. ICCL, New Mexico, USA (2018)
2. Lou, C., Liang, B., Gui, L., et al.: Affective dependency graph for sarcasm detection. In: Proc of SIGIR, pp.1844–1849. ACM, Virtual Event Canada (2021)
3. Wen, Z., Gui, L., Wang, Q., et al.: Sememe knowledge and auxiliary information enhanced approach for sarcasm detection. Inf. Process. Manage. **59**(3), 102883 (2022)
4. Zhang, Y., Ma, D., Tiwari, P., et al.: Stance level sarcasm detection with BERT and stance-centered graph attention networks. ACM Trans. Internet Technol. (TOIT). **23**(2), 1–20 (2022)
5. Cai, Y., Cai H., Wan X.: Multi-modal sarcasm detection in twitter with hierarchical fusion model. In: Proc of ACL, pp. 2506–2515. Association for Computational Linguistics, Florence, Italy (2019)
6. Wang, X., Sun, X., Yang, T., et al.: Building a bridge: a method for image-text sarcasm detection without pretraining on image-text data. In: Proc of NLPBT, pp.19–29. Association for Computational Linguistics, online (2020)
7. Pan, H., Lin, Z., Fu, P., et al.: Modeling intra and inter-modality incongruity for multi-modal sarcasm detection. In: Proc of EMNLP, pp.1383–1392. Association for Computational Linguistics, online (2020)
8. Liang, B., Gui, L., He, Y., et al.: Fusion and Discrimination: a multimodal graph contrastive learning framework for multimodal sarcasm detection. IEEE Trans. Affect. Comput. **15**(4), 1874–1888 (2024)
9. Xu, N., Zeng, Z., Mao W.: Reasoning with multimodal sarcastic tweets via modeling cross-modality contrast and semantic association. In: Proc of ACL, pp.3777–3786. Association for Computational Linguistics, online (2020)
10. Farabi, S., Ranasinghe, T., Kanojia, D., et al.: A survey of multimodal sarcasm detection. In: Proc of JCAI, pp. 8020–8028. The IJCAI organization, Jeju, South Korea (2024)
11. Zhang, C., Chen, J., Li, Q., et al.: Deep contrastive learning: a survey. Acta Automatica Sinica. **49**(1), 15–39 (2023)
12. Chen, T., Kornblith, S., Norouzi, M., et al.: A simple framework for contrastive learning of visual representations. In: Proc of ICML, pp. 1597–1607. ACM, Vienna, Austria (2020)
13. Gao, T., Yao, X., Chen D.: SimCSE: simple contrastive learning of sentence embeddings. In: Proc of EMNLP, pp. 6894–6910. Association for Computational Linguistics, Online and Dominican Republic (2021)
14. Khosla, P., Teterwak, P., Wang, C., et al.: Supervised contrastive learning. Adv. Neural. Inf. Process. Syst. **33**, 18661–18673 (2020)
15. Mai, S., Zeng, Y., Zheng, S., et al.: Hybrid contrastive learning of tri-modal representation for multimodal sentiment analysis. IEEE Trans. Affect. Comput. **14**(3), 2276–2289 (2022)
16. Devlin, J., Chang, M-W., Lee, K., et al.: BERT: Pre-training of deep bidirectional transformers for language understanding. In: Proc of NAAC, pp. 4171–4186. Association for Computational Linguistics, Minnesota, USA (2019)
17. Reimers, N., Gurevych I.: Sentence-BERT: Sentence embeddings using SIAMESE BERT-networks. In: Proc of EMNLP-IJCNLP, pp. 3982–3992. Association for Computational Linguistics, Hong Kong, China (2019)
18. Lu, J., Batra, D., Parikh, D., et al.: Vilbert: Pretraining task-agnostic vision linguistic representations for vision-and-language tasks. In: Proc of NeurIPS, pp. 32:13–23. Curran Associates Inc, Vancouver BC Canada (2019)
19. Tay, Y., Tuan, L., Hui, S., et al.: Reasoning with sarcasm by reading in-between. In: Proc of ACL, pp. 1010–1020. Association for Computational Linguistics, Melbourne, Australia (2018)

20. Xiong, T., Zhang, P., Zhu H, et al.: Sarcasm detection with self-matching networks and low-rank bilinear pooling. In: Proc of WWW, pp. 2115–2124. ACM, San Francisco USA (2019)
21. Dosovitskiy A., Beyer, L., Kolesnikov A., et al.: An image is worth 16x16 words: transformers for image recognition at scale. In: Proc of ICLR, pp. 1–21. World Academy of Science, Engineering and Technology, Virtual Conference (2021)
22. Wei, Y., Duan, M., Zhou, H., et al.: Towards multimodal sarcasm detection via label-aware graph contrastive learning with back-translation augmentation. Knowl. Based Syst. **300**(C) (2024)
23. Jia, M., Xie, C., Jing L.: Debiasing multimodal sarcasm detection with contrastive learning. In: Proc of AAAI, pp. 18354–18362. Association for the Advancement of Artificial Intelligence, Vancouver, Canada (2024)

Ensemble ToT and Its Application to Automatic Grading

Yuki Ito[1](✉) and Qiang Ma[2]

[1] Graduate School of Informatics, Kyoto University, Social Informatics Course Yoshidahonmachi, Sakyo-ku, Kyoto-shi, Kyoto 606-8501, Japan
ito.yuki.w26@kyoto-u.jp
[2] Graduate School of Science and Technology, Kyoto Institute of Technology Matsugasakihashikamicho, Sakyo-ku, Kyoto 606-8585, Japan
qiang@kit.ac.jp

Abstract. Providing students with detailed and timely grading feedback is essential for self-learning. While existing LLM-based grading systems are promising, most of them rely on one single model, which limits their performance. To address this, we propose **Ensemble Tree-of-Thought (ToT)**, a framework that enhances LLM outputs by integrating multiple models. Using this framework, we develop a grading system. Ensemble ToT follows three steps: (1) analyzing LLM performance, (2) generating candidate answers, and (3) refining them into a final result. Based on this, our grading system first evaluates the grading tendencies of LLMs, then generates multiple results, and finally integrates them via a simulated debate. Experimental results demonstrate our approach's ability to provide accurate and explainable grading by effectively coordinating multiple LLMs.

Keywords: Large Language Models · EdTech · Ensemble · ToT

1 Introduction

Since the release of ChatGPT in 2022, generative AI has gained significant popularity, with many students increasingly adopting it as a learning aid for various purposes. A survey on students' use of AI for learning [4] identified several common applications, such as generating ideas, summarizing readings, and so on. Among these, AI's ability to provide personalized and immediate learning support is particularly valued. Notably, some students in the survey reported using AI to obtain grading and its explanations on their completed homework. These grading comments help enhance these students' depth of thinking and understanding. This highlights the importance of providing students with detailed, point-by-point feedback on their homework to facilitate self-learning.

Many automated grading methods have been proposed to provide quick and accurate feedback to a large number of students. For example, Gobbo et al. introduced GradeAid, a framework that predicts grading scores of student answers by extracting both lexical and semantic features [9]. Similarly, Huang et al. developed a grading system for reading comprehension answers, which automatically generates hints to guide students to the correct answers [12]. However, there are some challenges with these methods. For example, GradeAid does not have a function to explain the reason for grading, making it difficult for students to understand how to revise their answers. Although Huang et al.'s system provides feedback, they state that the quality of the feedback is not yet at a satisfactory level [12]. These issues are major obstacles to supporting self-learning.

A promising approach to overcoming these challenges is integrating multiple LLMs into grading systems. Existing methods typically rely on a single LLM. In contrast, a previous study [19] demonstrated that using ensembles of pre-trained BERT models for scoring tasks significantly improves prediction accuracy. This highlights the advantages of combining multiple models. However, for LLMs, especially for language generation models, methods for integrating the outputs of multiple models have not yet been well-studied.

In this paper, inspired by ensemble learning [24] and Tree-of-Thought [23], we propose a novel framework called **Ensemble ToT**, which combines multiple LLMs. Based on this framework, we then develop a grading system named **Graders by Ensemble ToT (GET)** to support self-learning.

Ensemble ToT combines the strengths of multiple language generation models by integrating ensemble learning techniques with the Tree-of-Thought (ToT) approach. Figure 1 shows the overview of this framework.

It operates in three steps:

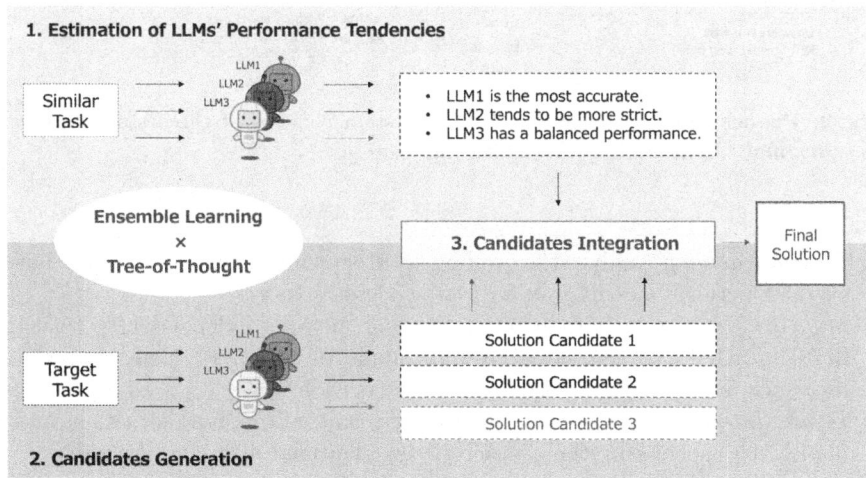

Fig. 1. Overview of Ensemble ToT Framework: The framework integrates ensemble learning techniques with the Tree-of-Thought (ToT) approach. It identifies individual LLM performance tendencies and synthesizes multiple candidate solutions generated by LLMs into a single refined result.

1. **Estimation of LLM performance tendencies**: Inspired by ensemble learning principles, we estimate the performance of individual LLMs on tasks similar to the target task.
2. **Candidates Generation**: Drawing on the ToT, multiple LLMs independently generate candidate solutions for the target task.
3. **Candidates Integration**: These candidate solutions are synthesized into a single, refined result, considering the performance tendencies.

The GET system is an automatic grading system based on Ensemble ToT (Fig. 2). It operates in the following three steps.

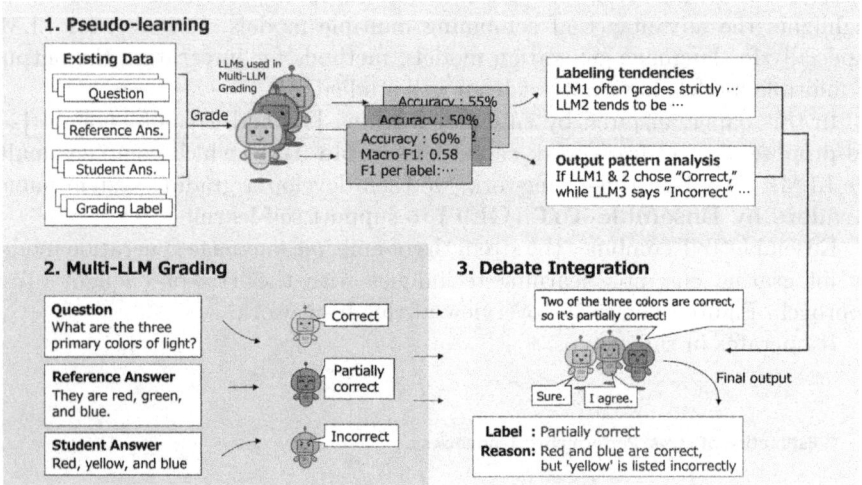

Fig. 2. Process Diagram of GET: The system consists of three stages: *pseudo-learning*, *multi-LLM grading*, and *debate integration*.

1. *Pseudo-learning*: Analyze the grading accuracy and tendencies of LLMs based on their performance in grading past student answers.
2. *Multi-LLM Grading*: Grade target student answers independently with the LLMs. Each LLM generates grading labels (Correct, Partially correct, or Incorrect) and provides reasons for their decisions.
3. *Debate Integration*: Integrate the grading results through a discussion among the LLMs, considering the characteristics identified in *pseudo-learning*.

The major contributions of this work are as follows:

- Propose Ensemble ToT, a novel framework to collaborate multiple LLMs effectively (Sect. 3.1).
- Propose an automated grading system, GET, to support students' self-learning (Sect. 3.2 - Sect. 3.5).

- Conduct a comparison evaluation on the grading performance of the GET system with a public dataset (Sect. 4 - Sect. 5).
  - In the experiment, design a new automated evaluation method to assess the quality of grading reasons using multiple LLMs.

2 Related Work

2.1 Frameworks for Improving the Output of a Single LLM

Several frameworks have been developed to enhance LLM performance. For example, Chain-of-Thought (CoT) [21] improves problem-solving by guiding LLMs to generate intermediate reasoning steps before arriving at a final answer.

The Tree-of-Thoughts (ToT) [23] extends CoT by enabling LLMs to explore multiple reasoning paths. It decomposes problems into intermediate steps, generating multiple candidate solutions per step. These candidates are evaluated through heuristics or voting to identify the best options. This allows the model to explore diverse paths in parallel, enhancing decision-making in complex tasks.

Graph-of-Thoughts (GoT) [2] further extends ToT by modeling reasoning as a graph, where nodes represent thoughts and edges capture dependencies. This facilitates integrating multiple ideas and refining solutions through feedback loops.

In this paper, we propose Ensemble ToT, which extends ToT by incorporating multiple LLMs through ensembling. This approach leverages the strengths of multiple models, improving performance beyond what a single LLM can achieve.

2.2 Ensembling Multiple Models

Beyond single-model frameworks, several methods for combining multiple LLMs have been explored. Ensemble learning [24] is a widely used technique in machine learning. It enhances performance by aggregating the outputs from multiple models [13]. One popular method, Stacking [22], trains several models on the same dataset and uses an additional model to combine their predictions.

Another approach, Mixture of Experts (MoE) [3], focuses on LLMs by combining several specialized sub-models into a single resource-efficient model. MoE activates only the relevant sub-models based on input, with a gating mechanism directing the queries to the most appropriate expert.

While ensemble learning and MoE share similarities with our approach, they differ in key ways. Ensemble learning improves accuracy by combining multiple outputs, but it is not directly suited for combine outputs from LLMs which are natural language. While MoE is a method used in language generation models, its objective is to optimize computation within a single architecture rather than integrating multiple independent LLMs. Our approach, Ensemble ToT, applies ensemble learning principles to multiple LLMs, enhancing accuracy through collaborative decision-making across models.

2.3 Automatic Grading Systems

The application of LLMs in education has gained significant attention. Automated short-answer grading (ASAG) [11] is no exception. ASAG evaluates student answers to open-ended questions, typically a few sentences long. Existing grading methods can be broadly divided into two categories: score or label prediction and detailed feedback generation.

Many studies focus on assigning scores or labels to student answers. For example, GradeAid [9] assign numeric scores to student answers. It extracts lexical and semantic features from them and processes them through a regression model, which performs well across datasets. [19] also predicts grading scores by ensembling several BERT models, showing enhanced performance. Additionally, [5] focuses on predicting grading labels (pass/fail) for student answers using GPT models. They found that GPT-4 performed well when guided by a grading example from existing data.

These methods focus primarily on score or label prediction and do not provide the reasoning behind the grading process. This distinguishes them from the GET system, which generates both grading labels and the reasoning behind the grades.

Recently, some methods have aimed to generate detailed feedback alongside scores or labels. For example, Huang et al. fine-tuned a GPT model to evaluate student answers and generate hints to guide students toward the correct answers [12]. We have also proposed a method that combines an LLM with grammatical analysis to generate both grading labels and explanations [14].

In addition, Fateen et al. propose ASAS-F-RAG [7], a grading system that combines retrieval-augmented generation (RAG) with LLM. This method retrieves the 3 to 5 most similar past examples using the ColBERT [16] retriever. They are then fed to the LLM to guide the prediction of scores, labels (e.g., Correct, Partially correct, Incorrect), and feedback.

While most existing methods typically use a single LLM, our proposed framework, Ensemble ToT, integrates several LLMs by combining ToT with ensemble learning. This collaborative approach compensates for individual models' weaknesses, resulting in more accurate grading outcomes.

3 Proposed Framework and System

In this section, we first introduce the core idea, Ensemble ToT. Then, we define the problem that the GET system aims to tackle. Finally, we describe the system's overall structure in detail.

3.1 Ensemble ToT

Inspired by ensemble learning [24] (particularly Stacking [22]) and Tree-of-Though (ToT) [23], we propose a novel framework called Ensemble ToT for leveraging multiple LLMs. Ensemble ToT operates in three steps.

First, it identifies the performance tendencies of LLMs on tasks similar to the target task. This phase mirrors the training phase in Stacking. However, we do not actually train the LLMs, since they have large sizes, and fine-tuning them is resource-intensive. We design an alternative process called *pseudo-learning*, as described in Sect. 3.3.

Second, based on the ToT framework, Ensemble ToT generates multiple candidate solutions using several LLMs. While the original ToT uses a single LLM to generate multiple candidates, Ensemble ToT utilizes different LLMs for each candidate. This increases the diversity of the candidates, leading to a wider range of solutions. In the GET system, this is implemented as *multi-LLM grading*, detailed in Sect. 3.4.

Finally, Ensemble ToT combines the candidate solutions while considering the identified performance tendencies of the LLMs. This process corresponds both to ensemble learning, which combines the outputs of multiple models through a newly trained model, and to ToT, which merges thought candidates using heuristics or voting mechanisms. In the GET system, an LLM agent merges candidate solutions into the best possible solution, by referring performance tendencies. This is implemented as *debate integration*, explained in Sect. 3.5.

The following section provides a detailed explanation of the GET system, which implements the Ensemble ToT framework.

3.2 Problem Definition

The GET system's objective is to grade short student answers. The symbols used in this paper are explained in Table 1.

Table 1. Definition of Symbols

Symbol	Definition
$Q = \{q_1, q_2, \ldots, q_m\}$	Set of questions for students
$R = \{r_1, r_2, \ldots, r_m\}$	Set of reference answers for Q
$S_i = \{s_{i_1}, s_{i_2}, \ldots, s_{i_n}\}$	Set of n students' answers for q_i
$GL = \{\text{Correct}, \text{Partially Correct}, \text{Incorrect}\}$	Set of grading labels
GR	Set of grading reasons
$gl_{i_j} \in GL$	Grading label for s_{i_j}
gr_{i_j}	Grading reason for s_{i_j}

The problem addressed by the GET system is to grade student answers based on the question content and the corresponding reference answers. Specifically, for the input tuple (q_i, r_i, s_{i_j}), the system predicts a pair consisting of a grading label and a grading reason, (gl_{i_j}, gr_{i_j}).

Mathematically, this task can be represented as follows:

$$(gl_{i_j}, gr_{i_j}) = GET(q_i, r_i, s_{i_j}) \tag{1}$$

Following the Ensemble ToT procedure, the GET system tackles this task through three steps: *pseudo-learning*, *multi-LLM grading*, and *debate integration*.

3.3 Pseudo-Learning

Pseudo-learning aims to evaluate the capabilities and grading tendencies of LLMs employed in the system. It is conducted before grading student answers. The process comprises two steps: *grading past-data* and *tendencies identification*.

Grading Past-Data. In the *grading past-data* step, each LLM independently grades answers from an existing dataset D', such as historical student answers with grading results. Then it computes the LLMs' performance metrics.

Formally, each LLM agent predicts grading label and reason pairs (gl'_{i_j}, gr'_{i_j}) based on the input tuples $(q'_i, r'_i, s'_{i_j}) \in D'$. The predicted grading labels (denoted as pl'_{i_j}) are then evaluated against ground truth gl'_{i_j} using performance metrics, including accuracy and macro F1-score.

Tendencies Identification . *Tendencies identification* analyzes the strengths, weaknesses, and grading biases of each LLM based on the performance metrics calculated in the previous step. The analysis is conducted from two perspectives:

- **Labeling tendencies compared to other LLMs** (e.g., an LLM excels at identifying errors but is overly strict).
- **Output pattern analysis for the most likely label prediction** (e.g., if LLM1 and LLM2 assign *Correct* while only LLM3 assigns *Partially correct*, the ground truth is *Partially correct* in many cases).

For labeling tendencies, an LLM interprets the numeric metrics and explains its findings in natural language. Through this process, numerical data are transformed into text form. LLMs are generally more effective at processing natural language than numerical data. This transformation ensures the results of *pseudo-learning* can be utilized more effectively in the subsequent processes.

For output pattern analysis, the following algorithm determines the most likely grading label for a given combination of LLM output labels considering the *grading past-data* results and the data distributions.

1. **Count occurrences**: For each student answer, count the occurrences of LLM output label combination (pl'_1, pl'_2, pl'_3) and their associated ground truth labels gl', where $pl'_i \in GL$ is the predicted label of LLM i, and $gl' \in GL$ is the ground truth label.
2. **Calculate ratios**: For each label combination, calculate the ratio of observed label counts to the expected label counts from the prior probabilities (e.g., the percentage of the number of labels in the dataset).
3. **Assign most likely labels**: Output all labels whose ratios exceed a predefined threshold as most likely labels. If no label exceeds the threshold, the label with the highest frequency of occurrence is output.

3.4 Multi-LLM Grading

The *multi-LLM grading* is the phase to generate multiple grading result candidates of the actual student answers. It involves two steps: *few-shot selection* and *independent grading*, as shown in Fig. 3.

Few-Shot Selection. In *few-shot selection*, we select three appropriate grading examples for each student answer s_{i_j} from the past grading data D'. This process is based on the study by Fateen et al. [7]. They demonstrated that using ColBERT [16] to retrieve 3 to 5 highly relevant grading examples from past data and providing them to the LLM improves grading accuracy. Based on this, our system also searches for grading examples as follows:

1. **Retrieve relevant grading examples:**
   - If the question q_i has been posed to students in D', retrieve the corresponding grading examples: $(q_i, r_i, s'_{i_l}, gl'_{i_l}, gr'_{i_l})$.
   - Otherwise, gets all data $(q'_k, r'_k, s'_{k_l}, gl'_{k_l}, gr'_{k_l}) \in D'$.
2. **Prepare data for embedding:** Concatenate all items in each grading example into a single string.
3. **Generate embeddings:** Use ColBERT's encoders to generate token embeddings (vector representations capturing semantic meaning) for both the concatenated grading examples and the student answer s_{i_j}.
4. **Aggregate similarity scores:** Sum the maximum similarity scores across all tokens in s_{i_j}.
5. **Rank and select examples:** Rank the grading examples in the descending order of their total similarity scores and select the top three examples.

Independent Grading. In *independent grading*, three LLMs individually grade the student answer s_{i_j} by referencing the selected grading examples. Specifically, each LLM receives the question, reference answer, and student answer (q_i, r_i, s_{i_j}) as input. For questions q_i not present in the dataset D', additional information about the grading criteria is also provided to enhance grading accuracy for unseen questions. Based on this input, each LLM predicts both a grading label (gl_{i_j}) and a corresponding reason (gr_{i_j}). This process yields three independent grading assessments.

To enhance output diversity, each LLM is assigned a distinct role name explicitly stated in the prompt. This approach is based on the study by Ormerod et al. [19], which demonstrated that combining models with different characteristics can improve performance. In this system, based on the macro F1-score in *pseudo-learning*, the model with the highest score is designated as a *Skilled Expert Grader*, the second highest as a *University Teacher*, and the lowest as a *Student TA*. This assignment introduces different perspectives, thereby enhancing the diversity of grading results.

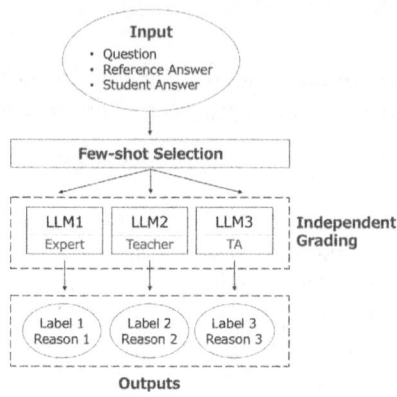
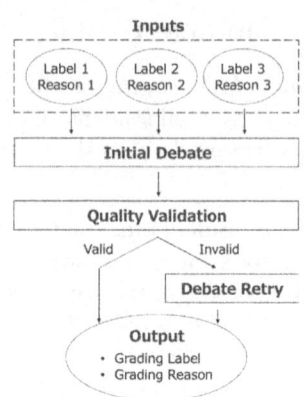

Fig. 3. Multi-LLM Grading Flow   **Fig. 4.** Debate Integration Flow

3.5 Debate Integration

In *debate integration* phase, grading results from *multi-LLM grading* are merged into a single result. This process has three steps: *initial debate*, *quality validation*, and *debate retry*, as shown in Fig. 4.

Initial Debate. In the *initial debate*, the LLM with the highest accuracy in *pseudo-learning* simulates a discussion among three graders to integrate their grading results. The LLM receives a debate template, where the graders' parts are initially left blank. Using the results from *pseudo-learning* and *multi-LLM grading*, it fills in the template by predicting what each grader would say in a real discussion to decide on the final grading result.

The debate follows four stages:

1. *Ice break*: Graders introduce themselves based on grading tendencies identified in *pseudo-learning*.
2. *Divergence*: Graders share their grading results from *multi-LLM grading*.
3. *Conversion*: Graders exchange opinions about each other's grading. They may emphasize their original opinion or, if they realize a grading mistake, adjust their stance.
4. *Voting*: Graders vote for the most likely grading label and reason to determine the grading conclusion.

This process is inspired by Dong et al.'s study [6], but adapted to fit our system.

In this debate, if the question q_i exists in the past dataset D', the LLM can access the three examples selected during *few-shot selection*. Otherwise, no examples are available. This ensures the system can grade new questions effectively without overly relying on past data.

Quality Validation. In *quality validation*, the LLM that conducted the debate reassesses the grading conclusion from the *initial debate* to ensure its validity. If the conclusion is deemed valid, it is finalized as the system output. Otherwise, the LLM suggests a revised grading result.

Debate Retry. The *debate retry* occurs only when the grading result is deemed invalid in the *quality validation* phase. During this phase, the LLM facilitates an additional discussion among the graders and selects the best result between the original grading from the *initial debate* and the revised result from *quality validation*. The selected result becomes the final output of the system.

4 Experimental Setup

We experimented to measure the performance of our system. In this section, we describe the experimental setup.

4.1 Dataset

For evaluation, we used the Short Answer Feedback (SAF) dataset [8]. This dataset focuses on a university-level computer networking course and includes questions, model answers, and student answers. Additionally, it contains grading labels (Correct, Partially correct, or Incorrect) and justification for these labels, both provided by human graders.

The dataset is divided into four subsets: train, validation, test-unseen-answers (UA), and test-unseen-questions (UQ). Each subset contains 1700, 427, 375, and 479 instances, respectively. There are two types of test sets: UA contains the same questions as the train set but different student answers, while UQ contains only questions that are different from those in the train set.

In this study, we used the train set for *pseudo-learning* and evaluated system performance using the UA and UQ test sets.

4.2 Evaluation Method

The output of the GET system is categorized into two tasks: grading label prediction and grading reason generation.

To evaluate the label prediction performance, we used accuracy and macro F1-score as metrics. On the other hand, to assess the quality of grading reasons, we newly introduced an automated evaluation using LLMs. In this method, LLMs compare the grading reasons recorded in the dataset with those generated by the system to verify their validity. The procedure is as follows:

1. The three LLMs used in the GET system receive the question, reference answer, student answer, the grading reason recorded in the dataset, and the reason generated by the system as input: $(q_i, r_i, s_{i_j}, gr_{i_j}, pr_{i_j})$.
2. Each LLM compares gr_{i_j} and pr_{i_j} to determine if gr_{i_j} is valid or invalid.

3. The number of grading reasons judged as valid by each LLM is counted, and their proportions to the total number of grading reasons are calculated.
4. The average of these proportions is computed to obtain the final result.

By averaging the evaluation results from the three LLMs, we aim to achieve an objective evaluation that does not rely on a specific model.

4.3 Model Selection and Parameter Settings

The GET system uses three LLMs. For the experiments, we selected these models considering ease of deployment and cost-effectiveness, keeping real-world educational applications in mind.

For *multi-LLM grading*, we used three models: Mixtral-8x22B [15], Gemini 1.5 Flash [20], and Meta-Llama-3-8B-Instruct [1]. These models were chosen because they either offer free API access or are open-source, making them suitable in terms of cost and usability.

For *pseudo-learning*, we adopted Gemini 1.5 Flash [20] to analyze the labeling tendencies of LLMs. According to benchmarks, among the three models used for *multi-LLM grading*, Gemini 1.5 Flash demonstrated the best performance in mathematics-related tasks [10,17,18] (Accessed: December 26, 2024). Since the labeling tendency analysis involves numerical metrics such as accuracy and F1-score, we considered Gemini to be suitable due to its strong mathematical processing capabilities.

In *pseudo-learning*, the output pattern analysis needs a threshold to determine the most likely label for each combination of outputs. Through multiple trials, we determined the threshold to be 1.2.

Additionally, the temperature for the three LLMs was set to 0.7, and the maximum number of newly generated tokens was unified at 8192, which matches the maximum input length of ColBERT [16]. All other parameters were set to their default values for each LLM.

4.4 Baseline Methods

The performance of the GET system is compared with the following baselines.

- **Criteria-based Grading System** [14]: A system proposed by the authors. It extracts grading criteria from each question based on predefined grammar rules, then evaluates student answers using a fine-tuned language model to generate grading labels and their corresponding reasons.
- **ASAS-F-RAG** [7]: A grading system combining LLMs and RAG, developed by Fateen et al. Based on the publicly available GitHub repository, we implemented this method using Mixtral-8x22b, which showed the most stable and high-grading performance among the three models during *pseudo-learning*.
- **Single LLM Grading**: Without using the Ensemble ToT framework, a single LLM (Mixtral-8x22b, Llama-3-8B-Instruct, Gemini 1.5 Flash) individually grades student answers. The grading process follows the same steps as multi-LLM grading.

- **Ensemble Single**: A variation of the GET system implemented with only a single LLM (Mixtral-8x22b) instead of three different LLMs. *Pseudo-learning* is omitted as it relies on comparing outputs from multiple LLMs. Additionally, to mimic *multi-LLM grading*, the same LLM is used three times with different role names to grade student answers.

5 Result and Discussion

5.1 Analysis of Grading Label Prediction

The accuracy and macro F1-score of grading label predictions are shown in Table 2. All metrics are rounded to the fifth decimal place.

The GET system achieved the highest accuracy on both the UA and UQ sets, up to 77.87%. It also recorded the highest F1 score on the UA set and the average of UA/UQ. This confirms the effectiveness of the Ensemble ToT framework, which combines the outputs of multiple models.

In particular, for the UQ set, Gemini 1.5 Flash showed the highest accuracy and F1 score among the baselines that only used multi-LLM grading. However, the model used in the debate integration phase was Mixtral-8x22b. Despite this, the proposed method outperformed Gemini, achieving higher performance. This suggests that the proposed system can identify the optimal grading candidates through pseudo-learning and debate, without relying on a specific model.

Furthermore, the proposed method achieved better performance than the Ensemble Single baseline. This indicates that utilizing diverse LLMs is more important than the individual performance of a single LLM.

Table 2. Evaluation Results of Grading Label Prediction

	Accuracy			Macro F1-score		
	UA	UQ	Avg.	UA	UQ	Avg.
Criteria-base	0.7387	0.5595	0.6491	0.5496	0.5525	0.5511
ASAS-F-RAG	0.7093	0.6534	0.6814	0.6315	**0.6289**	0.6302
Mixtral-8x22b	0.7680	0.6284	0.6982	0.6686	0.5820	0.6253
Gemini 1.5 Flash	0.7387	0.6472	0.6929	0.7110	0.6146	0.6628
Llama3-8b-it	0.6987	0.4760	0.5873	0.5809	0.4622	0.5215
Ensemble Single	0.7600	0.6326	0.6963	0.6539	0.5830	0.6184
GET	**0.7787**	**0.6701**	**0.7244**	**0.7128**	0.6268	**0.6698**

5.2 Analysis of Grading Reason Generation

Table 3 shows the LLM-based evaluation results for grading reasons.

The proposed system achieved the highest average percentage of valid grading reasons in both the UA and UQ datasets, up to 70.22%. This demonstrates our system's ability to generate more semantically accurate explanations than the baselines. Moreover, it confirms that Ensemble ToT is effective not only for label prediction (a classification task) but also for grading reason generation (a text generation task). This underscores the framework's versatility and its potential applicability to various tasks.

Additionally, the proposed system consistently exhibited high performance across two different datasets. This suggests that it can maintain stable performance even when the dataset changes, supporting the reliability of the GET system's grading capability.

Table 3. Evaluation Results of Grading Reason Generation

	UA	UQ	Avg.
Criteria-base	0.4267	0.1635	0.2951
ASAS-F-RAG	0.6355	0.6145	0.6250
Mixtral-8x22b	0.6969	0.6110	0.6539
Gemini 1.5 Flash	0.6960	0.6319	0.6639
Llama-3-8B-it	0.5902	0.3709	0.4806
Ensemble Single	0.6871	0.6284	0.6578
GET	**0.7022**	**0.6360**	**0.6691**

5.3 Limitations and Future Works

This study has several limitations. First, the grading performance of the system when using LLMs other than the three models tested in our experiments remains unverified. Additionally, we have not yet examined how students perceive and evaluate the grading results provided by the GET system.

Moving forward, we aim to further enhance the performance of the GET system. For this enhancement, user experiments to collect feedback from students and instructors will also provide valuable insights.

Ensemble ToT is not limited to grading tasks – it is a versatile framework applicable to other domains. In the future, we will explore its potential applications beyond the education field.

6 Conclusion

In this study, we propose **Ensemble ToT**, a framework to effectively integrate multiple LLMs. Ensemble ToT unifies the outputs of multiple LLMs in three steps: analyzing performance tendencies, generating candidate solutions, and integrating them into a single high-accuracy solution.

Based on this framework, we develop an automated grading system, **Graders by Ensemble ToT (GET)**. It also works in three steps: *pseudo-learning* (analyzing LLM grading tendencies), *multi-LLM grading* (generating labels and reasons), and *debate integration* (determining final results through simulated discussions). Through this collaborative process, the LLMs complement each other, aiming to enable more accurate grading.

Experiments showed that GET achieved a maximum grading accuracy of 77.87%, outperforming existing methods. Additionally, up to 70.22% of its generated grading reasons were judged as valid, demonstrating the effectiveness of integrating multiple LLMs.

For future work, we aim to enhance the performance of the proposed system and explore broader applications of Ensemble ToT beyond grading.

Acknowledgment. Part of this research was supported by Grant-in-Aid for Scientific Research (23K28094).

References

1. AI@Meta: Llama 3 model card (2024). https://github.com/meta-llama/llama3/blob/main/MODEL_CARD.md
2. Besta, M., et al.: Graph of thoughts: solving elaborate problems with large language models. In: Proceedings of the AAAI Conference on Artificial Intelligence, vol. 38, pp. 17682–17690 (2024)
3. Cai, W., Jiang, J., Wang, F., Tang, J., Kim, S., Huang, J.: A survey on mixture of experts. arXiv preprint arXiv:2407.06204 (2024)
4. Chan, C.K.Y., Hu, W.: Students' voices on generative AI: perceptions, benefits, and challenges in higher education. Int. J. Educ. Technol. High. Educ. **20**(1), 43 (2023). https://doi.org/10.1186/s41239-023-00411-8
5. Chang, L.H., Ginter, F.: Automatic short answer grading for Finnish with ChatGPT. In: Proceedings of the AAAI Conference on Artificial Intelligence, vol. 38, pp. 23173–23181 (2024)
6. Dong, Y., Ding, S., Ito, T.: An automated multi-phase facilitation agent based on LLM. IEICE Trans. Inf. Syst. **107**(4), 426–433 (2024). https://doi.org/10.1587/TRANSINF.2023IHP0011
7. Fateen, M., Wang, B., Mine, T.: Beyond scores: A modular rag-based system for automatic short answer scoring with feedback. IEEE Access **12**, 185371–185385 (2024)
8. Filighera, A., Parihar, S., Steuer, T., Meuser, T., Ochs, S.: Your answer is incorrect... would you like to know why? Introducing a bilingual short answer feedback dataset. In: Proceedings of the 60th Annual Meeting of the Association for Computational Linguistics (Volume 1: Long Papers), pp. 8577–8591. Association for Computational Linguistics, Dublin, Ireland (2022). https://doi.org/10.18653/v1/2022.acl-long.587,

9. del Gobbo, E., Guarino, A., Cafarelli, B., Grilli, L.: Gradeaid: a framework for automatic short answers grading in educational contexts - Design, implementation and evaluation. Knowl. Inf. Syst. **65**(10), 4295–4334 (2023). https://doi.org/10.1007/S10115-023-01892-9
10. Google DeepMind: Gemini. https://deepmind.google/technologies/gemini/. Accessed 26 Dec 2024
11. Haller, S., Aldea, A., Seifert, C., Strisciuglio, N.: Survey on automated short answer grading with deep learning: from word embeddings to transformers. CoRR abs/2204.03503 (2022). https://doi.org/10.48550/ARXIV.2204.03503
12. Huang, J., Lee, Y., Kwon, O.: DIRECT: toward dialogue-based reading comprehension tutoring. IEEE Access **11**, 8978–8987 (2023). https://doi.org/10.1109/ACCESS.2022.3233224
13. Ibomoiye, D.M., Sun, Y.: A survey of ensemble learning: concepts, algorithms, applications, and prospects. IEEE Access **10**, 99129–99149 (2022). https://doi.org/10.1109/ACCESS.2022.3207287
14. Ito, Y., Ma, Q.: Supporting student self-learning using generative AI. In: Proceedings of the 2024 the 16th International Conference on Education Technology and Computers 2024, pp. 97–103. ICETC 2024, Association for Computing Machinery, New York, NY, USA (2025). https://doi.org/10.1145/3702163.3702177,
15. Jiang, A.Q., Sablayrolles, A., Roux, A., Mensch, A., Savary, B., et al.: Mixtral of experts. arXiv preprint arXiv:2401.04088 (2024)
16. Khattab, O., Zaharia, M.: ColBERT: efficient and effective passage search via contextualized late interaction over BERT. In: roceedings of the 43rd International ACM SIGIR Conference on Research and Development in Information Retrieval, SIGIR 2020, pp. 39–48. ACM (2020). https://doi.org/10.1145/3397271.3401075,
17. Meta: Meta-llama-3-8b. https://huggingface.co/meta-llama/Meta-Llama-3-8B. Accessed 26 Dec 2024
18. Mistral AI: Mixtral 8x22b. https://mistral.ai/news/mixtral-8x22b/. Accessed 26 Dec 2024
19. Ormerod, C.: Short-answer scoring with ensembles of pretrained language models. arXiv preprint arXiv:2202.11558 (2022)
20. Reid, M., Savinov, N., Teplyashin, D., Lepikhin, D., Lillicrap, T.P., et al.: Gemini 1.5: unlocking multimodal understanding across millions of tokens of context. arXiv preprint arXiv:2403.05530 (2024)
21. Wei, J., et al.: Chain-of-thought prompting elicits reasoning in large language models. Adv. Neural. Inf. Process. Syst. **35**, 24824–24837 (2022)
22. Wolpert, D.H.: Stacked generalization. Neural Netw. **5**(2), 241–259 (1992). https://doi.org/10.1016/S0893-6080(05)80023-1
23. Yao, S., Yu, D., Zhao, J., Shafran, I., Griffiths, T., et al.: Tree of thoughts: Deliberate problem solving with large language models. In: Advances in Neural Information Processing System, vol. 36: Annual Conference Neural Information Processing System (NeurIPS 2023) (2023). http://papers.nips.cc/paper_files/paper/2023/hash/271db9922b8d1f4dd7aaef84ed5ac703-Abstract-Conference.html
24. Zhou, Z.H.: Ensemble Methods: Foundations and Algorithms. Chapman & Hall/CRC, 1st edn. (2012)

Improving Prompt-Based Learning Framework for Mental Health Aspect Detection from Social Media

Jia-Ling Koh[1](✉) ⓘ, Hsiao-Ting Huang[1], and Yin-Ju Lien[2]

[1] Department of Computer Science and Information Engineering, Taipei, Taiwan
jlkoh@csie.ntnu.edu.tw
[2] Department of Health Promotion and Health Education, National Taiwan Normal University, Taipei, Taiwan
yjlien@ntnu.edu.tw

Abstract. Mental health detection on social media is challenging due to limited labeled data, data imbalance, and informal text structures. This study proposes IS iPET, an incremental selection training strategy that enhances Pattern-Exploiting Training (PET) and iterative PET (iPET) by gradually incorporating and strategically selecting training samples for fine-tuning Masked Language Models (MLM). Additionally, a margin-based loss function improves class separability. Experiments on Chinese social media posts show IS iPET improves precision by 20% and F1-score by 10%, while maintaining strong performance with 50% less training data. In open-environment testing, IS iPET achieves 0.81 precision in help-seeking behavior detection, demonstrating its real-world applicability. These findings suggest IS iPET is an effective semi-supervised approach for mental health detection.

Keywords: Prompt-based Learning · Mental Health Aspect Detection · Social Media Data

1 Introduction

As modern society faces increasing emotional and psychological challenges, mental health literacy (MHL) has become a crucial area of study. MHL assessments evaluate individuals' understanding of mental disorders, available treatments, and willingness to seek help. These assessments help researchers analyze how different aspects of MHL influence help-seeking behaviors [8, 10]. Traditional methods rely on expert-scored questionnaires [2, 11], but these approaches are costly and lack scalability.

With the rise of social media, users openly share their mental health experiences, providing a valuable but unstructured data source for large-scale MHL analysis [14]. For instance, posts like *"I have seen a doctor and was diagnosed with slight depression"* indicate a self-reported mental illness, while *"I have an appointment next week"* shows help-seeking behavior. Conversely, *"Sometimes emotional breakdowns occur"* suggests emotional distress, and *"I dare not seek medical treatment"* reflects a resistance to seek help. Analyzing such posts enables automated detection of mental health aspects, helping researchers identify patterns of emotional distress and help-seeking behavior.

Detecting mental health aspects in social media posts is challenging due to the scarcity of annotated data, class imbalance, and the complexity of informal text. Labeling datasets requires expert verification, making large-scale collection expensive and time-consuming. Recently, encoder-based prompt learning has proven effective for text classification with fewer labeled samples [12, 13], making it a promising approach for this task.

The iPET framework applies semi-supervised learning by fine-tuning Masked Language Models (MLM) with prompt-based learning and expanding training data through pseudo-labeling [12]. However, it lacks data selection during fine-tuning, leading to model bias due to noisy pseudo-labeling, especially in imbalanced datasets with scarce positive samples. To address this, we introduce Incremental Selection iPET (**IS iPET**), which integrates systematic data selection and multi-round fine-tuning to improve training effectiveness.

This study detects three mental health aspects—mental illness, emotional distress, and help-seeking behavior—in Chinese social media posts. Our approach enhances pre-trained language model fine-tuning by strategically selecting training samples, incrementally incorporating new data, and generating high-quality pseudo-labels to expand the dataset. Inspired by Spaced Learning [1] and Feedback Learning [5], our model addresses the challenge of detecting subtle mental health indicators. Spaced Learning ensures gradual knowledge retention by incrementally fine-tuning the model with new data across multiple rounds, while Feedback Learning refines training data selection by prioritizing misclassified or uncertain samples for retraining. Additionally, we introduce margin loss [16, 17] into the MLM loss function, enhancing class differentiation. We evaluate IS iPET in three perspectives: enhancing PET and iPET performance through the IS strategy, testing robustness with reduced training samples, and generalizing to a simulated open environment.

This study makes the following key contributions:

1. Proposes IS iPET, an improved training strategy for PET and iPET, increasing precision by 20% and F1-score by 10%.
2. Introduces margin loss in MLM fine-tuning to enhance differentiation between mental health aspects.
3. Evaluates IS iPET in a in an open-environment dataset to show the practicality of our approach in detecting mental health aspects in social media.

The rest of this paper is structured as follows: Sect. 2 reviews related work, Sect. 3 details data preparation, Sect. 4 presents the IS iPET training framework, Sect. 5 reports results of experiments, and Sect. 6 concludes with future research directions.

2 Related Works

Mental Health Literacy (MHL), introduced by Tony Jorm in 1997 [4], refers to public understanding and the ability to address mental health issues. MHL emphasizes the recognition, management, and prevention of mental disorders, and has expanded to include attitudes toward stigma and help-seeking behaviors. The MHL scale includes five key components [2]: (1) mental health maintenance (M), (2) understanding mental disorders and treatments (R), (3) addressing stigma (S), (4) help-seeking efficacy (HE), and (5) help-seeking attitudes (HA). The HA aspect covers responses to mental health and emotional issues, while Help-Seeking Behavior (HB) refers to seeking professional help.

This paper focuses on detecting three aspects of mental health in social media posts: (1) mental illness, (2) emotional distress, and (3) help-seeking behavior. The first two tasks involve binary classification for identifying posts related to mental illness or emotional distress, while the third task classifies posts based whether the post demonstrating help-seeking behavior.

Neural networks, such as CNNs and RNNs, are used to extract semantic features from text. CNNs capture local features [6, 7], while RNNs model text as a sequence for semantic understanding. The Transformer model [15], introduced in 2017, improved upon these by using positional embeddings and a self-attention mechanism, enabling better context understanding.

Pre-trained large-scale language models have gained popularity for text classification [3], as they require fewer labeled datasets and computational resources. These models learn semantic representations and can be fine-tuned for specific tasks with smaller datasets, such as sentiment analysis using prompt-based learning.

The PET (Pattern-Exploiting Training) and iPET (iterative PET) frameworks [12] use semi-supervised learning, fine-tuning a pre-trained MLM with a small labeled dataset, then soft-labeling a larger unlabeled dataset. iPET enhances PET by incrementally expanding the pseudo-labeled dataset, refining predictions over multiple rounds. However, in imbalanced datasets, random sampling during fine-tuning may cause mislabeling and suboptimal performance due to biases in pseudo-labeled data.

3 Data Preparation for Mental Health Aspect Detection

3.1 Problem Definition

The dataset used in this study consists of posts from Taiwan's PTT (Bulletin Board System). Experts in mental health literacy label posts based on defined mental health aspects. If a post contains a segment matching a targeted aspect, it is labeled as *positive* and includes a judgment sentence. If not, the post is labeled *negative*. The dataset includes each post with its corresponding label and *judgment sentence*, if applicable.

[Mental Health Aspect Detection] Given a text d_i consisting of multiple sentences, $d_i = \langle s_{i,1}, s_{i,2}, \ldots \rangle$, where $s_{i,k}$ represents the k-th sentence in d_i, the task is to predict whether the text exhibits a mental health aspect, assigning either a *positive* or *negative* label.

This study detects three mental health aspects: *having a mental illness*, *experiencing emotional problems*, and *exhibiting help-seeking behavior*, corresponding to separate classification tasks.

Most content in a post is unrelated to the targeted aspects, such as daily life descriptions. For example, a post may say, "Last week I was scolded by my boss, even though I had worked hard and thought I had done well before the report, but still didn't meet his expectations..." while relevant indicators like "The doctor said I have moderate depression" are often found in a single sentence. Training on entire posts may introduce noise, while analyzing individual sentences may lose context. To address this, we preprocess posts by concatenating consecutive sentences until they reach a specified length, forming *segments* that provide more meaningful training units. Instead of classifying entire posts, we focus on whether any segment exhibits the targeted aspect. If any segment is positive, the entire post is labeled as positive; otherwise, the post is labeled negative.

3.2 Sentence Segment and Labeling Process

This section describes how post content is transformed into sentence segments through three steps: sentence segmentation, removal of special characters and numbers, and sentence merging.

1. Sentence Segmentation.

Social media posts often contain punctuation marks such as commas, semicolons, periods, colons, question marks, exclamation marks, spaces, and line breaks, which indicate sentence boundaries. In the first step, posts are segmented based on these punctuation marks, resulting in multiple sentences of varying lengths, some of which may contain noise.

2. Character and Number Removal.

To ensure the model correctly interprets text semantics, non-Chinese characters, special symbols (e.g., @, ~, 「, 」, etc.), and numbers are removed. In this study, the 'bert-base-chinese' model is used as the pre-trained language model.

3. Sentence Merging

Short sentences may lack sufficient semantic information. To address this, consecutive sentences are merged until they reach a predefined length threshold l, forming sentence segments. Given a text $d_i = \langle s_{i,1}, s_{i,2}, \ldots, s_{i,n} \rangle$, the process starts with $s_{i,1}$.

If $s_{i,1}$ meets the length threshold, it becomes a segment $seg_{i,1}$.

If not, additional sentences $s_{i,2}, \ldots, s_{i,k}$ are appended (where $k \geq 2$) until the total length meets or exceeds l, forming $seg_{i,1}$.

This process repeats for the next sentence $s_{i,2}$, generating subsequent segments $seg_{i,2}, seg_{i,3}, \ldots$, ensuring that the entire text is converted into meaningful segments: $d_i = \{seg_{i,1}, seg_{i,2}, \ldots, seg_{i,m}\}$.

Label Assignment.

Each sentence segment $seg_{i,j}$ (where $j = 1, 2, \ldots, m$) is assigned a label based on the post's classification:

- If $d_i.label =$ 'positive', we check whether the judgment sentence $d_i.label\_sentence$ is contained within the segment:
- If $seg_{i,j}$ includes $d_i.label\_sentence$, it is labeled 'positive'.
- Otherwise, it is labeled 'none', meaning it does not explicitly contain the judgment sentence but may still be relevant.
- If $d_i.label =$ 'negative', all segments in d_i are labeled 'negative'.

For a text d_i, the sentence segment sets are defined as follows:

- **Positive segments**:

$$L_p(d_i) = \{seg_{i,j} | seg_{i,j} \in d_i \wedge seg_{i,j}.label = \text{'}positive\text{'}\} \tag{1}$$

- **Negative segments**:

$$L_n(d_i) = \{seg_{i,j} | seg_{i,j} \in d_i \wedge seg_{i,j}.label = \text{'}negative\text{'}\} \tag{2}$$

- **Unlabeled segments**:

$$U(d_i) = \{seg_{i,j} | seg_{i,j} \in d_i \wedge seg_{i,j}.label = \text{'}none\text{'}\} \tag{3}$$

For a dataset D, the union of all segments across all texts is defined as:
$L_p(D) = \bigcup_{d_i \in D} L_p(d_i)$, $L_n(D) = \bigcup_{d_i \in D} L_n(d_i)$, and $U(D) = \bigcup_{d_i \in D} U(d_i)$.
The complete labeled set, denoted as $L(D)$, consists of all positive and negative segments. The full set of all segments, denoted as $Seg(D)$, includes both labeled and unlabeled segments.

4 Training Aspect Detectors Using Prompt Learning

4.1 Prompt Learning Training Setup

Prompt learning requires a pre-trained language model (MLM), a prompt template, and a category verbalizer [9]. This study uses BERT as the MLM model. For each detection task, a manually designed prompt template is applied. For instance, in the "having a mental illness" detection task, the prompt template is: "I [MASK] have a men-tal illness." The category verbalizer maps 'Positive' to "do" and 'Negative' to "don't".

The MLM predicts [MASK] by assigning:

- A higher probability to "do" for positive samples.
- A higher probability to "don't" for negative samples.

In Chinese, both "do" and "don't" correspond to single-word tokens.

4.2 IS iPET Training Framework

The PET and iPET frameworks fine-tune MLM models by randomly selecting an equal number of samples from each class. However, when class distributions are imbalanced, this approach may not yield optimal performance. To address this, we propose the IS training strategy, which gradually incorporates balanced batches of positive and negative samples over multiple fine-tuning rounds while strategically selecting new training samples in each round. By combining incremental data addition with strategic selection, the MLM model progressively adapts to the target task, improving fine-tuning effectiveness.

The IS training strategy is compatible with both PET and iPET frameworks. When applied to iPET, it forms the IS iPET training framework (Fig. 1), where training samples are incrementally and strategically selected during Step (1) MLM-Tuning. The strategy supports multiple fine-tuning rounds ($t = 0, 1, \ldots, I$), with selection criteria and loss

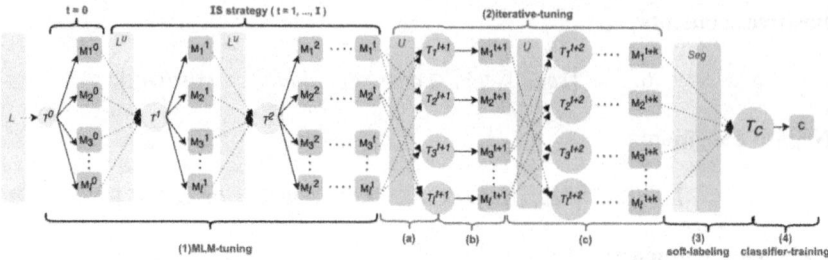

Fig. 1. IS iPET Training Framework

functions detailed in later. Similarly, IS PET incorporates the strategy within the PET framework.

The semi-supervised learning phase consists of three key steps. In Step (2) Iterative Tuning, the fine-tuned MLM models generate pseudo-labels for unlabeled data, which are then incrementally incorporated into training. Step (3) Soft-Labeling refines this process by generating soft labels not only for unlabeled data but also for labeled data, ensuring a larger and more informative training set. Finally, in Step (4) Classifier Training, a BERT-based sequence classifier is trained using these soft-labeled samples. While these steps follow the original iPET design, minor modifications are introduced to improve training effectiveness.

4.3 Initial Training Rounds for Fine-Tuning the MLM Model

This section describes the selection of training data in the first round of fine-tuning and its integration into prompt templates for aspect detection. Based on different randomly initialized training samples, as detailed in this section, the pretrained MLM is fine-tuned by prompt learning to obtain multiple MLM models, denoted as $M_1^0, ..., M_l^0$, forming an ensemble. In this study, we set $l = 4$.

- Initial Fine-Tuning Round.

In the first round, n positive and n negative samples are randomly selected from the labeled sentence segment set L, forming n training pairs. We set $n = 20$. These $2n$ samples constitute L^s (denoted as T^0), while the remaining segments form L^u for future rounds. Each sentence segment seg is defined as: $seg = \langle w_1^{seg}, ..., w_{len(seg)}^{seg} \rangle$. Let a prompt template $Pat = \langle w_1^{Pat}, ..., [MASK], ..., w_{len(Pat)}^{Pat} \rangle$. The sentence segment is concatenated with a prompt template Pat to form $Pat(seq)$ as follows:

$$Pat(seq) = \text{concatenate}([CLS]; seg; Pat) \quad (4)$$

The MLM model encodes the semantic representation and extracts the [MASK] position representation, denoted as $e_{[MASK]}$. A fully connected layer predicts the class label l from label set L, where $l \in \{positive, negative\}$:

$$s_{MLM}(l|seg) = \text{MLM}([MASK] = v(l)|Pat(seq)) \quad (5)$$

A *softmax* function normalizes scores to compute probabilities for each label l:

$$q_{MLM}(l|seg) = \frac{\exp(S_{MLM}(cl|seg))}{\sum_{l' \in L} \exp(S_{MLM}(l'|seg))} \quad (6)$$

The final predicted class l is chosen based on the highest probability:

$$l_{predict} = \underset{l \in L}{argmax} \, q_{MLM}(l|seg) \quad (7)$$

- MLM Model Loss Function.

To enhance classification performance, we adopt a margin-based ranking loss that enforces a prediction score margin between positive and negative samples (seg_i^+, seg_i^-). The binary cross-entropy (BCE) loss measures the difference between the predicted probability $\hat{y}$ and the ground truth y, where $y = 1$ for positive samples and $y = 0$ for negative samples:

$$L_{CE}(seg_i^*) = -(y * log(\hat{y}) + (1-y) * log(1-\hat{y})) \quad (8)$$

where $\hat{y} = q_{MLM}(c = 1|seg)$.

To further improve class separation, we introduce margin loss L_{pair}, which penalizes small differences between the predicted probabilities of positive and negative samples:

$$L_{pair}(seg_i^+, seg_i^-) = max(0, margin - (\hat{y}_p - \hat{y}_n)) \quad (9)$$

where $\hat{y}_p$ and $\hat{y}_n$ are the predicted probabilities for positive and negative samples. If $\hat{y}_p$ exceeds $\hat{y}_n$ by at least the margin (set to 0.5 in our experiments), the loss is 0; otherwise, the difference is penalized.

The final loss function combines BCE and margin loss, weighted by α (default: 0.5):

$$L = \frac{1}{n} \sum_{i=1}^{n} (\alpha(L_{CE}(seg_i^+) + L_{CE}(seg_i^-)) + (1-\alpha)L_{pair}(seg_i^+, seg_i^-)) \quad (10)$$

4.4 Multi-round Fine-Tuning of the MLM Model

The IS training strategy progressively incorporates labeled data over multiple rounds to fine-tune the MLM model. Each round retains previously used data, introduces new samples, and updates model parameters.

Training data is added incrementally using either random selection or strategic selection. The latter evaluates unselected samples (L^u) based on predictions from the previously fine-tuned MLM model, selecting those that require further learning for inclusion in the next round. By strategically adding training data, the model strengthens its ability to distinguish features while retaining previously used data for repeated training, preventing it from overly adapting to newly added samples.

- Random Incremental Selection of Training Data.

In each training round, n positive and n negative samples are randomly selected from L^u without replacement and added to the training data. The newly added samples in round t are denoted as $T^t_{positive}$ and $T^t_{negative}$. The training set is updated as follows:

$$T^t = T^{t-1} \cup T^t_{positive} \cup T^t_{negative} \tag{11}$$

The selected samples are then removed from L^u:

$$L^u = L^u - T^t_{positive} - T^t_{negative} \tag{12}$$

This process continues until one class in L^u has fewer than n samples. At that point, all remaining samples from that class are selected, while n samples from the other class are randomly chosen to complete the final training set T^t, marking the completion of the MLM fine-tuning process.

- Strategy-Based Selection of Training Data.

This approach uses the previously fine-tuned MLM model $\mathcal{M}^{(t-1)}$ to predict pseudo-labels for unselected sentence segments in L^u, forming the set L^u_{pseudo}. Since multiple MLM models are trained, the ensemble is represented as:

$$\mathcal{M}^{(t-1)} = \{M_1^{(t-1)}, \ldots, M_{len(\mathcal{M})}^{(t-1)}\} \tag{13}$$

The prediction scores for a given sentence segment seg are aggregated based on each model's relative importance. The weighted score for each class label l is computed as:

$$s_{\mathcal{M}^{(t-1)}}(l|seg) = \frac{1}{Z} \sum_{M_i^{(t-1)} \in \mathcal{M}^{(t-1)}} w(M_i^{(t-1)}) \bullet s_{M_i^{t-1}}(l|seg) \tag{14}$$

where $w\left(M_i^{(t-1)}\right)$ represents the weight of model $M_i^{(t-1)}$, determined by its accuracy on $T^{(t-1)}$ before fine-tuning, and Z is the sum of all model weights. A *softmax* function then converts the prediction scores into probabilities for each label l:

$$q_{\mathcal{M}^{(t-1)}}(l|seg) = \frac{\exp(S_{\mathcal{M}^{(t-1)}}(l|seg))}{\sum_{l' \in L} \exp(S_{\mathcal{M}^{(t-1)}}(l'|seg))} \tag{15}$$

Negative Sample Selection Strategy.

In round t, negative samples for the training set $T^t_{negative}$ are selected as follows:

1. Identify misclassified negatives: Extract samples from L^u_{pseudo} where the model incorrectly predicted a negative sample as positive.
2. Prioritize high-confidence misclassifications: Misclassified negatives are grouped into five probability bins: [0.9, 1.0), [0.8, 0.9), [0.7, 0.8), [0.6, 0.7) and [0.5, 0.6).
3. Select from the highest probability bin first: Since higher predicted probabilities indicate stronger prediction bias, selection starts with the [0.9,1.0) bin, accumulating n samples. If fewer than n samples are available, continue selecting from the next highest bin until the required count is met.

4. Handle cases with fewer misclassified negatives: If the total number of misclassified negatives is $\leq n$, all available samples are selected.

Positive Sample Selection Strategy.
In round t, positive samples for the training set $T^t_{positive}$ are selected as follows:

1. Identify correctly classified positives: Extract samples from L^u_{pseudo} where the model's predicted label matches the true label.
2. Prioritize low-confidence positives: Select the n samples with the lowest predicted probability, as they are likely near the classification boundary and require further training to enhance model robustness.

Final Round of Fine-Tuning Using the IS Training Strategy.
After multiple rounds of strategic selection, when misclassified negative samples are $\leq n$, they form the final negative training set $T^t_{negative}$. The final positive training set $T^t_{positive}$ is selected using the positive sample selection strategy, ensuring it matches the size of $T^t_{negative}$. is then selected to match its size. Once this final fine-tuning round is completed, the MLM fine-tuning phase under the IS training strategy concludes.

4.5 Semi-Supervised Learning Phase

The semi-supervised learning phase in the IS iPET training framework consists of three steps: iterative tuning, soft-labeling, and classifier training, as shown in Fig. 1.

Iterative Tuning.
Fine-tuned MLM models $M^t_1, ..., M^t_l$ generate pseudo-labels for the unlabeled dataset U. Following the iPET method, these pseudo-labeled samples are progressively incorporated into training, refining the model into $\mathcal{M}^{(t+1)}$.

Soft-Labeling.
The original iPET framework generates soft labels for U using iteratively fine-tuned MLM models $M^{(t+k)}_1., M^{(t+k)}_l$. Since this study includes a labeled dataset L, we extend soft-labeling to incorporate both labeled and unlabeled data, expanding the training set and mitigating model overconfidence. The soft-labeled dataset is drawn from seg, which includes both U and L. This improved soft-labeling strategy is referred to as $Strategy\_{(soft-labelingdata)}$.

Classifier Training.
The final step involves training a BERT-based binary classifier with a fully connected classification layer. Each training sample in T_c consists of a text segment seg and its corresponding soft label $q_\mathcal{M}(l|seg)$, representing the probability of a positive or negative classification.

To optimize classifier learning, we employ Kullback-Leibler (KL) divergence as the loss function, where $q_\mathcal{M}(l|seg)$ serves as the ground truth and is compared against the classifier's predicted probability $q_C(l|seg)$. The KL divergence loss is computed as:

$$L_{KL}(seg) = \text{KL}(q_\mathcal{M}(l|seg)\|q_C(l|seg)) = \sum_{l \in L} q_\mathcal{M}(l|seg) \log \frac{q_\mathcal{M}(l|seg)}{q_C(l|seg)} \quad (16)$$

This ensures that the classifier effectively learns from the soft-labeling probabilities, enhancing generalization and improving classification accuracy.

5 Performance Evaluation

5.1 Dataset Description

Labeled Dataset.
This study performs mental health aspect detection on social media posts across three tasks: (1) detecting **M**ental **D**isorders (MD), (2) identifying **E**motional **D**istress (ED), and (3) recognizing **H**elp-seeking **B**ehavior (HB). The dataset consists of Chinese-language posts from Taiwan's electronic bulletin board, PTT, which were manually selected and labeled according to these aspects. The labeled datasets are referred to as DB_{MD}, DB_{ED}, and DB_{HB}. Table 1 provides their statistical details.

Test Dataset for Simulating an Open Environment.
To evaluate real-world detection performance, 5,500 posts were collected from three PTT boards: *Prozac*, *Psychiatry*, and *Women'Talk*. Models trained on the labeled dataset were used to analyze these posts, with predictions validated by domain experts. Since most predictions were negative, all positive predictions were reviewed, while negative predictions were randomly sampled for verification. After expert validation, a final test dataset of 2,122 posts, referred to as DB_{open}, was established. Table 1 presents the distribution of positive and negative cases across the three mental health aspects.

5.2 Experimental Design and Results

Experiments on the labeled dataset are designed to evaluate the following:

1. Effectiveness of the IS training strategy in improving PET and iPET performance.
2. Impact of reduced training samples on model robustness.
3. Contribution of individual strategies within the IS iPET framework.

These experiments use the three labeled datasets and apply 5-fold cross-validation to compare different training frameworks for mental health aspect detection. Additionally, two experiments train models on the labeled dataset and evaluate their performance on the open-environment test dataset. The evaluation metrics include accuracy(A), precision(P), recall(R), and F1-score(F1) for detecting positive samples.

[Exp. 1] Effectiveness of the IS Training Strategy in PET and iPET.
Table 2 compares the performance of PET, iPET, IS PET and IS iPET in mental health aspect detection. The results indicate that iPET slightly underperforms PET in Precision and F1-score across all three tasks. This aligns with our observations that fine-tuning the MLM model only once with labeled data may be insufficient. Although iPET benefits from iterative fine-tuning with pseudo-labeled data, mislabeling introduce biases that ultimately degrade model performanc. Consequently, iPET-trained models perform worse than PET-trained models in this setting.

In contrast, incorporating IS training significantly improves Precision in both PET and iPET. IS training reduces false positives by reinforcing positive samples near the classification boundary while preventing negative samples from being misclassified.

This leads to higher Precision but a slight drop in Recall, as some borderline positives may be excluded. Despite this trade-off, IS training substantially boosts F1-score, with IS iPET outperforming IS PET in both Precision and Recall, resulting in an overall superior performance.

Table 1. Statistical information of the data sets.

	DB_{MD}	DB_{ED}	DB_{HB}	DB_{open}		
				MD	ED	HB
# posts of positive samples	318	381	320	283	827	220
# posts of negative samples	490	328	483	1839	1295	1902

Table 2. Performance of PET, iPET, IS PET, and IS iPET.

	DB_{MD}				DB_{ED}				DB_{HB}			
	A	P	R	F1	A	P	R	F1	A	P	R	F1
PET	0.748	0.612	0.981	0.754	0.777	0.706	1	0.828	0.72	0.59	0.972	0.734
iPET	0.726	0.592	0.984	0.739	0.758	0.69	1	0.816	0.707	0.578	0.988	0.729
IS PET	0.869	0.808	0.874	0.84	0.856	0.793	0.989	0.881	0.863	0.811	0.856	0.833
IS iPET	0.871	0.807	0.884	0.844	0.862	0.797	0.995	0.885	0.872	0.822	0.866	0.843

Table 3. Performance at different training data sampling ratios.

Sampling ratios	DB_{MD}		DB_{ED}		DB_{HB}	
	IS iPET	iPET	IS iPET	iPET	IS iPET	iPET
100%	0.807	0.592	0.796	0.69	0.822	0.578
75%	0.783	0.583	0.761	0.687	0.808	0.552
50%	0.757	0.581	0.75	0.674	0.779	0.543
25%	0.708	0.543	0.736	0.663	0.747	0.485
10%	0.643	0.548	0.692	0.635	0.668	0.452

[Exp. 2] Impact of Reduced Training Data on IS iPET.
This experiment examines how reducing the training sample size affects iPET and IS

Table 4. Performance impact of ablation.

	DB_{MD}			DB_{ED}			DB_{HB}		
	P	R	F1	P	R	F1	P	R	F1
IS iPET	0.807	0.884	0.844	0.796	0.995	0.884	0.822	0.866	0.843
w/o $Strategy_{selection}$	0.604	0.987	0.749	0.686	1	0.814	0.57	0.975	0.72
w/o $Strategy_{margin\ loss}$	0.738	0.928	0.822	0.768	0.995	0.867	0.786	0.897	0.838
w/o $Strategy_{soft-labeling data}$	0.747	0.893	0.814	0.758	0.995	0.86	0.787	0.866	0.824

Table 5. Performance on DB^{open} with and without aspect representative keyword filtering.

	MD				ED				HB			
	A	P	R	F1	A	P	R	F1	A	P	R	F1
DB^{open}	0.96	0.84	0.88	0.86	0.88	0.79	0.92	0.85	0.9	0.77	0.76	0.76
$DB^{open}_{w/keywords}$	0.85	0.85	0.8	0.83	0.72	0.85	0.76	0.8	0.66	0.81	0.58	0.67

iPET. Training sets were randomly reduced to 75%, 50%, 25%, and 10% of the original dataset, following the 5-fold cross-validation setup from Experiment 1. Precision was chosen as the primary metric.

Table 3 presents the results. As training data decreases, both frameworks exhibit a decline in Precision. However, IS iPET maintains a Precision drop within 5% at 50% of the data and within 10% at 25%. Remarkably, IS iPET trained on just 10% of the data still outperforms iPET trained on the full dataset. When using only 10% of the original data, the number of positive sentence segments drops below 100, highlighting the effectiveness of IS training in few-shot scenarios. By selecting informative samples, IS iPET significantly improves Precision despite data scarcity.

[Exp. 3] Effectiveness of Individual Strategies in IS iPET.

To assess the contribution of individual strategies, three modified versions of IS iPET were tested:

- w/o $Strategy_{selection}$: Replaces strategic selection with random sampling.
    w/o $Strategy_{margin\ loss}$: Removes margin loss term, relying solely on prediction accuracy.
- w/o $Strategy_{soft-labeling}$: Generates pseudo soft-labels only from unlabeled data, excluding labeled data.

Table 4 compares these versions against full IS iPET framework. w/o $Strategy_{selection}$ results in the most significant drop in Precision, confirming strategic selection as critical for performance.

Without strategy selection, Recall approaches 1, suggesting that randomly expanding positive-class data increases coverage but also introduces false positives. Conversely, strategic selection effectively reduces misclassified negatives, significantly improving Precision. Although removing soft-labeling has a smaller impact, incorporating labeled data into soft labeling boosts Precision by over 4%, confirming its importance when sufficient labeled data is available.

[Exp. 4] Evaluating IS iPET in an Open-Environment Test Dataset.
To assess real-world applicability, models trained on the labeled dataset were tested on DB_{open}. First, MD and ED detectors were evaluated separately, achieving 0.8 Precision.

For HB detection, an additional filtering step was introduced to align with mental health literacy definitions. Posts in DB_{open} were first screened using the MD and ED detectors, and only those classified as mental disorder or emotional distress were passed to the HB detector. This step reduced false positives in HB detection and improved Precision by 10%, reaching 0.77, while maintaining Recall.

[Exp. 5] Detection Performance with Aspect Representative Keyword Filtering.
Since mental health-related posts are rare, most social media content does not contain targeted aspects. To improve detection efficiency in open datasets, a preliminary filtering mechanism was introduced.

Representative keywords for each target aspect A_j were identified using term frequency (TF) analysis. Positive samples' judgment sentences were concatenated into D_p, and negative samples into D_n. Words appearing in D_p were collected in $D.term$, and TF ratios between D_p and D_n were computed. The top-N words with the highest relative scores formed the representative keyword set for A_j. If these keywords appear in a post, it is likely a positive sample. The keyword filtering mechanism pre-selects candidate posts before applying aspect detection models.

Table 5 compares results with and without keyword filtering. Precision improves across all three aspects, though Recall decreases slightly. However, in open-environment testing, Precision is the more critical metric, as the actual number of positive cases is unknown. Keyword filtering effectively removes irrelevant posts, enhancing detection efficiency and real-world applicability.

6 Conclusion

This study introduces the IS training strategy to enhance PET and iPET by incrementally adding training samples and refining model biases through multi-round fine-tuning. It also incorporates margin-based loss to improve class differentiation. Experiments show that IS training significantly boosts classification performance, increasing Precision by 20% to over 0.8 across three mental health datasets. Even with a 50% reduction in training data, IS iPET maintains at least 0.75 Precision, demonstrating robustness. In open-environment tests, filtering posts with MD and ED detectors before HB analysis achieves 0.81 Precision, confirming the framework's applicability to social media-based mental health detection. Future work may focus on improving negative sample diversity through candidate filtering, clustering, and resampling, while active learning could optimize labeling by prioritizing high-uncertainty predictions for expert annotation, reducing manual effort while expanding the dataset.

References

1. Cepeda, N.J., Pashler, H., Vul, E., Wixted, J.T., Rohrer, D.: Distributed practice in verbal recall tasks: a review and quantitative synthesis. Psychol. Bull. **132**(3), 354–380 (2006)
2. Chao, H., Lien, Y., Kao, Y., Tasi, I., Lin, H., Lien, Y.: Mental health literacy in healthcare students: an expansion of the mental health literacy scale. Int. J. Environm. Res. Public Health **17** (2020)
3. Devlin, J., Chang, M., Lee, K., Toutanova, K.: BERT: pre-training of deep bidirectional transformers for language understanding. In Proceedings of the Conference of the North American Chapter of the Association for Computational Linguistics: Human Language Technologies. Association for Computational Linguistics, Minneapolis, Minnesota, 4171–4186 (2019)
4. Jorm, A.F., Korten, A.E., Jacomb, P.A., Christensen, H., Rodgers, B., Pollitt, P.A.: Mental health literacy: a survey of the public's ability to recognize mental disorders and their beliefs about the effectiveness of treatment. Med. J. Aust. **166** (1997)
5. Kluger, A.N., Denisi, A.S.: The effects of feedback interventions on performance: a historical review, a meta-analysis, and a preliminary feedback intervention theory. Psychol. Bull. **119**, 254–284 (1996)
6. Kim, Y.: Convolutional neural networks for sentence classification. In: Conference on Empirical Methods in Natural Language Processing (EMNLP'14) (2014)
7. Kalchbrenner, N., Grefenstette, E., Blunsom, P.: A convolutional neural network for modelling sentences. In Proceedings of the 52nd Annual Meeting of the Association for Computational Linguistics. Association for Computational Linguistics, pp. 655–665 (2014)
8. Kutcher, S., Wei, Y., Coniglio, C.: Mental health literacy: past, present, and future. In The Canadian Journal of Psychiatry / La Revue canadienne de psychiatrie **61**(3), 154–158 (2016)
9. Liu, P., Yuan, W., Fu, J., Jiang, Z., Hayashi, H., Neubig, G.: Pre-train, prompt, and predict: a systematic survey of prompting methods in natural language processing. ACM Comput. Surv. **55**, 1–35 (2021)
10. Mojtabai, R., Evans-Lacko, S., Schomerus, G., Thornicroft, G.: Attitudes toward mental health help seeking as predictors of future help-seeking behavior and use of mental health treatments. Psychiatr. Serv. **67**(6), 650–657 (2016)
11. O'Connor, M., Casey, L.M.: The mental health literacy scale (MHLS): a new scale-based measure of mental health literacy. Psychiatry Res. **229**, 511–516 (2015)
12. Schick, T., Schütze, H.: Exploiting cloze-questions for few-shot text classification and natural language inference. In: Proceedings of the 16th Conference of the European Chapter of the Association for Computational Linguistics: Main Volume (EACL'21), Paola Merlo, Jörg Tiedemann, and Reut Tsarfaty (Eds.). Association for Computational Linguistics, pp. 255–269 (2020)
13. Scao, T.L., & Rush, A.M.: How many data points is a prompt worth?. In: Proceedings of the Conference of the North American Chapter of the Association for Computational Linguistics: Human Language Technologies. Association for Computational Linguistics, pp. 2627–2636 (2021)
14. Turcan, E., McKeown, K.: Dreaddit: A reddit dataset for stress analysis in social media. In: Proceedings of the Tenth International Workshop on Health Text Mining and Information Analysis (LOUHI 2019), pp. 97–107, Hong Kong. Association for Computational Linguistics (2019)

15. Vaswani, A., et al.. Attention is all you need. In: Advances in Neural Information Processing Systems, pp. 5998–6008 (2017)
16. Gao, J., Pantel, P., Gamon, M., He, X., Deng, L.: Modeling interestingness with deep neural networks. In Proceedings of the 2014 Conference on Empirical Methods in Natural Language Processing, pp. 2–13 (2014)
17. Goldberg, Y.: A primer on neural network models for natural language processing. J. Artif. Intell. Res. **57**, 345–420 (2016)

DInos: A Deep Reinforcement Learning Approach to Generalizable Autoscaling in Stateless Cloud Applications

Constantinos Bitsakos[1]($\boxtimes$)[ID], Dimitrios Tsoumakos[2][ID], Ioannis Konstantinou[3][ID], and Nectarios Koziris[1][ID]

[1] CSLAB, National Technical University of Athens (NTUA), Athens, Greece
{kbitsak,nkoziris}@cslab.ece.ntua.gr
[2] DBLAB, National Technical University of Athens (NTUA), Athens, Greece
dtsouma@mail.ntua.gr
[3] Department of Informatics and Telecommunications, University of Thessaly, Volos, Greece
ikons@uth.gr

Abstract. Efficient autoscaling in Kubernetes (K8s)-managed in-memory systems like Redis remains a critical challenge, especially under highly dynamic workloads. Traditional threshold-based mechanisms (e.g., HPA) often fail to anticipate sudden demand surges, leading to poor performance and inefficient resource use.

We introduce **DInos**, a Deep Reinforcement Learning (Deep RL) agent enhanced with LSTM layers and transfer learning, designed for proactive and adaptive autoscaling in Kubernetes. As an evolution of our earlier agent DERP, DInos leverages temporal workload modeling and pre-trained policies to generalize across deployments with minimal retraining. DInos utilizes a customizable reward function balancing throughput, latency, resource usage, and pod efficiency.

DInos achieves up to **17.3× higher rewards** in simulation and a **5.5× improvement** in real-world K8s-Redis deployments by forecasting spikes, optimizing pod counts and maintaining low latency, providing a robust autoscaling solution for volatile, cloud-native environments.

Keywords: Kubernetes · Deep Reinforcement Learning · LSTMs · Autoscaling Redis

1 Introduction

Kubernetes (K8s) is the de facto standard for orchestrating containerized applications, offering reactive autoscaling through mechanisms like the Horizontal Pod Autoscaler (HPA) [37]. However, such threshold-based approaches struggle with dynamic, time-sensitive workloads, often leading to overprovisioning or increased latency.

Redis, a stateless in-memory key-value database, is widely adopted in latency-critical cloud services [29]. Under traffic spikes, reactive autoscaling approaches frequently fail to respond well enough, resulting in degraded performance.

To overcome these limitations, our previous work introduced *DERP* [10], a Deep Reinforcement Learning (Deep RL) autoscaler. DERP significantly outperformed traditional threshold-based autoscalers and earlier RL-based autoscaling methods that used Q-tables or decision trees. In this paper, we further enhance DERP by introducing a new deep neural network architecture, extending it to utilize additional performance metrics such as CPU and memory usage, as well as pod count.

Building on these improvements, we propose our main contribution: **DInos**, an advanced LSTM-based Deep RL autoscaling agent that integrates transfer learning for rapid generalization across deployments [39]. DInos leverages temporal modeling and policy reuse to proactively anticipate workload changes and optimize resource allocation. Both DERP and DInos are trained with a customizable reward function that aims to balance throughput, latency, resource usage (CPU and memory), and pod efficiency [38], enabling versatile adaptation to diverse system goals.

DInos achieves up to a 17× improvement over DERP in simulation and a 5.5× gain over its non-transfer variant in real Kubernetes-Redis environments, clearly demonstrating the significant advantages of temporal modeling and policy transfer in proactive autoscaling. Furthermore, its transfer learning capability enables DInos to quickly and effectively adapt to diverse environments, new applications, different frameworks, and varying workloads without extensive retraining.

2 Related Work

Kubernetes (K8s) has emerged as a leading platform for cloud-native orchestration [4], driving substantial research in autoscaling. Traditional scaling mechanisms include the Horizontal Pod Autoscaler (HPA) [37], Vertical Pod Autoscaler (VPA) [6], and Cluster Autoscaler [3], relying on reactive CPU and memory thresholds. Such methods often struggle under rapidly changing workloads, causing resource inefficiencies and latency spikes.

KEDA [19], an event-driven Kubernetes autoscaler, improves flexibility using external metrics sources such as Kafka or Prometheus. However, its lack of predictive intelligence and adaptive learning limits its effectiveness with highly dynamic workloads.

Previous autoscalers employed reinforcement learning (RL) or heuristic-based approaches, notably Tiramola [34] and VCONF [28], initially targeting NoSQL or VM-based environments. Classic RL methods employing Q-tables or decision trees [24, 27, 36] had inherent scalability limitations and insufficient generalization.

Our earlier work, DERP [10], significantly outperformed these threshold-based and classical RL methods by employing Deep Q-Learning with neural networks. DERP introduced fine-grained autoscaling policies based on throughput, latency, and resource metrics. Despite these improvements, DERP lacked temporal modeling and knowledge transfer mechanisms, restricting its predictive accuracy and generalizability.

Other ML-based autoscalers, such as Imdoukh et al. [16] and the Bi-LSTM system by Dang-Quang and Yoo [11], enhanced workload forecasting but lacked an integrated adaptive RL framework, limiting responsiveness.

Research by Toka et al. [33] and Koukis et al. [18] explored predictive scaling and network overhead impacts, respectively. Studies like SCADIS [17], Harmonia [40], and performance benchmarks [9,22,30] provided insights into Redis-specific scenarios. Additional work investigated node-level elasticity [32], extreme-scale workloads [21], and dynamic cloud scaling strategies [7,25].

Advancements in LSTM network optimization [20,39], and Deep RL methodologies [15,38] underscored the importance of combining temporal forecasting with RL. Transfer learning has also been recognized as critical for rapid generalization across deployments [26]. Kubernetes-based ML management frameworks, like Scanflow-K8s [23] and performance studies by Fogli et al. [12], further emphasize the need for intelligent autoscaling.

Our proposed solution, **DInos**, significantly surpasses even the enhanced DERP architecture by integrating LSTM-based temporal forecasting, Double Deep Q-Learning, and transfer learning into a single agent. Unlike previous methods that treat forecasting and scaling separately, DInos learns proactive and predictive scaling strategies simultaneously. DInos demonstrates a substantial cumulative reward improvement of 17.3× over DERP in simulations and achieves a 5.5× reward gain compared to a non-transfer learning baseline in real Kubernetes/Redis deployments. These improvements result directly from DInos' capability to forecast resource needs, optimize provisioning proactively, and adapt rapidly to previously unseen workloads.

In summary, DInos represents an adaptive, intelligent autoscaling framework capable of leveraging historical workload patterns and generalizing across environments, establishing a robust solution for dynamic cloud-native autoscaling.

3 System Model and Background

3.1 Kubernetes and Redis Setup

Our system is deployed on a Kubernetes (K8s) cluster running a stateless Redis environment (Fig. 1). Redis, being an in-memory key-value store, is well-suited for dynamic autoscaling due to its statelessness and fast response times. Kubernetes orchestrates Redis pods and enables elastic scaling in response to workload fluctuations.

Metrics such as CPU-memory usage and latency are retrieved directly from the Kubernetes Metrics Server [5] and provided to our Deep RL agents for scaling decisions. Prometheus [2] and Grafana [1] were used only for monitoring and visualization. Our custom autoscaler interacts with the control plane to apply scaling actions via the kube-api-server.

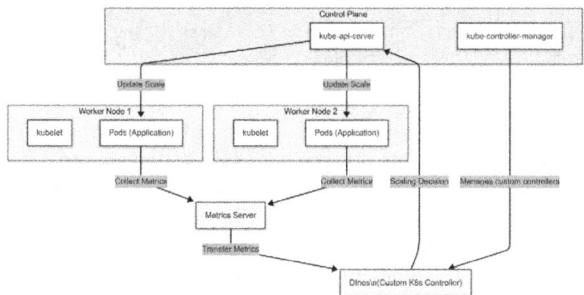

Fig. 1. System architecture of our proposed DInos autoscaling agent within a Kubernetes cluster. The agent interacts with the Kubernetes control plane by receiving time-series metrics from Redis pods via the Metrics Server. Based on these inputs, DInos issues proactive scaling decisions–adding or removing pods–by communicating with the kube-api-server. The agent learns workload patterns using deep reinforcement learning, enabling intelligent resource allocation that balances performance and cost across dynamic environments.

3.2 Deep Reinforcement Learning Agents

We compare two Deep RL-based agents: a dense-layer agent (enhanced DERP) and an LSTM agent enhanced with transfer learning (DInos). Both agents take a state vector of system metrics–load, CPU usage, memory usage, latency, and pod count–and return scaling actions.

DERP Agent (Dense). DERP enchance our previous work [10] by using two dense layers (128 units each) with ReLU activations. It is trained using a Double DQN [8] to avoid overestimation of Q-values and provides a baseline for comparison, This architecture is shown in Fig. 2.

- **Input:** System state vector
- **Hidden:** Two dense layers with 128 neurons
- **Output:** Q-values for scaling actions

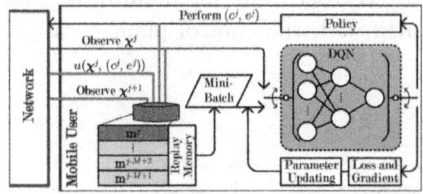 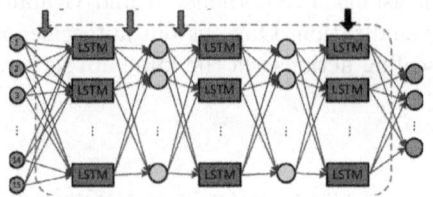

Fig. 2. Double Deep Q-Network (Double DQN) architecture used by both DERP and DInos for stable learning and decision-making. It separates action selection (online network) from value estimation (target network), mitigating Q-value overestimation and improving training stability. This architecture supports discrete action spaces, such as scaling decisions for pod count.

Fig. 3. LSTM-based neural network architecture employed by DInos. The LSTM layers are designed to capture long-term temporal dependencies in workload metrics, including periodic trends and abrupt spikes. This enables the agent to proactively scale resources by forecasting future system states rather than reacting to current load alone.

DInos Agent (LSTM + Transfer Learning). DInos extends the dense-layer-based DERP variant with LSTM layers and transfer learning. It leverages these LSTM layers to effectively model temporal patterns in workloads. Initially, a base model is extensively trained in a simulation environment whose parameters and load scenarios have been carefully formalized based on extensive real-world experiments and observations (Fig. 7, right). This pretrained model thus captures generalized scaling policies applicable to various workload patterns. DInos is subsequently initialized with this pretrained model when deployed in new environments, enabling it to rapidly adapt to different workloads with minimal additional training. Fine-tuning occurs on the target workload using a smaller batch size of 16, significantly accelerating adaptation and ensuring robust performance across diverse real-world deployment scenarios.

- **Input:** Sequence of past states
- **LSTM:** Memory layers with fine-tuning
- **Output:** Q-values for scaling decisions

The Spike Factor: While DERP reacts to general workload trends, it struggles with sudden spikes. DInos' LSTM layers anticipate these patterns using temporal memory, adjusting resources proactively–particularly effective for flash-sale or seasonal traffic surges (Figs. 3, 4).

3.3 Reward Function

The reward function balances five objectives:

$$\begin{aligned}reward = {} & \text{load\_factor} \cdot \text{next\_load} - \text{latency\_factor} \cdot \text{next\_latency} \\ & - \text{pods\_penalty} \cdot \text{num\_pods} - \text{cpu\_penalty} \cdot \text{cpu} - \text{memory\_penalty} \cdot \text{memory}\end{aligned} \quad (1)$$

This reward formulation is designed to highlight the versatility and capability of our agents in learning optimal scaling strategies under multiple objectives. While this particular reward function focuses on balancing throughput, latency, and resource consumption, it is fully customizable. System operators or cloud vendors can easily redefine the reward structure to reflect specific operational goals or application priorities–such as cost minimization, latency guarantees, or energy efficiency. DInos and enhanced DERP retain their effectiveness across such formulations, showcasing their adaptability to diverse autoscaling policies and deployment environments.

3.4 Simulation Environment

Workload patterns are generated using a sinusoidal base, Gaussian noise, and a sharp spike between timesteps 1500âĂŞ1600 to emulate bursty real-world traffic.

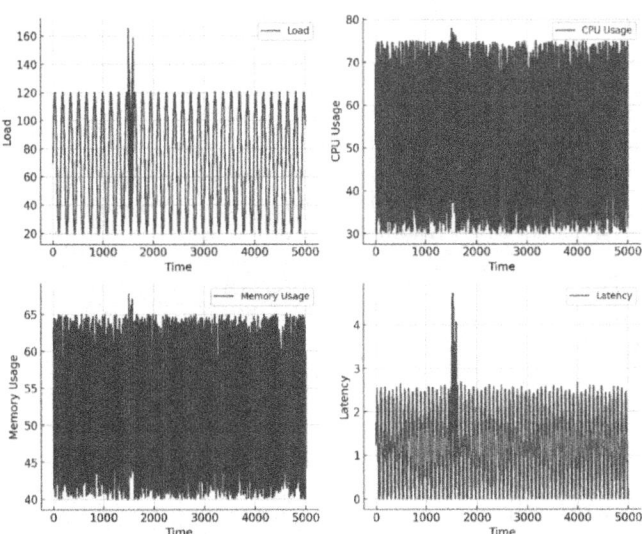

Fig. 4. Input metrics generated by the simulation environment: load, CPU usage, memory usage, and latency over 5000 timesteps. The load follows a sinusoidal pattern with added Gaussian noise and a sharp spike between timesteps 1500 and 1600, simulating a burst workload. These metrics form the state input for both enhanced DERP and DInos, testing their ability to maintain performance under dynamic and unpredictable conditions.

CPU, memory, and latency metrics are modeled accordingly after observing real-world behaviors (see Fig. 4 and the right subfigure of Fig. 7).

3.5 Training Details

Both agents are trained using the Huber loss function [14] for stability. A Double DQN framework (Fig. 2) is used to decouple action selection from value estimation and improve learning in noisy environments.

3.6 DInos Autoscaling Algorithm

Algorithm 1 DInos: Transfer Learning Autoscaler

1: **Input:** Pretrained model (optional), current system state s_t
2: Initialize Q-network (pretrained if using DInos)
3: Initialize target network and replay buffer R
4: **for** each timestep t **do**
5:    Observe s_t
6:    **if** exploration condition met **then**
7:       Select random action a_t
8:    **else**
9:       Select action $a_t = \arg\max_a Q(s_t, a)$
10:    **end if**
11:    Apply a_t, observe r_t, s_{t+1}
12:    Store (s_t, a_t, r_t, s_{t+1}) in R
13:    Sample minibatch from R
14:    **for** each (s_t, a_t, r_t, s_{t+1}) in minibatch **do**
15:       Compute target: $y = r_t + \gamma \max_a \hat{Q}(s_{t+1}, a)$
16:       **if** transfer learning is enabled **then**
17:          $Q(s_t, a_t) \leftarrow (1-\alpha)Q(s_t, a_t) + \alpha y$
18:       **else**
19:          $Q(s_t, a_t) \leftarrow Q(s_t, a_t) + \eta(y - Q(s_t, a_t))$
20:       **end if**
21:    **end for**
22:    Periodically update target network $\hat{Q}$
23: **end for**

---

4 Implementation

4.1 Training Setup

We trained two Deep RL agents–an enhanced version of DERP (with dense layers) and DInos (with LSTM and transfer learning)–using system metrics collected from a Redis cluster orchestrated by Kubernetes. The input state for each agent includes:

- Load (request rate to Redis),
- CPU usage,
- Memory usage,
- Latency,
- Number of pods.

The environment simulated dynamic workloads with oscillating patterns and abrupt traffic spikes (see Fig. 4), allowing agents to learn both reactive and proactive autoscaling behavior.

Agents selected one of three actions–add, remove, or maintain pods–every 15 s. The reward was computed at each timestep using the following weighted linear function:

$$\text{reward} = 3.0 \cdot \text{next\_load} - 5.0 \cdot \text{next\_latency} \\ - 10.0 \cdot \text{num\_pods} - 0.5 \cdot \text{next\_cpu\_usage} - 0.5 \cdot \text{next\_memory\_usage} \tag{2}$$

This formulation promotes high system throughput (via the positive weight on `load`) and low latency, while discouraging over-provisioning and excessive resource usage. The specific weights were determined through empirical tuning and sensitivity analysis during simulation experiments. These coefficients are not fixed and can be adapted to suit different infrastructure priorities, such as cost minimization or latency-critical performance.

4.2 Kubernetes Integration

To enable real-time scaling by the agents, we disabled the default Horizontal Pod Autoscaler (HPA) and created a custom Python environment class that directly interfaces with the Kubernetes API. This class executes scaling actions issued by the agent and fetches current system metrics from the Kubernetes Metrics Server [5].

The cluster is configured with stateless Redis pods, ensuring that pod addition or removal can be handled gracefully without service disruption.

4.3 Challenges and Observations

Several technical challenges emerged during development:

- **Latency sensitivity:** Capturing realistic latency patterns required simulating spikes during load surges, guided by latency trends observed in the real-world environment. This helped ensure that the simulation environment reflected realistic system behavior and reinforced the agents ability to respond appropriately.
- **Temporal forecasting:** DERP, even the enchanced version, was unable to anticipate future workload shifts. In contrast, DInos successfully leveraged LSTM layers to recognize and prepare for recurring spikes, improving responsiveness (see Fig. 3).

- **Cost-performance tradeoff:** The pod penalty in the reward function helped balance throughput against overprovisioning, though it required careful tuning to avoid excessive scaling.

Despite these challenges, both agents were successfully trained in the simulated environment. DInos demonstrated superior performance by adapting to future workload conditions through transfer learning and temporal modeling. Full results are discussed in Sect. 5.

5 Experimental Results

5.1 Experimental Setup

Experiments were conducted on a Kubernetes cluster running on a large Ubuntu 20.04 VM with the following specs:

- **RAM:** 191 GiB, **CPU:** 32 vCPUs (Skylake, 2 GHz), **GPU:** Cirrus Logic, **Disk:** 750 GiB (EXT4)

Using Kubernetes on such a high-resource VM allowed the setup to emulate a full-scale cluster. We used Python 3.8 and PyTorch, and recorded load, latency, CPU-memory usage and pod count for evaluation.

5.2 Autoscaling Agents Compared

We evaluate two Deep RL-based autoscaling agents:

- **Enhanced DERP (Dense):** A Deep Q-learning agent using dense layers
- **DInos (LSTM + Transfer):** An advanced LSTM-based agent that leverages transfer learning for faster adaptation

5.3 Simulation Results

In the simulation environment our agents achieved:

- **DERP:** 2.2×10^5 reward
- **DInos:** 3.8×10^6 reward – a 17.3× improvement

Figure 4 illustrates the simulation environment, which features sinusoidal workloads combined with random noise and a sharp demand spike. Under these challenging conditions, DInos demonstrates a significant advantage over DERP, both in terms of accumulated rewards and system responsiveness. This superiority arises from DInos' effective forecasting and proactive scaling strategies. Detailed performance metrics and comparative analyses can also be found in Figs. 5 and 6. Specifically, Fig. 6 (right) clearly illustrates DInos' adaptive scaling capabilities, as it dynamically adjusts pod counts in response to workload fluctuations. The figure demonstrates how DInos efficiently balances resource usage and latency by proactively predicting workload patterns and swiftly responding to spikes, showcasing the adaptability and robustness of our agent under volatile simulated conditions.

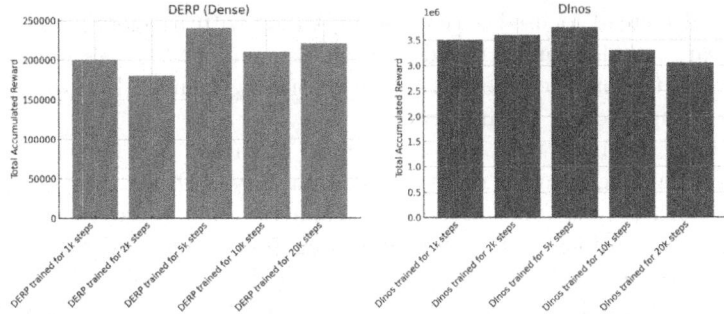

Fig. 5. Comparison of total accumulated rewards between DERP and DInos at different training checkpoints (1k, 2k, 5k, 10k, and 20k steps). While DERP's performance improves with training, it plateaus early. In contrast, DInos, leveraging LSTM layers and transfer learning, achieves significantly higher rewards across all training durations, highlighting its superior adaptability and learning efficiency in dynamic environments.

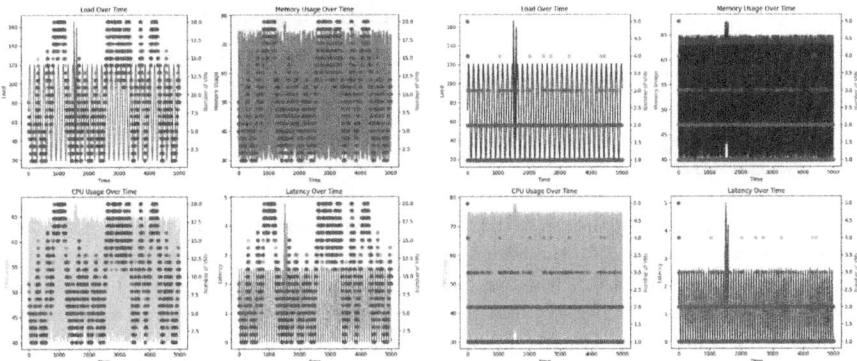

Fig. 6. Evaluation of DERP (left) and DInos (right) based on runtime metrics over 5000 timesteps. DERP responds reactively, often under- or over-provisioning during spikes. DInos, using LSTM layers and transfer learning, shows superior stability by anticipating load fluctuations and adapting pod count accordingly. The result is lower average latency, better CPU and memory efficiency, and higher throughput under varying load.

5.4 Real-World Results

In real Kubernetes/Redis deployments, DInos achieved:

- **DInos:** 308,000 reward
- **Baseline LSTM agent (no transfer):** 56,000 reward

In real Kubernetes/Redis deployments, DInos achieved a cumulative reward of 308,000, significantly outperforming its LSTM-based baseline without transfer learning, which attained a reward of only 56,000. Due to DERP's relatively lower performance in the simulation scenarios, we excluded it from real-world evaluations and instead focused on comparing DInos with its closest non-transfer-learning variant. This comparison demonstrates a remarkable 5.5× improvement, underscoring the practical benefits of incorporating transfer learning and temporal modeling into our RL framework.

Figure 7 (right) provides detailed metrics from the live Kubernetes/Redis deployment, highlighting the real-world adaptability of DInos. The figure explicitly demonstrates how DInos proactively manages resource provisioning by forecasting load variations and promptly scaling pod counts, maintaining stable resource utilization and consistently low latency despite noisy and spiky traffic. This confirms that DInos can rapidly adapt to diverse production workloads with minimal retraining effort, thereby validating the practical effectiveness and robustness of our proposed solution in dynamic cloud-native environments.

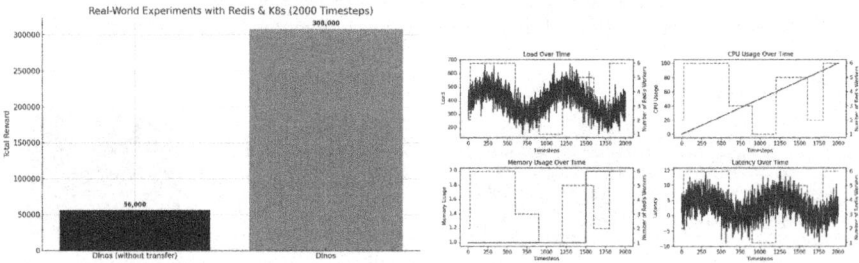

Fig. 7. Real-world evaluation of DInos in a live Kubernetes/Redis deployment. **Left:** Cumulative reward over 2000 timesteps, comparing DInos against its LSTM-only baseline. DInos achieves a total reward of 308,000, outperforming the LSTM-only version by 5.5×, showcasing the effectiveness of transfer learning in accelerating adaptation to production environments. **Right:** Temporal system behavior of DInos, demonstrating how the agent dynamically adjusts the number of Redis pods in response to noisy and spiky load patterns. The figure shows that DInos maintains low latency and stable resource usage while proactively scaling based on predicted workload trends.

5.5 Key Insights

- **Load Forecasting:** DInos anticipates spikes and scales preemptively via LSTM memory, unlike DERP which reacts post-factum.
- **Transfer Learning:** DInos requires minimal retraining and adapts quickly to new workloads, outperforming baseline LSTM agents in both accuracy and speed.
- **Resource Efficiency:** DInos minimizes pod count while maintaining low latency and high throughput.

5.6 Performance Summary

Figure 8 highlights how **DInos** consistently outperforms the enchanced **DERP** agent across all key metrics–accumulated reward, CPU usage, memory usage, and system latency–in both simulated and real-world Kubernetes/Redis deployments. These metrics collectively demonstrate DInos' superior adaptability, forecasting ability, and efficiency under volatile workloads.

In simulation, DInos achieved a cumulative reward of 3.8×10^6, compared to just 2.2×10^5 for DERP, representing a **17.3× improvement**. This drastic gain reflects DInos' capacity to anticipate spikes and proactively allocate resources using its LSTM-enhanced decision-making. In real-world experiments, DInos outperformed a non-transfer LSTM baseline with a reward of 308,000 versus 56,000, demonstrating a **5.5× gain** in environments with real noise and system complexity.

Notably, these improvements are achieved while maintaining lower average CPU and memory consumption and significantly reducing latency fluctuations. This confirms that DInos does not simply overprovision resources to gain performance but instead learns to scale intelligently, minimizing cost while maximizing responsiveness.

The underlying reward function, which balances system throughput, latency, and resource usage, serves as a demonstration of the agent's capability to optimize under multi-objective conditions. However, this formulation is intentionally flexible: system operators or cloud vendors can customize the reward structure to reflect domain-specific policies–such as strict latency bounds, energy-aware scaling, or monetary budget constraints. Both DERP and DInos maintain robust performance across such variations, underscoring their **versatility and generalizability** in real-world autoscaling scenarios.

Method	Environment	Results (Accumulated Reward)
DERP	Simulation	≈ 220,000
DInos (with transfer)	Simulation	≈ 3,800,000
DInos (without transfer)	Real World	≈ 56,000
DInos (with transfer)	Real World	≈ 308,000

Fig. 8. Summary table comparing DERP and DInos across simulation and real-world deployments. Metrics include accumulated reward, average CPU and memory usage, and system latency. DInos consistently outperforms DERP in both environments, demonstrating better generalization, resource efficiency, and responsiveness to workload changes–especially in scenarios involving sharp demand spikes.

6 Discussion and Conclusion

Our experimental results confirm that Deep Reinforcement Learning (Deep RL) can outperform traditional autoscaling strategies in Kubernetes environments.

While DERP (dense-layer-based) improved over threshold-based baselines, it lacked the foresight needed for handling workload volatility. In contrast, DInos, our LSTM-based agent enchanced transfer learning, learned to proactively scale resources by identifying temporal workload patterns and leveraging knowledge from prior deployments.

In simulation, DInos achieved a 17.3× improvement in accumulated reward over DERP, and in real-world Redis deployments, it delivered 5.5× better performance compared to a baseline LSTM agent without transfer learning. These gains reflect more efficient resource utilization, lower latency, and better responsiveness under unpredictable workloads.

Future Directions

Future research can explore extending DInos to more complex environments:

- **Stateful Applications:** Applying DInos to stateful systems like Cassandra introduces new challenges such as data replication, consistency, and safe pod identity management due to the constraints of StatefulSets. Transfer learning can accelerate adaptation in such environments by leveraging prior knowledge to reduce retraining overhead.
- **Expanded Metrics:** Incorporating disk I/O, network bandwidth, or storage usage could provide a more holistic view of system health and enable even finer-grained scaling decisions.
- **Advanced RL Architectures:** Exploring algorithms like Dueling DQN [35], PPO [31], or SAC [13] may improve stability or allow continuous action spaces, especially in multi-objective environments.

Key Takeaways

DInos significantly improves upon prior Deep RL agents by combining LSTM forecasting with transfer learning. This enables faster convergence, better generalization, and minimal retraining across different deployments. Our approach demonstrates that predictive, learning-based autoscaling can lead to lower costs, higher performance, and robust adaptability–making it a compelling choice for modern, cloud-native systems.

These findings open new avenues for generalizing Deep RL-based autoscalers beyond Redis, across cloud platforms, and into more complex distributed workloads.

Competing Interests. The authors declare no competing interests relevant to this work.

References

1. Grafana: The open platform for analytics and monitoring (2021). https://grafana.com. Accessed 21 Oct 2024
2. Prometheus: Monitoring system and time series database (2021). https://prometheus.io. Accessed 21 Oct 2024
3. Cluster autoscaler (2024). https://github.com/kubernetes/autoscaler/blob/master/cluster-autoscaler/README.md. Accessed 22 Oct 2024
4. Horizontal pod autoscaler (2024). https://kubernetes.io/docs/tasks/run-application/horizontal-pod-autoscale/. Accessed 22 Oct 2024
5. Kubernetes metrics server (2024). https://github.com/kubernetes-sigs/metrics-server. Accessed 21 Oct 2024
6. Vertical pod autoscaler. https://kubernetes.io/docs/concepts/workloads/pods/pod-lifecycle/#vertical-pod-autoscaling (2024). Accessed 22 Oct 2024
7. Ardagna, C., et al.: A competitive scalability approach for cloud architectures. IEEE Trans. Serv. Comput. (2014). https://doi.org/10.1109/TSC.2014.2372786
8. Bell-Thomas, A.H.: Exploring variational deep Q networks. arXiv preprint arXiv:2004.05615 (2020). https://arxiv.org/abs/2004.05615
9. Ben Seghier, N., Kazar, O.: Performance benchmarking and comparison of NoSQL databases: Redis vs MongoDB vs Cassandra using YCSB tool. IEEE (2021). https://doi.org/10.1109/ICRAMI52622.2021.9585956
10. Bitsakos, C., Konstantinou, I., Koziris, N.: Derp: A deep reinforcement learning cloud system for elastic resource provisioning. In: 2018 IEEE International Conference on Cloud Computing Technology and Science (CloudCom), pp. 21–29. IEEE (2018)
11. Dang-Quang, N.M., Yoo, M.: Deep learning-based autoscaling using bidirectional long short-term memory for Kubernetes. Appl. Sci. **11**(9) (2021). https://doi.org/10.3390/app11093835
12. Fogli, M., et al.: Performance evaluation of Kubernetes distributions in federated cloud infrastructure. In: IEEE International Conference on Cloud Computing (2021). https://doi.org/10.1109/CLOUD.2021.00073
13. Haarnoja, T., Zhou, A., Abbeel, P., Levine, S.: Soft actor-critic: off-policy maximum entropy Deep Reinforcement Learning with a stochastic actor. In: Proceedings of the 35th International Conference on Machine Learning (ICML), pp. 1861–1870. JMLR.org (2018)
14. Huber, P.J.: Robust estimation of a location parameter. Ann. Math. Stat. **35**(1), 73–101 (1964)
15. Ikemoto, J., et al.: Application of deep reinforcement learning to control problems. In: International Symposium on Control, Automation, and Systems (2019). https://doi.org/10.23919/ICCAS.2019.8912116
16. Imdoukh, M., Ahmad, I., Alfailakawi, M.G.: Machine learning-based auto-scaling for containerized applications. Neural Comput. Appl. **32**(13), 9745–9760 (2019). https://doi.org/10.1007/s00521-019-04507-z
17. Kimm, H., Li, Z., Kimm, H.: Scadis: supporting reliable scalability in Redis replication on demand. IEEE (2017). https://doi.org/10.1109/SmartCloud.2017.9
18. Koukis, G., Skaperas, S., Kapetanidou, I.A., Mamatas, L., Tsaoussidis, V.: Performance evaluation of kubernetes networking approaches across constraint edge environments (2024). https://arxiv.org/abs/2401.07674
19. Kubernetes-based Event Driven Autoscaling: KEDA: Kubernetes-based event driven autoscaling. https://keda.sh/ (2024). Accessed 31 Mar 2024

20. Kuchaiev, O., Ginsburg, B.: Factorization tricks for LSTM networks. arXiv preprint arXiv:1703.10722 (2017). https://arxiv.org/abs/1703.10722
21. Lankes, S.: Hermitcore: a unikernel for extreme scale computing. ACM (2016). https://doi.org/10.1145/2931088.2931093
22. Li, P., Luo, B., Zhu, W., Xu, H.: Cluster-based distributed dynamic cuckoo filter system for Redis. Taylor & Francis (2019). https://doi.org/10.1080/17445760.2019.1599889
23. Liu, P., et al.: Scanflow-k8s: Agent-based framework for autonomic management and supervision of ml workflows in Kubernetes clusters. In: International Conference on Machine Learning and Applications (2020). https://doi.org/10.1109/ICMLA.2020.00041
24. Lolos, K., Konstantinou, I., Kantere, V., Koziris, N.: Elastic management of cloud applications using adaptive reinforcement learning. In: 2017 IEEE International Conference on Big Data (Big Data), pp. 203–212 (2017). https://api.semanticscholar.org/CorpusID:19567764
25. Malhotra, M., et al.: Dynamic scaling of web services for xen based virtual cloud environment. Int. J. Cloud Comput. (2020). https://doi.org/10.1504/IJCC.2020.10034234
26. Pan, S.J., Yang, Q.: A survey on transfer learning. IEEE Trans. Knowl. Data Eng. **22**(10), 1345–1359 (2010)
27. Puterman, M.L.: Markov Decision Processes: Discrete Stochastic Dynamic Programming. Wiley, New York, NY, USA (1994)
28. Rao, J., Bu, X., Xu, C., Wang, L., Yin, G.: VCONF: a reinforcement learning approach to virtual machines auto-configuration. In: Proceedings of the 6th International Conference on Autonomic Computing (ICAC 2009), pp. 137–146. ACM (2009)
29. Sanfilippo, S., Stancliff, M.: Redis: A high-performance, in-memory, key-value store. Redis Labs (2013). https://redis.io/
30. Sanka, A.I., Chowdhury, M., Cheung, R.: Efficient high-performance FPGA-REDIS hybrid NoSQL caching system for blockchain scalability. Elsevier (2021). https://doi.org/10.1016/j.comcom.2021.01.017
31. Schulman, J., Wolski, F., Dhariwal, P., Radford, A., Klimov, O.: Proximal policy optimization algorithms. arXiv preprint arXiv:1707.06347 (2017)
32. Thurgood, B., Lennon, R.G.: Cloud computing with Kubernetes cluster elastic scaling. ACM (2019). https://doi.org/10.1145/3341325.3341995
33. Toka, L., Dobreff, G., Fodor, B., Sonkoly, B.: Machine learning-based scaling management for kubernetes edge clusters. IEEE Trans. Netw. Serv. Manage. **18**(1), 958–972 (2021). https://doi.org/10.1109/TNSM.2021.3052837
34. Tsoumakos, D., Konstantinou, I., Boumpouka, C., Sioutas, S., Koziris, N.: Automated, elastic resource provisioning for NoSQL clusters using TIRAMOLA. In: 13th IEEE/ACM International Symposium on Cluster, Cloud and Grid Computing (CCGrid), pp. 34–41. IEEE (2013)
35. Wang, Z., Schaul, T., Hessel, M., van Hasselt, H., Lanctot, M., de Freitas, N.: Dueling network architectures for Deep Reinforcement Learning. In: Proceedings of the 33rd International Conference on Machine Learning (ICML), pp. 1995–2003. JMLR.org (2016)
36. Watkins, C.J.C.H., Dayan, P.: Q-learning. Mach. Learn. **8**(3), 279–292 (1992)
37. Yeom, Y.J., Kim, T., Park, D.H., Kim, S.: Horizontal pod autoscaling in Kubernetes for elastic container orchestration. Sensors **20**(16), 4621 (2020)
38. Yu, K., Liu, Y., Wang, Q.: Review of deep reinforcement learning. IEEE Access (2020). https://doi.org/10.1109/ACCESS.2020.2979650

39. Yu, Y., Si, X., Hu, Z., Zhang, J.: A review of recurrent neural networks: LSTM cells and network architectures. Neural Comput. (2019). https://doi.org/10.1162/neco_a_01199
40. Zhu, H., Bai, Z., Li, J.: Harmonia: Near-linear scalability for replicated storage with in-network conflict detection. ACM (2019). https://doi.org/10.14778/3368289.3368301

Influential Slot and Tag Selection in Billboard Advertisement

Dildar Ali, Suman Banerjee[✉], and Yamuna Prasad

Indian Institute of Technology Jammu, Jammu 181221, Jammu and Kashmir, India
{2021rcs2009,suman.banerjee,yamuna.prasad}@iitjammu.ac.in

Abstract. The selection of influential billboard slots remains an important problem in billboard advertisements. Existing studies on this problem have not considered the case of context-specific influence probability. To bridge this gap, in this paper, we introduce the CONTEXT DEPENDENT INFLUENTIAL BILLBOARD SLOT SELECTION PROBLEM. First, we show that the problem is NP-hard. We also show that the influence function holds the bi-monotonicity, bi-submodularity, and non-negativity properties. We propose an orthant-wise Stochastic Greedy approach to solve this problem. We show that this method leads to a constant-factor approximation guarantee. Subsequently, we propose an orthant-wise Incremental and Lazy Greedy approach. In a generic sense, this is a method for maximizing a bi-submodular function under the cardinality constraint, which may also be of independent interest. We analyze the performance guarantee of this algorithm as well as time and space complexity. The proposed solution approaches have been implemented with real-world billboard and trajectory datasets. We compare the performance of our method with several baseline methods, and the results are reported. Our proposed orthant-wise stochastic greedy approach leads to significant results when the parameters are set properly with reasonable computational overhead.

Keywords: Billboard Advertisement · Billboard Database · Trajectory Database · Influence Probability · Bi-submodularity

1 Introduction

In recent times, *Billboard Advertisement* has emerged as an effective out-of-home advertisement technique due to multiple reasons such as being easy to adopt, ensuring a return on investment[1]. If we have the location information of a group of people over different time stamps and locations of a set of billboards then appropriate advertisement contents could be displayed on the billboards, and it may lead to an influence among the people. In billboard advertisements, the billboards are owned by some billboard owners (e.g., Lamar, Sigtel, etc.), and different commercial houses approach a billboard owner for a number of billboard slots depending on their budget. Given a trajectory database, a billboard

---

[1] https://www.thebusinessresearchcompany.com/report/billboard-and-outdoor-advertising-global-market-report.

database, and a positive integer k, which k billboard slots should be chosen to maximize the influence? This problem has been referred to as the TOP-k INFLUENTIAL BILLBOARD SLOT SELECTION PROBLEM [2], and a few solution methodologies are available. The influence probability between a billboard slot and a trajectory has been considered in all these studies [2,3,14] is the same and does not vary. However, in practice, a low-income person will be more influenced toward a low-cost product rather than a high-cost product. Hence, the influence probability is dependent on context, and this notion is captured as a tag-dependent influence probability. In recent times, tag-based influence maximization has gained significant attention, and most of the studies on this topic are concerned with social networks [5,11]. In both studies, the authors have proposed a bi-set function and an incremental greedy approach that exploits the submodularity property of the influence function. However, such studies have not been done in the context of billboard advertisement, although actual influence is dependent on both slot and tag. Now, the question is that given two positive integers k and ℓ, which k influential slots and ℓ influential tags should be chosen such that the influence is maximized. To the best of our knowledge, such a problem has not been addressed in billboard advertisement settings. However, some studies focus on the influence maximization in the presence of tags in social networks. The first study by Ke et al. [7], where they studied the problem of finding k seed nodes and r influential tags to maximize the influence in the network. Subsequently, other solution methodologies exist in the literature, e.g., the community-based approach [4] that exploits the bi-submodularity of the influence function. This paper bridges this gap by studying the influential billboard slot selection problem in tag-specific influence probability settings. We have posed this problem as a maximization of the bi-submodular set function [10]. In the literature, several practical problems have been modeled as a maximization of a bi-submodular function, such as influence maximization in social networks [11], drug-drug interaction detection [6], and many more. In particular, we make the following contributions in this paper:

- We study this problem in the tag-specific influence probability setting, where the goal is to select influential slots and tags to maximize the influence.
- We establish several important properties of the influence function and exploit them to design efficient algorithms to solve this problem.
- We propose an efficient Orthent-wise Stochastic Greedy maximization algorithm and subsequently introduce Incremental and Lazy Greedy algorithms.
- We analyze the algorithm to understand its time and space complexities, performance guarantee, and conduct experiments with real-world trajectory datasets to exhibit the effectiveness and efficiency of the proposed approach.

Rest of the paper is organized as follows. Section 2 describes the required background and defines our problem formally. The proposed solution approaches have been described in Sect. 3. Section 4 describes the experimental evaluations of the proposed solutions. Finally, Sect. 5 concludes this study and provides future research directions.

2 Background and Problem Definition

2.1 Trajectory and Billboard and Tag Database

A trajectory database contains location information of moving objects over time. In this problem context, the trajectory database $\mathcal{D}$ contains tuples of the form $(\mathcal{U}', \text{loc}, [t_1, t_2])$, signifies the set of people $\mathcal{U}'$ was at the location loc for the duration $[t_1, t_2]$. For any tuple $p \in \mathcal{D}$, let p_u denote the set of people associated with it. Let $\mathcal{U} = \{u_1, u_2, \ldots, u_n\}$ denote the set of people covered by the trajectory database, and hence $\mathcal{U} = \bigcup_{p \in \mathcal{D}} p_u$. This is defined as the people for which there exists at least one tuple that contains the people, i.e., $\mathcal{U} = \{u_i : \exists (\mathcal{U}', \text{loc}, [t_1, t_2]) \in \mathcal{D} \text{ and } u_i \in \mathcal{U}'\}$. Similarly, $\mathcal{L}$ denotes the set of locations that are covered by the trajectory database $\mathcal{D}$, i.e., $\mathcal{L} = \{\text{loc}_i : (\mathcal{U}', \text{loc}_i, [t_j, t_k]) \in \mathcal{D}\}$. Let $[T_1, T_2]$ be the duration for which the trajectory database $\mathcal{D}$ contains the movement data. The billboard database $\mathcal{B}$ stores information about billboards across a city. Each entry is a tuple $(b_{id}, \text{loc}, \text{slot\_duration}, \text{cost})$, where b_{id} is the billboard ID, loc is the location, slot_duration is the slot duration, and cost is the associated cost. Assume all billboards operate over the period $[T_1, T_2]$, with each slot having duration Δ. A billboard slot is represented as a tuple of billboard ID and slot duration. The set of all billboard slots, denoted as $\mathcal{BS}$, is defined as: $\mathcal{BS} = \{(b_i, [t_j, t_j + \Delta]) : i \in [m] \text{ and } t_j \in \{1, \Delta + 1, 2\Delta + 1, \ldots, T_2 - \Delta + 1\}\}$. The tag database contains information about tags (i.e., advertisement content) from the commercial clients. The tag database $\mathcal{T}$ contains a tuple of the form (tag_id, tag_cost), which signifies each tag contains its corresponding unique tag ID and cost. The influence providers allocate slots to commercial clients to maximize product influence. The key question becomes: how can we quantify the influence of a set of billboard slots? This is addressed in Definition 1.

Definition 1 (Influence of Billboard Slots). *Given a trajectory database $\mathcal{D}$, and a subset of billboard slots $\mathcal{S} \subseteq \mathcal{BS}$, the influence of $\mathcal{S}$ can be defined as the expected number of trajectories is influenced can be computed using Eq. 1.*

$$\phi(\mathcal{S}) = \sum_{u_i \in \mathcal{U}} [1 - \prod_{b_j \in \mathcal{S}} (1 - Pr(b_j, u_i))] \quad (1)$$

Here, ϕ is the influence function that maps each subset of the billboard slots to its expected influence, hence $\phi : 2^{\mathcal{BS}} \longrightarrow \mathbb{R}_0^+$ and $\phi(\emptyset) = 0$. The influence model stated in Definition 1 has been widely accepted in the existing studies [2,13] on billboard advertisement. Assuming that a person u_i crosses a billboard slot, bs_i, at a time t_x. Now, assume that the advertisement content of an E-Commerce house is displayed on that billboard in the slot $[t_i, t_j]$, and $t_x \in [t_i, t_j]$. Then u_i is likely to be influenced by the advertisement content with a certain probability. The billboard b_i will influence the user u_i with probability $Pr(bs_i, u_i)$. One of the way to calculate this value as, $Pr(bs_i, u_i) = \frac{Size(bs_i)}{\max_{bs_i \in \mathcal{BS}} Size(bs_i)}$ where $Size(bs_i)$ is the billboard panel size. We adopt this probability setting in our experiments as well. Although it can be calculated in several ways depending on the needs of applications [12–14]. As mentioned in the literature [7], whether a people will be influenced towards a brand or not is always context dependent. In this study we consider that every people $u_i \in \mathcal{U}$, for a billboard slot $b_j \in \mathcal{BS}$ and every relevant tag $c \in \mathcal{H}'$, there exists a non-zero tag specific influence probability and it is denoted by $Pr(u_i, b_j | c)$. This signifies the influence probability of the people u_i when he/she looks at some advertisement content containing the tag c at the billboard slot b_j. For any person, $u \in \mathcal{U}$, for any set of given tags $\mathcal{H}' \subseteq \mathcal{H}$, it is an important question how to calculate the aggregated influence of the tags in $\mathcal{H}'$. Assume that $\mathcal{H}'(u)$ denotes the subset of the tags used in the advertisement content which are visible to u. Now, the aggregated influence will be dependent on how the tags are aggregated. In this study, we use the independent tag aggregation, which has been stated in Definition 2.

Definition 2 (Independent Tag Aggregation). *For a subset of given slots* $\mathcal{S} \subseteq \mathcal{BS}$, *tags* $\mathcal{H}'$, *as per independent tag aggregation the aggregated influence probability of u can be computed using Eq. 2.*

$$Pr(u, \mathcal{S} | \mathcal{H}') = 1 - \prod_{(b,c) \in f} (1 - Pr(u, b | c)) \qquad (2)$$

Here, f denotes the tag assignment function. Now, the aggregated influence for a given subset of billboard slots $\mathcal{S}$, a set of given tags $\mathcal{H}'$ is denoted by $\Phi(\mathcal{S}, \mathcal{H}')$ and stated in Definition 3.

Definition 3 (Aggregated Influence). *The aggregated influence for a given subset of billboard slots* $\mathcal{S}$, *a set of given tags* $\mathcal{H}'$ *is defined as the sum of the influence probabilities of all the persons as stated in Eq. 3.*

$$\Phi(\mathcal{S}, \mathcal{H}') = \sum_{u \in \mathcal{U}} Pr(u | \mathcal{H}') \qquad (3)$$

Here, $\Phi(.,.)$ is a bi-set function which is a mapping from $2^{\mathcal{BS}} \times 2^{\mathcal{H}'}$ to the set of positive real number including 0, i.e., $\Phi : 2^{\mathcal{BS}} \times 2^{\mathcal{H}'} \longrightarrow \mathbb{R}_0^+$. It can be observed that for any subset of slots $\mathcal{S}$, if the tag set is $\emptyset$ then the aggregated influence will be 0. Hence, $\Phi(\mathcal{S}, \emptyset) = 0$ for all $\mathcal{S} \subseteq \mathcal{BS}$.

2.2 Problem Definition

As previously mentioned, selecting both tags and billboard slots is important. However, obtaining a required billboard slot from an influence provider is subject to payment, and the e-commerce house doing this advertisement will have budget constraints. So, the goal here is to select k many billboard slots and ℓ many tags to maximize the influence. We call this problem the CONTEXT DEPENDENT INFLUENTIAL BILLBOARD SLOT SELECTION PROBLEM, which asks for given k and ℓ, in which k influential slots and tags should be chosen respectively to maximize the influence. We state the problem in Definition 4.

Definition 4 (Context Dependent Influential Billboard Slot Selection Problem). *Given a trajectory database $\mathcal{D}$, a billboard database $\mathcal{B}$, and two positive integers k and ℓ, this problem asks to choose k influential billboard slots and ℓ influential tags such that the influence is maximized. Mathematically, this problem can be expressed as follows:*

$$(\mathcal{S}^{OPT}, \mathcal{H}^{OPT}) \longleftarrow \underset{\mathcal{S} \subseteq \mathcal{BS} \wedge \mathcal{H}' \subseteq \mathcal{H}}{argmax} \phi(\mathcal{S}|\mathcal{H}') \tag{4}$$

It is reasonable to consider that even if there is no tag, some default tag h' still exists, which can be used even if no tag is selected. We want to select ℓ many tags on top of the default tag. In Equation No. 4, $\mathcal{S}^{OPT}$ and $\mathcal{H}^{OPT}$ denote the optimal slot subset of k and the optimal tag subset of ℓ. It can be easily observed that the problem introduced in Definition 4 is the generalization of the INFLUENTIAL BILLBOARD SLOT SELECTION PROBLEM [1,2] where the context-dependent influence probability is not considered. Hence, Theorem 1 holds.

Theorem 1. *For a given k and ℓ, finding the optimal slot and tag set for the Context-Dependent Influential Billboard Slot Selection Problem is **NP-hard**.*

3 Proposed Solution Approach

Exhaustive Search Approach. In this approach, we enumerate all k-sized subsets of the set of billboard slots and ℓ-sized subsets of tags. Considering all the billboards are running for the duration $[T_1, T_2]$ and the slot duration of Δ time units, hence the number of billboard slots will be $\frac{T_2 - T_1 + 1}{\Delta} \cdot m$. So, the number k-sized subsets will be $\binom{\frac{T_2 - T_1 + 1}{\Delta} \cdot m}{k}$ and the number of ℓ sized subsets of $\mathcal{H}$ will be $\binom{|\mathcal{H}|}{\ell}$. Subsequently, we create all possible k-sized slot subsets and ℓ-sized tag subsets pairs, and for every possible slot-tag pair, we compute the influence and choose the one that gives the maximum influence and return it.

Orthant-Wise Greedy Maximization Algorithm. In this approach, we start with a default slot s' and default tag h', and our approach is as follows. First, we fix the tag set to $\{h'\}$ and apply an incremental greedy algorithm [8,9] that works based on marginal gain computation to obtain the k size slot set $\mathcal{S}'$. Now, fixing the slot set to $\mathcal{S}' \cup \{s'\}$, we apply incremental greedy algorithm to obtain the ℓ size tag set $\mathcal{H}'$. Next, we do the same thing; however first fix the slot set to the default slot and apply the incremental greedy algorithm to choose an ℓ size tag set $\mathcal{H}''$, and then we fix the tag set to $\mathcal{H}'' \cup \{h'\}$ and apply the incremental greedy algorithm to obtain the k size slot set $\mathcal{S}''$. So we have two slot-tag pair $(\mathcal{S}', \mathcal{H}')$ and $(\mathcal{S}'', \mathcal{H}'')$. We return one that leads to the maximum influence. This method consists of the following four optimization problems.

$$\mathcal{S}' \longleftarrow \underset{\mathcal{S} \subseteq \mathcal{BS} \wedge |\mathcal{S}| = k}{argmax} \phi(\mathcal{S} \cup \{s'\}, \{h'\}) \tag{5}$$

$$\mathcal{H}' \longleftarrow \underset{H \subseteq \mathcal{H} \wedge |H|=\ell}{argmax} \phi(\mathcal{S}', H \cup \{h'\}) \tag{6}$$

$$\mathcal{H}'' \longleftarrow \underset{H \subseteq \mathcal{H} \wedge |H|=\ell}{argmax} \phi(\{s'\}, H \cup \{h'\}) \tag{7}$$

$$\mathcal{S}'' \longleftarrow \underset{\mathcal{S} \subseteq \mathcal{BS} \wedge |\mathcal{S}|=k}{argmax} \phi(\mathcal{S} \cup \{s'\}, \mathcal{H}'') \tag{8}$$

Lazy Greedy Algorithm. This approach involves excessive influence function evaluations, leading to high execution time. However, it can be implemented efficiently, with fewer evaluations in most practical cases, though the worst-case scenario matches the incremental greedy algorithm. The key idea is to consider the first for loop and its first iteration. We compute the marginal gain for all slots with respect to the empty tag set, which is equivalent to computing their influence value. Subsequently, we sort the slots based on this value in descending order, and the first slot is chosen. Now, in the second iteration, we compute the marginal gain of the slots in sorted order and consider the following situation. Suppose the marginal gain of the i-th slot is less than that of the $(i+1)$-th slot. Now, applying the submodularity property, it can be ensured that even if we compute the marginal gain of the slots, it can not be more than the marginal gain of the i-th slot. Hence, from the $(i+1)$-th slot onward, there is no need to compute their marginal gains, and they can be skipped safely. This improves the execution time, though the worst-case time complexity will remain the same.

Stochastic Greedy Algorithm. In this approach [9] in each iteration instead of computing the marginal gains of all the remaining elements, we sample $\frac{n}{k} \log \frac{1}{\epsilon}$ many elements from the ground set for slot selection and $\frac{n}{\ell} \log \frac{1}{\epsilon}$ many elements for tag selection. The marginal gain is computed only for the sampled elements. Here, we mention that ϵ is a control parameter that controls the trade-off between the quality of the solution and the execution time. Algorithm 1 describes this process as pseudo-code.

Complexity Analysis. Now, we analyze the time and space requirements for Algorithm 1. Initialization at Line No. 1 and 2 will take $\mathcal{O}(1)$ time. To sample out $\frac{a}{k} \log \frac{1}{\epsilon}$ many element it will take $\mathcal{O}(a.\log \frac{1}{\epsilon})$ time. Now, for any billboard slot $s \in \mathcal{BS}$ and $l \in \mathcal{H}$, calculating influence using Eq. 1 will take $\mathcal{O}(t)$ time, in which t is the number of tuple in the trajectory database. In Line No. 5 computing marginal gain will take $\mathcal{O}(2.a.\log \frac{1}{\epsilon}.t)$ time and Line No. 6 will execute for $\mathcal{O}(k)$ time. So, Line No. 3 to 6 will take $\mathcal{O}(a.\log \frac{1}{\epsilon} + 2.a.\log \frac{1}{\epsilon}.t + k)$ time. In Line No. 7 to 10 will take $\mathcal{O}(b.\log \frac{1}{\epsilon} + 2.b.\log \frac{1}{\epsilon}.k.t + \ell)$ time and Line No. 11 to 14 will take $\mathcal{O}(b.\log \frac{1}{\epsilon} + 2.b.\log \frac{1}{\epsilon}.\ell.t + \ell)$. In the fourth greedy time taken by Line No. 15 to 18 is of $\mathcal{O}(a.\log \frac{1}{\epsilon} + 2.a.\log \frac{1}{\epsilon}.\ell.t + k)$. Finally, Line No. 19 to 22 will take $\mathcal{O}(2.k.\ell.t)$ time for final comparison. Hence, total time requirement of Algorithm 1 will be $\mathcal{O}(a.\log \frac{1}{\epsilon}.\ell.t + b.\log \frac{1}{\epsilon}.k.t + k.\ell.t)$. Now, the additional space requirement to store the lists $\mathcal{S}', \mathcal{S}'', \mathcal{H}', \mathcal{H}''$ and $\mathcal{R}$ will be $\mathcal{O}(k), \mathcal{O}(k), \mathcal{O}(\ell)$, $\mathcal{O}(\ell)$ and $\mathcal{O}(max(a.\log \frac{1}{\epsilon}, b.\log \frac{1}{\epsilon}))$ respectively. Hence, total space requirement for Algorithm 1 will be of $\mathcal{O}(max(a.\log \frac{1}{\epsilon}, b.\log \frac{1}{\epsilon}) + 2k + 2\ell)$.

Now, we analyze this methodology and prove some theoretical results.

Algorithm 1: Stochastic Greedy Algorithm for the Influential Slots and Tags Selection Problem

Data: The Trajectory Database $\mathcal{D}$, The Billboard Database $\mathcal{B}$, Context Specific Influence Probabilities, Two Positive Integers k and ℓ.
Result: $\mathcal{S} \subseteq V(G)$ with $|\mathcal{S}| = k$ and $\mathcal{H}' \subseteq \mathcal{H}$ with $|\mathcal{H}'| = \ell$ such that $\phi(\mathcal{S}, \mathcal{H}')$ is maximized.

1   $\mathcal{S}' \longleftarrow \{s'\}, \mathcal{S}'' \longleftarrow \{s'\}, \mathcal{H}' \longleftarrow \{h'\}, \mathcal{H}'' \longleftarrow \{h'\}$;
2   $\mathcal{R} \longleftarrow \emptyset, \epsilon \longleftarrow 0.01$;
3   **for** $i = 1$ *to* k **do**
4      $\mathcal{R} \longleftarrow$ Sample $\frac{a}{k} \log \frac{1}{\epsilon}$ many elements from $\mathcal{BS} \setminus \mathcal{S}'$;
5      $s^* \longleftarrow \underset{s \in \mathcal{R}}{argmax} \; \phi(\mathcal{S}' \cup \{s\}, \{h'\}) - \phi(\mathcal{S}', \mathcal{H}')$;
6      $\mathcal{S}' \longleftarrow \mathcal{S}' \cup \{s^*\}$;
7   **for** $i = 1$ *to* ℓ **do**
8      $\mathcal{R} \longleftarrow$ Sample $\frac{b}{\ell} \log \frac{1}{\epsilon}$ many elements from $\mathcal{H} \setminus \mathcal{H}'$;
9      $h^* \longleftarrow \underset{h \in \mathcal{R}}{argmax} \; \phi(\mathcal{S}', \mathcal{H}' \cup \{h\}) - \phi(\mathcal{S}', \mathcal{H}')$;
10     $\mathcal{H}'' \longleftarrow \mathcal{H}'' \cup \{h^*\}$;
11   **for** $i = 1$ *to* ℓ **do**
12     $\mathcal{R} \longleftarrow$ Sample $\frac{b}{\ell} \log \frac{1}{\epsilon}$ many elements from $\mathcal{H} \setminus \mathcal{H}''$;
13     $h^* \longleftarrow \underset{h \in \mathcal{R}}{argmax} \; \phi(\mathcal{S}'', \mathcal{H}'' \cup \{h\}) - \phi(\mathcal{S}'', \mathcal{H}'')$;
14     $\mathcal{H}'' \longleftarrow \mathcal{H}'' \cup \{h^*\}$;
15   **for** $i = 1$ *to* k **do**
16     $\mathcal{R} \longleftarrow$ Sample $\frac{a}{k} \log \frac{1}{\epsilon}$ many elements from $\mathcal{BS} \setminus \mathcal{S}''$;
17     $s^* \longleftarrow \underset{s \in \mathcal{R}}{argmax} \; \phi(\mathcal{S}'' \cup \{s\}, \mathcal{H}'') - \phi(\mathcal{S}'', \mathcal{H}'')$;
18     $\mathcal{S}'' \longleftarrow \mathcal{S}'' \cup \{s^*\}$;
19   **if** $\phi(\mathcal{S}', \mathcal{H}') > \phi(\mathcal{S}'', \mathcal{H}'')$ **then**
20   return $\mathcal{S}'$ and $\mathcal{H}'$;
21   **else**
22   return $\mathcal{S}''$ and $\mathcal{H}''$

Lemma 1. *The number of influence function evaluations by Algorithm 1 will be equal to* $4(a+b) \log \frac{1}{\epsilon}$, *i.e.,* $\mathcal{O}((a+b) \log \frac{1}{\epsilon})$.

Lemma 2. *Consider the first* for loop *of Algorithm 1 and assume that after the execution of its i-th iteration, the solution set is $\mathcal{S}'_i$. The expected influence gain of Algorithm 1 in the $(i+1)$-th will be at least* $\frac{1-\epsilon}{k} \sum_{s^* \in \mathcal{S}^{OPT} \setminus \mathcal{S}'} \phi(\mathcal{S}' \cup \{s^*\}, \{\mathcal{H}'\}) - \phi(\mathcal{S}', \mathcal{H}')$.

Theorem 2. *Let $\mathcal{S}^{OPT}$ and $\mathcal{H}^{OPT}$ be an optimal k-sized and an ℓ-sized slot and tag set, respectively. Also assume $\mathcal{S}^A$ and $\mathcal{H}^A$ are the k-sized and an ℓ-sized slot and tag set returned by Algorithm 1. Then* $\phi(\mathcal{S}^A, \mathcal{H}^A) \geq (1 - \frac{1}{e} - \epsilon)^2 \cdot \phi(\mathcal{S}^{OPT}, \mathcal{H}^{OPT})$. *In other words, Algorithm 1 gives* $(1 - \frac{1}{e} - \epsilon)^2$ *factor approximation guarantee.*

Proof. It can be observed that any one of the following two cases may happen.
Case I: $\mathcal{S}^A = \mathcal{S}', \mathcal{H}^A = \mathcal{H}'$ and **Case II:** $\mathcal{S}^A = \mathcal{S}'', \mathcal{H}^A = \mathcal{H}''$

Let, $\mathcal{S}'_i = \{s_1, s_2, s_3, \ldots s_i\}$ defines the solutions at each step returns by first `For Loop` in Algorithm 1 after i^{th} iteration. Now, from lemma 2 we can write,

$$E[\Delta(s_{i+1}|\mathcal{S}', \mathcal{H}')|\mathcal{S}', \mathcal{H}'] \geq \frac{1-\epsilon}{k} \sum_{s^* \in \mathcal{S}^{OPT} \setminus \mathcal{S}'} \Delta(s^*|\mathcal{S}', \mathcal{H}') \qquad (9)$$

Using the submodularity property, we can obtain,

$$\sum_{s^* \in \mathcal{S}^{OPT} \setminus \mathcal{S}'} \Delta(s^*|\mathcal{S}', \mathcal{H}') \geq \Delta(\mathcal{S}^{OPT}|\mathcal{S}'_i, \mathcal{H}')$$

$$\geq \phi(\mathcal{S}^{OPT}, \mathcal{H}') - \phi(\mathcal{S}'_i, \mathcal{H}')$$

Now, if we put these results in Eq. (14), we get,

$$E[\phi(\mathcal{S}'_{i+1}, \mathcal{H}') - \phi(\mathcal{S}'_i, \mathcal{H}')|\mathcal{S}'_i] \geq \frac{1-\epsilon}{k} \phi(\mathcal{S}^{OPT}, \mathcal{H}') - \phi(\mathcal{S}'_i, \mathcal{H}')$$

Now, if we take expectation over $\mathcal{S}'_i$, we can obtain,

$$E[\phi(\mathcal{S}'_{i+1}, \mathcal{H}') - \phi(\mathcal{S}'_i, \mathcal{H}')] = \frac{1-\epsilon}{k} \phi(\mathcal{S}^{OPT}, \mathcal{H}') - \phi(\mathcal{S}'_i, \mathcal{H}')]$$

If we apply induction to it,

$$E[\phi(\mathcal{S}'_k, \mathcal{H}')] \geq (1 - (1 - \frac{1-\epsilon}{k})^k) \cdot \phi(\mathcal{S}^{OPT}, \mathcal{H}^\mathcal{A})$$

$$\geq (1 - \frac{1}{e} - \epsilon) \cdot \phi(\mathcal{S}^{OPT}, \mathcal{H}^\mathcal{A}) \qquad (10)$$

Now, in a similar way, for the second, third, and fourth `For Loop`, we can write:

$$E[\phi(\mathcal{S}'_k, \mathcal{H}'_\ell)] \geq (1 - \frac{1}{e} - \epsilon)^2 \cdot \phi(\mathcal{S}^{OPT}, \mathcal{H}^{OPT}) \qquad (11)$$

$$E[\phi(\mathcal{S}'', \mathcal{H}''_\ell)] \geq (1 - \frac{1}{e} - \epsilon) \cdot \phi(\mathcal{S}^\mathcal{A}, \mathcal{H}^{OPT}) \qquad (12)$$

$$E[\phi(\mathcal{S}''_k, \mathcal{H}''_\ell)] \geq (1 - \frac{1}{e} - \epsilon)^2 \cdot \phi(\mathcal{S}^{OPT}, \mathcal{H}^{OPT}) \qquad (13)$$

4 Experimental Evaluations

This section describes the experimental evaluations of the proposed solution approaches. Initially, we start by describing the datasets used in our experiments.

Dataset Description. We use two widely studied datasets for our experiments [2,13]. The first dataset includes 227,428 check-in records from New York City[2], collected over ten months (April 12, 2012–February 16, 2013), with details like timestamps, GPS coordinates, and user IDs. The second dataset, VehDS-LA[3], contains 74,170 vehicle records from 15 streets in Los Angeles, featuring street names, GPS coordinates, and timestamps. Additionally, billboard data from LAMAR[4] includes billboard ID, venue ID, GPS coordinates, timestamps, and panel size. The New York City dataset has 716 billboards (1,031,040 slots), and Los Angeles has 1,483 billboards (2,135,520 slots).

Key Parameters. All the parameters are summarized in Table 1, including the number of billboard slots k and tags ℓ to be picked. The user-defined parameter ϵ defines the size of random subsets. The distance threshold, λ, determines the maximum distance a billboard can influence the trajectories. In each experiment, we fixed one parameter value and varied the other parameter values. All codes are executed in Python using Jupyter Notebook in an HP Z4 workstation with 64 GB of memory and an Xeon(R) 3.50 GHz processor.

Table 1. Parameter Settings

Parameter	Values
k	25, 50, 100, 150, 200
ℓ	10, 20, 30, 40, 50
ϵ	0.01, 0.05, 0.1, 0.15, 0.2
λ	25 m, 50 m, 75 m, 100 m, 125 m

Baseline Methodologies. We compared our proposed solutions with the following baseline methods:

Random Slot and Random Tag (RSRT): In this method, k many random slots and ℓ many random tags are chosen and returned as solution.

Random Slot and High-Frequency Tag (RSHFT): Tag frequency is defined by the number of associated people. We count and sort tags by frequency, then return ℓ tags and k random slots from the sorted list.

Maximum Coverage Slot and Random Tag (MAXSRT): The coverage of a billboard slot is the number of people passing by it. We compute and sort the coverage for all slots, then return k slots from the sorted list and select ℓ tags uniformly at random.

Top-k Slot and Top-ℓ Tag (TSTT): This method calculates the individual influence of each billboard slot and tag, then sorts both in descending order. From the sorted lists, we select the Top-k billboard slots and Top-ℓ tags.

Top-k Slot and Random ℓ Tag (TSRT): In this method, the influence of each billboard slot is calculated, and the slots are sorted in descending order, with the Top-k selected. For tags, ℓ is randomly chosen from the unsorted list.

---

[2] https://www.nyc.gov/site/tlc/about/tlc-trip-record-data.page.
[3] https://github.com/Ibtihal-Alablani.
[4] http://www.lamar.com/InventoryBrowser.

Random k Slot and Top-ℓ Tag (RSTT): This method is the reverse of the TSRT approach. First, the influence of each billboard slot and tag is calculated. Tags are then sorted in descending order by influence, and the Top-ℓ tags are selected. From the billboard slots, k are randomly chosen.

Goals of our Experiments. In this study, we address the following Research Questions (RQ).

- **RQ1**: How does the influence value increase if we increase the number of slots and tags to be selected?
- **RQ2**: If we increase the number of slots and tags, how do the computational time requirements of the proposed and the baseline methods change?
- **RQ3**: If we increase the size of the trajectory, how do the proposed method's influence value and computational time requirement change?
- **RQ4**: For the stochastic greedy algorithm, if we change the value of ϵ, how do the computational time and the quality of the solution change?

4.1 Experimental Results with Discussions

In this section, we describe the experimental results and answer each research question posed in this work.

Budget (k, ℓ) Vs. Influence. Budget and influence are critical factors in billboard advertisement decisions. In our experiment, we analyzed the influence of varying billboard slots $(25, 50, 100, 150,$ and $200)$ for different tag values ℓ, shown in Fig. 1. The influence probability of 'tags' in the NYC dataset is unevenly distributed, with a few tags being highly influential while most are not. This distribution favors algorithms like 'Lazy Greedy', 'Stochastic Greedy', and baseline methods such as 'TSTT', 'RSTT', and 'RSHFT'. In contrast, 'MAXSRT' and 'TSRT' underperform due to random tag selection. Conversely, the LA dataset exhibits a more balanced influence distribution, leading to better performance for 'TSRT' and 'RSTT'. In the LA dataset among the baseline methods, 'TSTT' has almost equal influence to 'Stochastic Greedy'. On the other hand, in the NYC dataset, the influence probability of billboard slots is well distributed, and the influence difference between 'Stochastic Greedy' and 'TSTT' is differentiable, as shown in Fig. 1 (a, b, c, d, e). Now, when we increase the number of billboard slot from 25 to 200 with a fixed value of $\ell = 10$, $\epsilon = 0.01$, the influence value of 'Lazy Greedy', 'Stochastic Greedy', 'TSTT', 'RSTT', 'RSHFT', 'MAXSRT', and 'TSRT' are increases from 353.74, 353.36, 352.27, 339.50, 265.43, 50.93, 19.71 to 437.55, 434.21, 414.16, 376.49, 297.14, 84.25, 56.19 respectively. Similarly, if we fixed the number of billboard slot, $k = 200$ and vary ℓ from 10 to 50 then the influence value of 'Lazy Greedy', 'Stochastic Greedy', 'TSTT', 'RSTT', 'RSHFT', 'MAXSRT', and 'TSRT' increase from 437.55, 434.21, 414.16, 376.49, 297.14, 84.25, 56.19 to 651.45, 641.69, 617.26, 583.42, 577.81, 338.21, 294.46 respectively. Similar types of observations were also observed in the LA dataset. Therefore, among the proposed two methods, 'Lazy Greedy' gives more influence compared to 'Stochastic Greedy' because of the randomized element selection behavior of 'Stochastic Greedy', and 'TSTT' gives maximum influence among other baseline methods for both LA and NYC datasets as reported in Fig. 1.

Budget (k, ℓ) Vs. Time. To understand the time requirement for proposed and baseline methods, we vary different k, and ℓ values with respect to time. From

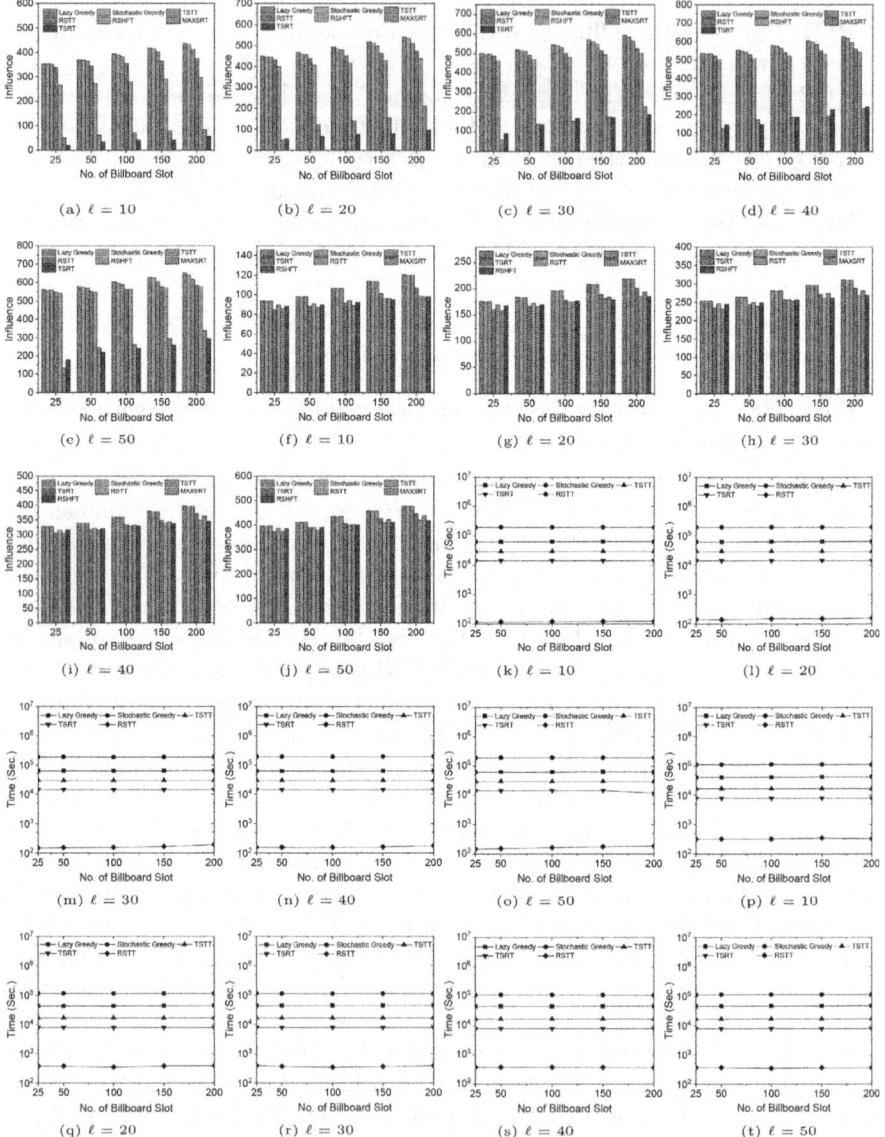

Fig. 1. (1) Influence varying ℓ, when $k = 25$ to 200, $\epsilon = 0.01$, (a, b, c, d, e) for NYC, and (f, g, h, i, j) for LA Dataset. (2) Time varying ℓ, when $k = 25$ to 200, $\epsilon = 0.01$, (k, ℓ, m, n, o) for NYC, and (p, q, r, s, t) for LA Dataset.

Fig. 1, it is observed that with a fixed value of ℓ, when k increases, the time requirement also increases. For example, in the LA dataset, when we fixed the value of $\ell = 10, \epsilon = 0.01$, and varied k value from 25 to 200, the time requirement in seconds for 'Lazy Greedy', 'Stochastic Greedy', 'TSTT', 'TSRT', and 'RSTT' increases from 40571, 112259, 16667, 7929, 323 to 43193, 114119, 16885, 7984, 332 respectively. Here, we observed that small changes in time between $k = 25$ and $k = 200$ happen, and this occurs due to marginal gain computation for each proposed method as well as the baseline method. Similarly, when we set $\ell = 50, \epsilon = 0.01$, and vary $k = 25$ to $k = 200$, the time requirement for 'Lazy Greedy', 'Stochastic Greedy', 'TSTT', 'TSRT', and 'RSTT' also increases from 46002, 113872, 16855, 7945, 360 to 47652, 115215, 16907, 7993, 365 respectively. One point needs to be noted that the experimental results of 'Lazy Greedy' are reported in Fig. 1, which is the best case time requirements, and in the worst case, it will take the same run time as 'Incremental Greedy'. However, when the dataset is large, 'Lazy Greedy' may not be the right choice. Now, in the case of 'Stochastic Greedy', its computational time is always far better than the 'Incremental Greedy' method as it is independent of the size of k, and ℓ as discussed in Lemma 1. In the case of the NYC dataset, a similar behavior is observed as of the LA dataset for the proposed and baseline methods. We have not reported the time requirements for the 'RSHFT', 'MAXSRT', and 'RSRT' methods as these methods take less than 10 s of computational time.

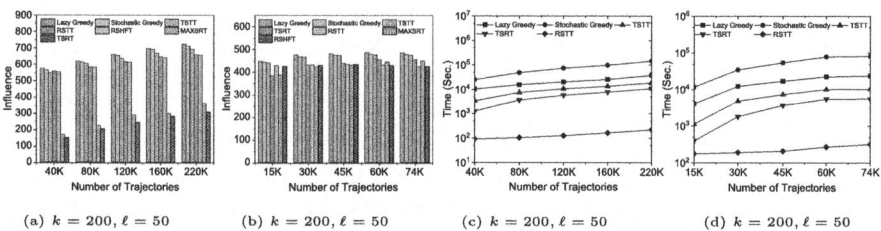

Fig. 2. (1) Influence varying trajectory size, when $k = 200$, $\ell = 50$, and $\epsilon = 0.01$ (a) for NYC, and (b) for LA Dataset. (2) Time varying trajectory size, when $k = 200$, $\ell = 50$, and $\epsilon = 0.01$ (c) for NYC, and (d) for LA Dataset.

Trajectory Size Vs. Influence, Time. Figure 2 shows the impact of varying trajectory size on influence and run time. We observe: (1) the influence of all proposed and baseline methods increases with the increment of trajectory size because more users can be influenced. (2) In the NYC and LA datasets, the influence of 'Lazy Greedy', and 'Stochastic Greedy' is consistently better than the baseline methods. We take $k = 200$, and $\ell = 50$, and vary trajectory size $40k$ to $200k$ for the NYC, and $15k$ to $74k$ for the LA dataset as shown in Fig. 2(a), 2(b). (3) In the NYC dataset, when trajectory sizes are $40k, 80k, 120k, 160k, 200k$, and their corresponding unique users encountered are 924, 969, 1017, 1064, 1083, respectively. In the LA dataset, when the trajectory varies between $15k$

to 74k, the number of unique users encountered is 2000. (4) Figs. 2(c), and 2(d) shows computational time for the NYC and LA dataset. We observe that 'Lazy Greedy', and 'Stochastic Greedy' scale linearly w.r.t. trajectory size, consistent in our analysis in both the NYC and LA datasets. Although the growth in time requirement in 'Stochastic Greedy' is faster than 'Lazy Greedy', e.g., when trajectory size varies from 40k to 220k, and 15k to 74k, the time requirement increases almost 6× and 6.5× times for NYC and LA datasets, respectively. However, in the 'Lazy Greedy', run time rises linearly in the best case as only one time marginal gain needs to be computed for all the elements, and from onwards only comparison operation needs to be executed. (5) Among the baseline methods, 'TSTT' takes the maximum time, and with the increase of trajectory size, the run time of all baseline methods increases linearly.

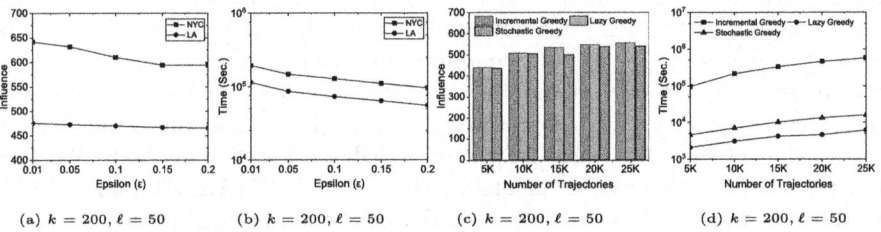

Fig. 3. (1) Influence varying ϵ (a), and Time Varying ϵ (b) when $k = 200$, $\ell = 50$ for NYC, LA Dataset. (2) Influence (c), Time (d), varying trajectory size, when $k = 200$, $\ell = 50$, and $\epsilon = 0.01$ for different Algorithms on NYC Dataset.

Epsilon (ϵ) Vs. Influence, Time. Figure 3(a), 3(b) shows the impact of varying ϵ values on 'Stochastic Greedy' w.r.t. influence, and time. We find: (1) when the ϵ value increases, the influence value decreases. The influence is decreasing more in the NYC dataset than in the LA dataset. (2) When the ϵ value varies from 0.01 to 0.2, the run time on both the NYC and LA datasets decreases linearly. In the 'Stochastic Greedy', we randomly pick a subset of elements, and the cardinality of the subset depends on the ϵ value. If the ϵ value decreases, then the subset size increases, and there is a minimal loss in influence compared to 'Incremental Greedy', however run time increases. For example, in the NYC dataset, when $k = 200, \ell = 50$ and vary ϵ for the value of 0.01 to 0.2, the influence values are 641.69, 631.60, 610.11, 594.27, 593.12 and the run-times are 193850, 147011, 127458, 110355, 95880 in seconds, respectively. A similar type of result was observed on the LA dataset as shown in Fig. 3(a), 3(b). So, the parameter ϵ, gives us the freedom to compromise either in influence or run time.

Additional Discussions. To find out the efficiency of 'Stochastic Greedy', we compare its performance with 'Incremental Greedy' and 'Lazy Greedy'. In our experiment, we fixed k, ℓ, ϵ value, which varies over different trajectory sizes. Our experiment shows that 'Incremental Greedy' and 'Lazy Greedy' achieve the same amount of influence; however, there is a huge difference when talking about

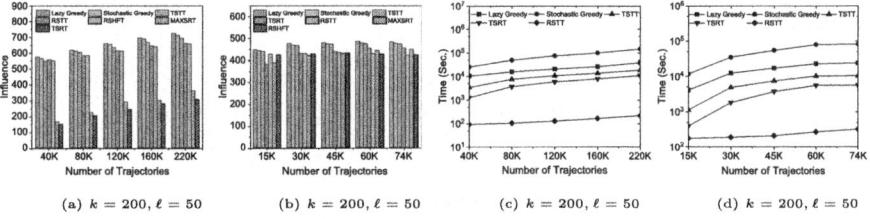

Fig. 4. (1) Influence varying Distance (λ) when $k = 200$, $\ell = 50$, and $\epsilon = 0.01$: (a) NYC Dataset, (b) LA Dataset. (2) Time varying Distance (λ) when $k = 200$, $\ell = 50$, and $\epsilon = 0.01$: (c) NYC Dataset, (d) LA Dataset.

run time. The 'Stochastic Greedy' achieves less influence than both 'Incremental Greedy' and 'Lazy Greedy' however it takes much less run time compared to 'Incremental Greedy'. As we previously discussed, in the worst case, 'Lazy Greedy' will take the same amount of time as 'Incremental Greedy' takes, and in our experiment, we, fortunately, got the best case results of 'Lazy Greedy' due to the nature of the datasets, as reported in Fig. 3(c), 3(d). When trajectory size increases from $5k$ to $25k$, the run time of 'Incremental Greedy', 'Lazy Greedy', and 'Stochastic Greedy' also increases from 96044, 2090, 4546 to 572512, 6144, 15905 s, i.e., 6x, 3x, 3.5x respectively. So, for trajectory size $25k$, 'Incremental Greedy' will take almost 36x more time than 'Stochastic Greedy', and we observe that for larger trajectory size, i.e., $200k$, the 'Incremental Greedy will not complete its execution with a reasonable computational time. We take $\lambda = 100$ m, and assume within the range of 100 m, a billboard slot can influence all trajectories with a certain probability as shown in Fig. 4. We have also experimented with varying λ values from 25 m to 125 m and observed that with the increment of λ value, the influence as well as run time increases because one billboard slot can influence more number of trajectories.

5 Conclusion

This paper has studied the problem of jointly selecting influential billboard slots and influential tags. First, we show that the influence function is non-negative, monotone, and bi-submodular. We show that the problem is **NP-hard** and propose an orthant-wise incremental greedy algorithm that gives a constant factor approximation algorithm. Though this method is simple to understand, it does not scale well when the trajectory dataset is large due to excessive marginal gain computations. To address this, we propose the orthant-wise Lazy and Stochastic Greedy approach, which executes fast while leading to more or less similar influence. Still, the problem is not solved on the ground because we must also report which tag will be displayed in which slot to maximize the influence. Developing more efficient techniques to address slot selection and allocation problems will remain an active area of research in the near future.

References

1. Ali, D., Banerjee, S., Prasad, Y.: Influential billboard slot selection using pruned submodularity graph. In: Chen, W., Yao, L., Cai, T., Pan, S., Shen, T., Li, X. (eds.) ADMA 2022. LNCS, vol. 13725, pp. 216–230. Springer, Cham (2022). https://doi.org/10.1007/978-3-031-22064-7_17
2. Ali, D., Banerjee, S., Prasad, Y.: Influential billboard slot selection using spatial clustering and pruned submodularity graph (2023)
3. Ali, D., Banerjee, S., Prasad, Y.: Multi-slot tag assignment problem in billboard advertisement. In: Chen, T., Cao, Y., Nguyen, Q.V.H., Nguyen, T.T. (eds.) ADC 2024. LNCS, vol. 15449, pp. 158–170. Springer, Singapore (2025). https://doi.org/10.1007/978-981-96-1242-0_12
4. Banerjee, S., Pal, B.: Budgeted influence and earned benefit maximization with tags in social networks. Soc. Netw. Anal. Min. **12**(1), 21 (2022)
5. Banerjee, S., Pal, B., Jenamani, M.: Budgeted influence maximization with tags in social networks. In: Huang, Z., Beek, W., Wang, H., Zhou, R., Zhang, Y. (eds.) WISE 2020. LNCS, vol. 12342, pp. 141–152. Springer, Cham (2020). https://doi.org/10.1007/978-3-030-62005-9_11
6. Hu, Y., Wang, R., Chen, F.: Drug-drug interactions (DDIs) detection from on-line health forums: bi-submodular optimization (BSMO). In: 2017 IEEE International Conference on Healthcare Informatics (ICHI), pp. 163–170. IEEE (2017)
7. Ke, X., Khan, A., Cong, G.: Finding seeds and relevant tags jointly: for targeted influence maximization in social networks. In: Das, G., Jermaine, C.M., Bernstein, P.A. (eds.) Proceedings of the 2018 International Conference on Management of Data, SIGMOD Conference 2018, Houston, TX, USA, 10–15 June 2018, pp. 1097–1111. ACM (2018)
8. Minoux, M.: Accelerated greedy algorithms for maximizing submodular set functions. In: Optimization Techniques: Proceedings of the 8th IFIP Conference on Optimization Techniques Würzburg, 5–9 September 1977, pp. 234–243. Springer, Cham (2005)
9. Mirzasoleiman, B., Badanidiyuru, A., Karbasi, A., Vondrák, J., Krause, A.: Lazier than lazy greedy. In: Proceedings of the AAAI Conference on Artificial Intelligence, vol. 29 (2015)
10. Schoot Uiterkamp, M.H.: A characterization of simultaneous optimization, majorization, and (bi-)submodular polyhedra. Math. Oper. Res. (2024)
11. Tekawade, A., Banerjee, S.: Influence maximization with tag revisited: exploiting the bi-submodularity of the tag-based influence function. In: Yang, X., et al. (eds.) ADMA 2023. LNCS, vol. 14176, pp. 772–786. Springer, Cham (2023). https://doi.org/10.1007/978-3-031-46661-8_51
12. Wang, L., Yu, Z., Yang, D., Ma, H., Sheng, H.: Efficiently targeted billboard advertising using crowdsensing vehicle trajectory data. IEEE Trans. Industr. Inf. **16**(2), 1058–1066 (2020). https://doi.org/10.1109/TII.2019.2891258
13. Zhang, P., Bao, Z., Li, Y., Li, G., Zhang, Y., Peng, Z.: Towards an optimal outdoor advertising placement: when a budget constraint meets moving trajectories. ACM Trans. Knowl. Discovery Data (TKDD) **14**(5), 1–32 (2020)
14. Zhang, Y., Li, Y., Bao, Z., Mo, S., Zhang, P.: Optimizing impression counts for outdoor advertising. In: Proceedings of the 25th ACM SIGKDD International Conference on Knowledge Discovery & Data Mining, KDD 2019, pp. 1205–1215. Association for Computing Machinery, New York, NY, USA (2019)

Speech-Scenario Generation Based on the Philosophy of a Prominent Leader Within a Small Community

Tetsuya Kitahata[✉], Kazuhiro Seki, and Akiyo Nadamoto

Konan University, 8-9-1 Okamoto, Higashinada-ku, Kobe, Hyogo, Japan
m2524012@s.konan-u.ac.jp, {seki,nadamoto}@konan-u.ac.jp

Abstract. Research about long text generation has been actively conducted with the advancement of large language models. However, generating long text that considers the unique philosophy within a small community, such as speeches used in graduation ceremonies or company inductions, remains challenging. The reason is that information and literature about prominent leaders in small communities are generally extremely limited compared to those about well-known prominent leaders. This study focuses on prominent leaders within small communities, such as university or company founders. It aims to generate speech scenarios that automatically share small communities' unique philosophies. In this paper, we target Hachisaburo Hirao, the founder of our university, and extract sentences containing his philosophies from his diaries, lectures, and autobiographies to create a quotations database. We then propose a method to generate speech scenarios based on the user's input theme using Retrieval-Augmented Generation (RAG) with the quotations database.

Keywords: Speech-scenario Generation · Philosophy · LLMs · RAG

1 Introduction

The philosophies exist within communities, such as schools and companies, that reflect the values and direction of the community. For example, Steve Jobs famously declared, "Innovation distinguishes between a leader and a follower." Similarly, Soichiro Honda, the founder of Honda, emphasized the importance of "The pursuit of one's dreams." Across different countries, there is a cultural emphasis on unique principles and values; however, in Japan, there is a particular tradition of cherishing the philosophies of the community's prominent leaders. In Japanese schools and companies, even within small communities, these philosophies often form the basis of various educational practices, serving as the foundation of the community and a pillar for long-term success. These philosophies are not mere words but act as a guide for education and management, deeply permeating the daily activities of students and employees as codes of conduct.

Since such philosophies play a crucial role in shaping the guidelines for the community's direction and the expected behavior of its members, sharing these principles within the community is of great importance. Moreover, these philosophies are often reflected in speeches given during graduation, entrance, or induction ceremonies. On the other hand, with the rapid development of generative AI, they can generate various types of content. With existing generative AI, the philosophies of famous prominent leaders such as Steve Jobs and Soichiro Honda have been learned, and it is possible to generate content that includes their philosophies to some extent. However, generative AI currently does not consider the principles of prominent leaders of small communities, such as small and medium-sized businesses and schools. This is likely due to the prominent leaders of such small communities being less well-known, making them less prominent in training datasets, as well as the limited amount of material available for learning. However, these philosophies have significant value even in small communities, and delivering speeches to community members who employ such philosophies is crucial for them. We believe that having these speeches presented by for example, avatars of prominent leaders can help members of the community quickly understand and internalize these values. Therefore, this paper proposes a method for generating speech scenarios that incorporate the philosophies of the prominent leaders of such small communities.

In this paper, we focus on Hachisaburo Hirao (1866–1945) (Hereafter "Hirao"), the founder (prominent leader) of Konan University, where the authors are affiliated, as the distinguished individual under study. Konan University is a medium-sized university, and Hirao is well known within the university but not widely recognized as a prominent leader in Japan. Thus, he is an appropriate subject for this study, which focuses on speech scenarios containing the philosophies of small community's prominent leaders. Konan University has educational philosophies such as "Prioritizing the cultivation of character and the promotion of health while respecting individuality and fostering the innate qualities of each person as a pioneer of character education." and "being gentlemen and ladies who are recognized worldwide." Hirao was an influential businessman, politician, and educator in Japan, and his ideals and words continue to be carried on at Konan University. However, it cannot be said that these philosophies and their background are thoroughly shared among the university's students and faculty members.

This research aims to generate speech scenarios sharing Hirao's philosophies using his dailies and written works. To this end, we use a retrieval-augmented generation (RAG) system which retrieves the sentences containing his principles and their contexts from his past writings and generates speech scenarios based on the retrieved texts. In this study, sentences or parts of sentences that reflect the ideas and principles of this distinguished individual are referred to as "quotations," while the original texts from which these quotations are extracted are referred to as "'episodes." We construct a database that consists of Hirao's quotations using literature related to him. We call this database "quotations database."

Figure 1 overviews the proposed RAG system. Our RAG system consists of a quotations database, query expansion, quotation search, and reranking.

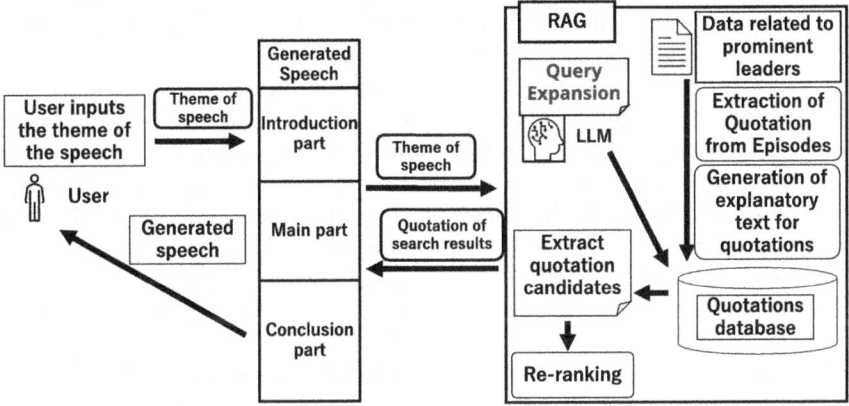

Fig. 1. Overview of our proposed system

The flow of generating a speech scenario with philosophies is as follows:

- The user inputs the theme of the speech.
- The system expands the query that is input by the user and extracts relevant quotations from the quotations database.
- It re-ranks the results and selects the quotations to generate a speech scenario.
- For each section of the speech structure, the system generates the speech scenario using the retrieved quotations and their associated episodes.

2 Related Work

There are many studies on text generation using large-scale language models and RAG. Lewis et al. [8] propose a basic framework of RAG for QA and language generation tasks. Cai et al. [1] propose a method for clarifying the logical structure of long-form QA responses and improving their actuality. Vassos et al. [16] developed an RAG system that extracts information from policy documents and provides answers in a QA format regarding policies. Sudhi et al. [13] propose a framework that explains the reasoning behind answers generated by large-scale language models in QA tasks. Tayal et al. [14] propose a dialog system that can respond appropriately to users' unclear questions by extending the user's input. These studies use RAG to improve the factuality of answers in QA and dialogue systems. In contrast, we generate speech scenarios based on the philosophy of prominent leaders from small communities where information is scarce based on RAG.

There are many studies that reproduce the personality and thoughts of specific people or characters using a method that prompts the user for persona. Serapio-García et al. [10] show that large-scale language models can reproduce human personality by incorporating persona descriptions into prompts. Lee et al. [7] propose a method for generating natural dialogue responses by retrieving persona information relevant to the intended output. Tsubota et al. [15] introduce a few-shot prompting technique that uses personality descriptions and representative statements to generate text characteristic to a specific individual. Kasahara et al. [6] propose a prompt-tuning method to build dialogue models that incorporate persona information. These studies input personas into prompts to reproduce the personality and tone of a specific person or character. However, simple persona descriptions or a few dialogue examples are insufficient to fully capture the depth of experiences, values, and philosophies of prominent figures. Therefore, in this study, we construct a quotations database from the books of prominent leaders.

Moreover, many studies refer to persona information extracted from dialogue histories to more accurately replicate specific individuals. Ma et al. [9] develop personalized chatbot by learning user profiles from individual dialogue histories. Zhong et al. [17] propose a framework called Memory Bank that constructs, updates, and retrieves memories from long-term user dialogue histories. Deng et al. [2] propose a method for personalized responses in e-commerce QA by extracting and integrating user preferences from post histories at three levels: knowledge, perspective, and vocabulary. These studies primarily extract persona information from dialogue histories. However, dialogue histories often contain significant noise, making it difficult to extract useful information from a target person's writings. Therefore, in this study, we construct a database of texts containing the philosophies and ideas of the target person and use it as a retrieval source in a RAG system. This enables the large-scale language model to generate speech scenarios based on the person's philosophy by incorporating this curated knowledge.

Many studies on text generation reproduce specific people through fine-tuning. Han et al. [4] introduce a contrastive learning to ensure persona consistency in generated dialogues. Hu et al. [5] propose a dialogue system that integrates knowledge graphs and large-scale language models distillation to maintain consistent persona characteristics. Shi et al. [12] proposed a method to build persona-consistent dialogue models by fine-tuning on conversations generated with specific character personas. Shao et al. [11] propose a method to reproduce conversations with a specific person, such as Beethoven, by collecting information about that person and using the dialogue data generated based on that information for fine tuning. These fine-tuning methods are difficult to accurately reproduce for people with a small amount of training data. Therefore, in this study, we construct a RAG using the target person's quotations to generate a speech scenario that includes the target person's philosophy.

3 Construction of the Quotations Database

3.1 Structure of the Quotations Database

In this study, we focus on small communities and aim to generate speech scenarios that reflect the principles of their prominent leaders. We input texts containing the leaders' thoughts and philosophies into a large language model as prompts. There are various methods for generating speech scenarios that embed such philosophies. This research takes an efficient approach by focusing on the quotations of prominent leaders, as such quotations are believed to inherently reflect their core philosophies. To produce diverse speech scenarios, it is essential to have a large collection of quotations. Hence, we manually constructed a quotations database using materials such as collections of speeches and autobiographies authored by the targeted leader, Hirao. The structure of the quotations database is illustrated in Fig. 2. The database used in this study consists of two tables: the quotations table and the episode table. The quotations table comprises the following elements:

- quotation: the extracted quotation.
- explanation_quotation: additional explanation information about the quotation.
- word_vector: vectorized explanation_quotation.
- episode_id: an identifier for the corresponding episode of the quotation.

The episode table consists of:

- episode_id: the identifier for the episode ID.
- episode_file: the content of the episode which includes the quotation.

This structured approach allows for efficient retrieval and utilization of the quotations and their contextual backgrounds, enabling the generation of meaningful and principled speeches tailored to the specific community.

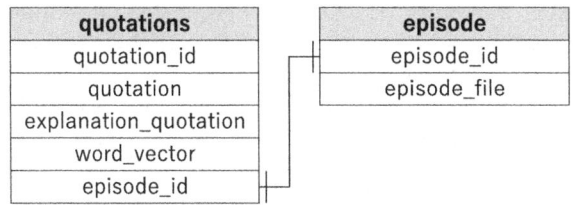

Fig. 2. Structure of quotations database

3.2 Identification of the Quotations

Converting Documents with Quotations into Text Data

The quotations used for speech-scenario generation are extracted from three books related to Hirao: "I think this way (In Japanese: Watashi wa Kou Omou)", "Hachisaburo Hirao Lecture Collection", and "Autobiography of Hachisaburo Hirao". These three works exist only in printed form, without any accompanying text data. Thus, they have to be converted into text (electronic) data for automatic processing. The following is an overview of these three books:

- I Think This Way:
  This book comprises 105 chapters and 337 pages and contains 24,563 sentences. In addition to the modern Japanese script used today, Japan had older character styles prior to World War II, and this book is entirely written in such older characters. It compiles Hirao's talks and free discussions that were published in newspapers and magazines.
- Hachisaburo Hirao Lecture Collection:
  This book consists of 37 chapters and 328 pages, with a total of 9,954 sentences. It contains transcripts of Hirao's lectures.
- Autobiography of Hachisaburo Hirao:
  Spanning 12 chapters and 482 pages, this autobiography contains 8,763 sentences. It details Hirao's life, from his childhood to the founding of Konan University.

First, we digitize the three books into text data using Optical Character Recognition (OCR). During this process, elements such as annotations and page numbers included in the text may cause layout distortions or inaccuracies, leading to parts that cannot be read accurately. We manually correct portions that have errors. Furthermore, we convert the older character into a modern character. In this study, we use the Python library kyujipy[1] for the conversion. We divide the data into appropriate units to record which quotations correspond to which episodes. Chapters are deemed suitable as the division units; however, while "I think this way" and "Hachisaburo Hirao Lecture Collection" have an average of 243 sentences per chapter, "Autobiography of Hachisaburo Hirao" contains an average of 730 sentences per chapter, resulting in significant differences in sentence counts per chapter. To address this, we divide "I think this way" and "Hachisaburo Hirao Lecture Collection" into each chapters. We divide "Autobiography of Hachisaburo Hirao" into section manually. As a result, there are 18 sections in "Autobiography of Hachisaburo Hirao". Through this process, we divide the text data from the three books into a total of 160 sections. The 160 sections become 160 episodes. On average, each episode contains 270 sentences and 9,139 characters.

---

[1] https://github.com/DrTurnon/kyujipy.

Extraction of Quotation from Episodes
We manually extract quotations from each episode. Specifically, annotators identify sentences they consider to be quotations within each episode. Three annotators are assigned to each episode. We determine a part of a quotation if two or more annotators have annotated it as such. In cases where the length of the extracted quotations differs among annotators, we determine the length of quotation based on the overlapping part annotated by two or more annotators. For example, if Annotator A annotates "Life is a long marathon race." as a quotation, and Annotator B annotates "Life is a long marathon race. The purpose of education is to build people who can endure this long race." as a quotation, we regard only the overlapping part "Life is a long marathon race." as a quotation. As a result of this process, we extracted 66 quotations from 160 episodes.

3.3 Generation of Explanatory Texts for Quotations

The extracted quotations are often concise, such as "Education cannot be mass-produced." Some quotations may also be ambiguous. Short or ambiguous quotations are less likely to be retrieved, even if they are important for the user's intended speech. To address this, we provide explanatory texts as supplementary information to clarify and make these quotations searchable.

We generate explanatory texts using LLMs. In this study, we utilize the GPT-4o text generation model and the text-embedding-3-large model via OpenAI's API[2,3]. To ensure the explanatory texts are relevant, we include the episode containing the quotation in the prompt. By inputting the quotation and its source episode into GPT-4o, we generate explanatory texts that reflect the thoughts and experiences of notable individuals.

We then vectorize the quotations and their explanatory texts using text-embedding-3-large and store them in a quotations database. Although LLMs can generate explanatory texts, they may produce hallucinations. To mitigate this, we input both the quotation and the episode to generate accurate background information. However, hallucinations may still occur due to unintended interpretations by the model. Such hallucinations remain to be addressed in future work. Figure 3 shows examples of quotations and their generated explanatory texts.

Quotation: "Education cannot be mass-produced."
Explanatory of the Quotation:
This quotation highlights the perspective that education cannot be conducted mechanically. The background information suggests that individuals possess unique talents and qualities, which necessitates personalized education to draw out these attributes. It argues against uniform education, emphasizing that the true purpose of education is to nurture individual talents.

Fig. 3. Example of Explanation of the quotation

---

[2] https://openai.com/index/gpt-4o-system-card.
[3] https://platform.openai.com/docs/guides/embeddings.

4 Quotation Search Method

This study aims to generate speech scenarios that incorporate the philosophies of prominent leaders by utilizing their notable quotes and the contextual episodes surrounding them. To achieve this, we retrieve an appropriate quotation from a quotations database based on the theme provided by the user. Ensuring consistency in the generated speech scenario requires the selection of quotes that align with the user's specified theme. For instance, if the theme is "learning at university," the quotes used in the speech should also pertain to education. However, user-provided themes are not always free from semantic ambiguity. To address this issue, we employ query expansion techniques to refine the meaning of the user's input theme. Subsequently, using the expanded query, we determine the most appropriate quotes by leveraging both the quotes themselves and their explanatory texts. To further enhance the relevance of the selected quotes, we perform re-ranking of the search results.

4.1 Query Expansion

User input themes may include keywords or short phrases that determine the topic of the generated speech scenario, such as "learning in university" or "mindset as a working professional." However, when such brief phrases are directly used as queries, vector search based on word similarity may fail to yield appropriate results.

To enhance the accuracy of quotations retrieval, we generate an explanatory description of the user-provided speech theme as a query expansion. This description is generated using an existing large language model (LLM), specifically GPT-4o in this study. The generated description is then utilized as an expanded query for searching the quotations database.

4.2 Re-ranking of Quotation Search Results

In search tasks where a single definitive result cannot be determined, such as in this study, re-ranking vector search results using a Cross-encoder can yield higher accuracy. Cross-encoders are commonly used for re-ranking and estimate similarity by simultaneously inputting the query and document into the model. Furthermore, Cross-encoders can evaluate the similarity between the query and the search target more accurately than simple vector searches. However, Cross-encoders cannot pre-store vectors of the search targets, leading to longer computation time when the number of search target documents increases. Therefore, in this study, we initially narrow down the quotation candidates to a certain number n using vector searches and then re-rank these candidates using a Cross-encoder. The following is the flow of re-ranking method.

1. We first perform searches using cosine similarity to extract quotation candidates for speech-scenario generation. The query is vectorized using text-embedding-3-large, and cosine similarity is calculated with vectors of all quotations and their explanations stored in the quotations database. The top n

candidates with the highest similarity are considered the final search result candidates. In this study, n is set to 10.
2. We re-rank the search result candidates using a Cross-encoder. The re-ranking model used is bge-reranker-large, developed by BAAI (Beijing Academy of Artificial Intelligence)[4,5]. This model recalculates the similarity between the query and the search result candidates, and the quotation with the highest similarity is considered the final search reseults.

5 The Method of Speech-Scenario Generation

5.1 Structure of Speech-Scenario Generation

This study aims to generate speech scenarios for ceremonies such as entrance and graduation. To achieve this, we generate the prompts considering the structure of the speech and then input them into GPT-4o for speech-scenario generation. Fukazawa et al. [3] analyze Japanese ceremonial speeches by dividing them into three parts: "introduction part," "main part," and "conclusion part." The introduction includes the opening greeting, expressions of gratitude to the audience, and the declaration of the speech's theme. The main part follows after presenting facts or the speaker's own opinions. This part often includes expressions of hope or expectation for people or events, such as calls to action like "Let us strive together." The conclusion part is where the speech scenario concludes with a closing declaration, such as "This concludes my address."

In this study, we generate speech scenarios for these three parts to create more authentic and coherent speech scenarios. In generating speech scenario, we employ a generative AI model for each part of the speech scenario. In this study, we utilize GPT-4o for this purpose.

5.2 Prompts for Speech-Scenario Generation

When using generative AI, the prompts are crucial. Generally, prompts consist of system prompts which are instructions to control the AI's behavior and response style, and user prompts which are instructions for specific tasks or questions directed at the AI. In this study, the system prompt provides overall role instructions for speech-scenario generation, while the user prompt provides role instructions for each part. Additionally, by using dynamic prompts with variables and templates for both types of prompts, we aim to generate speech scenarios that better satisfy user requirements.

Variables: We use variables in both system prompts and user prompts. The variables for user input consist of the speaker {person}, the situation {situation} in which the speech is given, the audience {audience}, and the theme of the speech {theme}. The system input editing consists of the quotation {quotation} used in the speech and the episode {episode} that provides background information for the quotation.

---

[4] https://www.baai.ac.cn/english.html.
[5] https://huggingface.co/BAAI/bge-reranker-large.

System Prompt: We use the system prompt to input information related to the overall speech-scenario generation. The system prompt consists of {person}, {situation}, and {audience}, the purpose of the speech, {theme}, {quotation}, and {episode}. Figure 4 shows an example of a system prompt. Here, we use a template for the purpose of the speech.

```
system_prompt
You are {person}.
Write a speech manuscript according to the following requirements.
### Speech Context
Occasion: {situation}
Audience: {audience}
### Purpose of the Speech
To share the philosophy of {person} with the audience.
### Theme of the Speech
"{theme}"
### Make sure to include the following quotation in the speech:
"{quotation}"
### Refer to the following episode as background for the quotation. Incorporate it as necessary:
{episode}
```

Fig. 4. Example of the system prompt

User Prompt: We generate the user prompt for each of the three components of the speech.

- Introduction part
  In general, the introduction is characterized by interpersonal considerations such as expressing gratitude or congratulations to the audience. Therefore, it is specified that expressions of gratitude or congratulations should always be generated. At the end of the introduction part, input the theme or quotation to state the theme of the speech to audiences clearly.
- Main part
  The prompt for the main part consists of text of the introduction part which is generated text in the introduction part, theme, quotation, and episode. By specifying text of the introduction part, we generate the main part as a continuation of the introduction, ensuring a smooth speech. The reason for specifying theme is to instruct the generation of a speech scenario directed at the audience in line with the theme and to incorporate quotations. The reason for specifying quotation is to ensure the inclusion of the philosophy of the prominent leader in the speech, making it most important to accurately and clearly explain the quotation to the user. Additionally, to suppress the generation of hallucinations when explaining the quotation, episode is specified as background information for the quotation. At the end of the main part, input instructions to include a message directed at the audience.
- Conclusion part
  In the conclusion, summarize the content of the main part concisely and reaffirm the key points as the closing of the speech. In generating the conclusion, input the sentence which is "re-emphasize the message to the audience"

to ensure the generation is audience-conscious. Finally, input instructions to make a closing declaration.

Figure 5 shows an example of a user prompt.

```
Introduction part
Write the introduction part of the speech according to the following conditions:
1.The tone must be appropriately formal for the opening address of {situation}.
2.The introduction part must include expressions of gratitude and congratulations directed toward the audience.
3. State the theme of speech succinctly and memorably.
Introduction part:

Main part
The following is the introduction of a speech. Write the main part that follows this introduction.
{Text of the Introduction}
Write the main part of the speech according to the following conditions:
1.Offer valuable advice or insight to audience based on the theme of "{theme}"
2.Naturally, incorporate "{quotation}" and provide a detailed explanation of its meaning.
3.Refer to the following "{episode}" and use specific examples or personal experiences to help illustrate the quotation.
4.Include a message to {audience} that encourages and empowers them to move forward with confidence and hope.
Main part:

Conclusion part
The following text contains the introduction part and main part of a speech.
Write the conclusion part that follows.
{Text of the Introduction part and Main part}
Write the conclusion part of the speech according to the following conditions:
1.Summarize the entire speech concisely and reaffirm its main points.
2.Once again, emphasize the message directed toward {audience}.
3.Conclude with a clear and appropriate closing remark suitable for {situation}.
Conclusion part:
```

Fig. 5. Each part of the user prompt

5.3 Experiments

We examined the validity of generated speech scenarios using the proposed method through human evaluation.

5.4 Condition of the Experiment

In this experiment, we use our proposed speech-scenario generation method incorporating Hirao's philosophy. The prominent leader in focus is Hirao, the founder of Konan University. Then, we input Hirao as the speaker and the audience as students of Konan University. Two types of speech scenarios are set: the entrance ceremony and the graduation ceremony. Seven themes are used for each scenario. The details of the themes used are shown in Table 1.

The subjects are 14 students from Konan University who understand Hirao's philosophy (9 males, 5 females; age range: 21–24 years). In our proposed method, we use a large language model to expand queries during the quotation search, which can result in different quotations being retrieved even for the same theme. Therefore, we generated three speech scenarios for each theme, resulting in 42 speech scenarios.

Theme: The Mindset of a Working Member of Society
Quotation from Search Results:
"Both the office and the workplace are battlefields. Do not indulge in excess nor seek comfort. As you graduate and enter society, I believe it is your responsibility to work hard and continue making relentless efforts without giving up."
Speaker:Hachisaburo Hirao
Occasion: Graduation Ceremony
Audience: Graduating students of Konan University
Speech Text:
Ladies and gentlemen, Thank you very much for attending today's graduation ceremony at Konan University. First, I would like to extend my heartfelt congratulations to all the graduates, their families, and the faculty and staff who have supported these students throughout their academic journey.
As you begin your path as members of society, I would like to share an important message with you. "Both the office and the workplace are battlefields. Do not indulge in excess or seek comfort. As you graduate and enter society, I believe it is your responsibility to work hard and continue making relentless efforts without giving up." Numerous challenges and hardships await you as you enter the working world. In these very moments, your true strength and determination will be tested. When I entered the business world, society underwent tremendous change and faced significant trials. Though I was neither a sociologist nor an academic, through my various business experiences, I came to believe that the spirit of business in service of the nation is essential for the youth who will lead the next generation. This spirit is not just about the pursuit of profit but about dedication to the greater good of the nation and society as a whole.
Precisely because we live in challenging times, I hope you will embrace this spirit, work sincerely, and contribute meaningfully to society. As members of the workforce, you are expected to fulfill your roles and responsibilities with a mindset that views the workplace as a battlefield and to commit yourselves to tireless effort. The ethos of "not indulging in excess nor seeking comfort" will serve as a powerful force to help you overcome adversity and foster your personal growth. I look forward to seeing the bright futures that each of you will create. Congratulations on your graduation.

Fig. 6. Example of generated speech

Table 2 shows the evaluation criteria for the generated speech scenario. Evaluation items 1 and 2 measure the consistency of the speeches, assessing whether the content aligns with the theme. Evaluation items 3 and 4 assess whether the speeches incorporate the community's philosophy. Evaluation item 5 evaluates our proposed quote retrieval method, specifically, whether Hirao's quotes are appropriately cited. Evaluation item 6 assesses the clarity of the speech structure we proposed. Evaluation item 7 measures the logical coherence of the speeches, checking for logical leaps or contradictions. Evaluation item 8 assesses the difficulty level of the speeches, evaluating whether the vocabulary and expressions are appropriately challenging. Evaluation item 9 measures the originality of the speeches, assessing whether they contain unique content or perspectives. Evaluation item 10 evaluates the message conveyed by the speeches.

We evaluated these 10 items using a 5-point Likert scale: 1. Strongly Disagree, 2. Disagree, 3. Neutral, 4. Agree, 5. Strongly Agree. We collected three evaluations from each of the 14 subjects, obtaining evaluations for a total of 42 generated speech scenarios. Figure 6 shows an example of a generated speech scenario.

5.5 Results and Discussion

Table 2 shows the results of experiment.

In Question 1, 66% of the subjects judged that the speeches used quotes that aligned with the theme. In Evaluation Item 2, 76% of the subjects judged that the speeches had content that aligned with the theme. The results show that the proposed method can generate speech scenarios with content that aligns with the theme. On the other hand, 12% of the subjects judged that the quotation

Table 1. Examples of speech themes for entrance and graduation ceremonies

No.	Theme
Speech theme of entrance ceremony	
1	A environment to learn freedom and responsibility
2	Educational opportunities for self-discovery and personal growth
3	Respecting diversity and growing together
4	Taking responsibility for one's choices
5	Finding your brilliance in a new environment
6	The importance of taking on challenges
7	A journey to find one's role
Speech theme of graduation ceremony	
8	The importance of continually learning by oneself and with others
9	Setting out toward your next goal
10	Contributing to society through knowledge
11	Small steps can change the future
12	The importance of thinking and acting integratively
13	How to deal with diverse values
14	How to make the most of yourself in a new environment

retrieved did not align with the theme. For example, in the entrance ceremony situation with the theme "Small efforts can change the future," the quotation is "It is dangerous to work without gaining people, and the resulting damage is immense. It cannot be compensated by frugality in clothing, food, and housing.". Although this quote appears to align with the theme, it is directed towards business people rather than students. Thus, we found that quotations sometimes do not match the situation, indicating the need to consider the situation.

In the question 3, 48% of the subjects judged that Hirao's philosophy was included, and in question 4, 59% of the subjects rated that Hirao's philosophy was included in the generated speech scenarios, which are positive results. On the other hand, a significant number of subjects rated 'neutral (3)' with 38% for question 3 and 36% for question 4. For example, in question 3, the subjects judged the generated speech scenario text of "The importance of advancing with appropriate personnel and cooperation brings more than mere efficiency. No matter how technologically advanced the world becomes, the fundamental flow is human understanding and cooperation." as 'neutral(3)'. This is because similar content is commonly made, making it difficult to determine if it is unique to Hirao's philosophy. This suggests that the philosophies of prominent leaders are not necessarily unique to those leaders. This problem needs to be addressed in future work.

In question 5, 77% of the subjects judged that Hirao's quotations were appropriately cited, while 9% of the subjects judged that the quotation were not appropriately cited. This is likely due to the retrieval of quotations that do not match the theme, resulting in unnatural citations. This is an issue to be addressed in the future. In question 6, 93% of the subjects judged that the structure of the speeches was clear. This indicates that most of the generated speech scenarios have a clear structure, as intended by the generation method.

In question 7, 76% of the subjects judged that there were no logical leaps or contradictions in the content, indicating that the proposed method can generate logical speeches. In question 8, 81% of the subjects judged that the vocabulary and expressions were of an appropriate difficulty level, showing that the generated speech scenarios are suitably challenging. However, some speeches included complex expressions and vocabulary. This was due to the use of difficult expressions quoted from episodes. While using episodes helps generate speech scenarios that reflect Hirao's experiences and thoughts, the results of questions 7 and 8 indicate that when the content of the episodes used is difficult, the generated speech scenarios can become harder to understand. The problem also needs to be addressed in future.

In question 9, 55% of the subjects judged that the speeches contained unique content or perspectives, while 40% judged 'neutral (3).' The reason for the high number of neutrals is that many of the generated speech scenarios reflect Hirao's philosophy, but this does not necessarily result in unique content. This issue is similar to that of Evaluation Items 3 and 4. In Evaluation Item 10, 88% of the subjects judged that the speeches had a strong message. This indicates that most of the generated speech scenarios possess a message based on Hirao's quotations and episodes.

Based on the above results, we found that the proposed method is appropriate in terms of the speech format, such as the clarity of the generated speech scenario structure and the absence of logical contradictions. On the other hand, we will consider a retrieval method that includes situations in quote searches.

Table 2. Questions of the Evaluation (%)

No.	Evaluation Item	5	4	3	2	1
1	Is the quotation used in the speech appropriate for the theme?	33	33	12	19	3
2	Is the content appropriate for the theme?	50	26	12	10	2
3	Does the speech include the philosophy of Konan University?	24	24	38	14	0
4	Does the speech include the philosophy of Hirao?	26	33	36	5	0
5	Does the speech appropriately quote the quotation of Hirao?	41	36	14	7	2
6	Is the structure of the speech (introduction, main part, conclusion) clear?	41	52	5	2	0
7	Is the content of the speech logical and consistent?	33	43	19	5	0
8	Are the vocabulary and expressions of an appropriate level of difficulty?	33	48	9	10	0
9	Are content and viewpoint unique?	29	26	40	5	0
10	Does the speech have a message?	57	31	7	5	0

Additionally, the challenge of handling cases where Hirao's philosophy is not unique but shares the same meaning as general philosophies was highlighted.

6 Conclusion

This study proposed a method for generating speech scenarios to share the philosophy of a community. By creating a quotations database from many books about prominent leaders and combining retrieval and prompt methods, we generated speech scenarios that include the philosophy. In this paper, we targeted on Hirao who are founder of our university. The evaluation of the proposed method showed that it can generate speech scenarios that include the prominent leaders' philosophy. However, issues were identified, such as the need to consider the context during quote retrieval and how to handle cases where Hirao's philosophy is not unique but shares the same meaning as general philosophies.

Future challenges include improving retrieval accuracy and verifying whether the method can generate speech scenarios that include the philosophy when applied to individuals other than Hirao. Additionally, by automating quote extraction, speech evaluation, and hallucination detection, we aim to enhance the versatility of the system.

Acknowledgements. This work was partially supported by Konan Digital Twin Research Center, the Research Institute of Konan University, JSPS KAKENHI Grant Numbers 24K03044 and 22K12280, and MEXT, Japan.

References

1. Cai, T., et al.: FoRAG: factuality-optimized retrieval augmented generation for web-enhanced long-form question answering, KDD 2024, pp. 199–210. Association for Computing Machinery (2024)
2. Deng, Y., Li, Y., Zhang, W., Ding, B., Lam, W.: Toward personalized answer generation in e-commerce via multi-perspective preference modeling. ACM Trans. Inf. Syst. **40**(4) (2022)
3. Fukasawa, N., Hillman Kobayashi, K.: Components and development patterns of Japanese Shikiji speeches. J. Tech. Jan. Educ. **14**, 27–34 (2012)
4. Han, Z., Zhang, S., Zhang, X.: Persona consistent dialogue generation via contrastive learning. In: Companion Proceedings of the ACM Web Conference 2023, WWW 2023 Companion, pp. 196–199. Association for Computing Machinery (2023)
5. Hu, L., Zhang, X., Song, D., Zhou, C., He, H., Nie, L.: Efficient and effective role player: a compact knowledge-grounded persona-based dialogue model enhanced by LLM distillation. ACM Trans. Inf. Syst. **43**(3) (2025)
6. Kasahara, T., Kawahara, D., Tung, N., Li, S., Shinzato, K., Sato, T.: Building a personalized dialogue system with prompt-tuning. In: Ippolito, D., Li, L.H., Pacheco, M.L., Chen, D., Xue, N. (eds.) Proceedings of the 2022 Conference of the North American Chapter of the Association for Computational Linguistics: Human Language Technologies: Student Research Workshop, pp. 96–105. Association for Computational Linguistics (2022)

7. Lee, J., Oh, M., Lee, D.: P5: plug-and-play persona prompting for personalized response selection. In: Bouamor, H., Pino, J., Bali, K. (eds.) Proceedings of the 2023 Conference on Empirical Methods in Natural Language Processing, pp. 16571–16582. Association for Computational Linguistics (2023)
8. Lewis, P., et al.: Retrieval-augmented generation for knowledge-intensive NLP tasks. In: NIPS 2020. Curran Associates Inc. (2020)
9. Ma, Z., Dou, Z., Zhu, Y., Zhong, H., Wen, J.R.: One chatbot per person: creating personalized chatbots based on implicit user profiles. In: Proceedings of the 44th International ACM SIGIR Conference on Research and Development in Information Retrieval, SIGIR 2021, pp. 555–564. Association for Computing Machinery (2021)
10. Serapio-García, G., et al.: Personality traits in large language models (2025)
11. Shao, Y., Li, L., Dai, J., Qiu, X.: Character-LLM: a trainable agent for role-playing. In: Bouamor, H., Pino, J., Bali, K. (eds.) Proceedings of the 2023 Conference on Empirical Methods in Natural Language Processing, pp. 13153–13187. Association for Computational Linguistics (2023)
12. Shi, H., Niu, K.: Enhancing persona consistency with large language models. In: Proceedings of the 2024 5th International Conference on Computing, Networks and Internet of Things, CNIOT 2024, pp. 210–215. Association for Computing Machinery (2024)
13. Sudhi, V., Bhat, S.R., Rudat, M., Teucher, R.: RAG-Ex: a generic framework for explaining retrieval augmented generation. In: Proceedings of the 47th International ACM SIGIR Conference on Research and Development in Information Retrieval, SIGIR 2024, pp. 2776–2780. Association for Computing Machinery (2024)
14. Tayal, A., Tyagi, A.: Dynamic contexts for generating suggestion questions in rag based conversational systems. In: Companion Proceedings of the ACM Web Conference 2024, WWW 2024, pp. 1338–1341. Association for Computing Machinery, New York, NY, USA (2024)
15. Tsubota, Y., Kano, Y.: Text generation indistinguishable from target person by prompting few examples using LLM. In: Kano, Y. (ed.) Proceedings of the 2nd International AIWolfDial Workshop, pp. 13–20. Association for Computational Linguistics (2024)
16. Vassos, S., et al.: Now I know! Empowering voters with rag-enabled LLMs to eliminate political uncertainty. In: Proceedings of the 13th Hellenic Conference on Artificial Intelligence, SETN 2024. Association for Computing Machinery (2024)
17. Zhong, W., Guo, L., Gao, Q., Ye, H., Wang, Y.: MemoryBank: enhancing large language models with long-term memory. In: Proceedings of the AAAI Conference on Artificial Intelligence, vol. 38, no. 17, pp. 19724–19731 (2024)

VarCGAN: Variational Cyclic Generative Adversarial Network For Music Genre Style Transfer

Pooja Singh[(✉)], Dhruv Mishra[id], and Ankita Khandelwal

Shiv Nadar Institution of Eminence, Greater Noida, India
{pooja.singh,dm409,ak844}@snu.edu.in

Abstract. Variational Autoencoders (VAEs) and Generative Adversarial Networks (GANs) are two prominent generative models that excel in different aspects of generative tasks of creating new content, such as images, text, or music using data and patterns. VAEs are known for their capability to learn smooth and probabilistic latent representations, which enable structured generation, but they often produce blurry and unrealistic outputs due to their inherent loss function. On the other hand, GANs generate sharp and visually appealing results by utilizing a discriminator to refine output but suffer from mode collapse, limiting their diversity. This paper proposes a novel hybrid architecture, the Variational Cyclic Generative Adversarial Network (VarCGAN), to transfer the musical style of a song from one genre to another. VarCGAN combines the latent space modeling capabilities of VAEs with the adversarial optimization of GANs to overcome the limitations of each approach.

The model introduces a cyclic adversarial loss, which ensures consistency and realism in style-transferred outputs while preserving the original song's musical essence. Furthermore, the hybrid design enables the generation of diverse variations of genre-transferred songs, capturing subtle stylistic features of the target genre. The proposed approach is evaluated on the GTZAN dataset, focusing on style transfers between classical, jazz, hip-hop, and rock genres. To the best of our knowledge, VarCGAN is the first framework to utilize this hybrid methodology for music genre style transfer, presenting a significant advancement in music composition and genre transformations.

Keywords: Music Genre Transfer · Machine Learning · Variational Autoencoders · Generative Adversarial Networks

1 Introduction

Music genres represent categorizations based on the stylistic and structural elements of musical compositions [4]. Different genres exhibit unique characteristics, such as rhythmic patterns, harmonic progressions, and instrumentation. Additionally, music often incorporates diverse languages and cultural elements, making it challenging to represent and even more difficult to translate between genres. However, translating a song from one genre to another can be of immense value in the music composition industry, enabling creative exploration and audience engagement [25].

Traditionally, genre translation was considered time-consuming and computationally infeasible. Recent advancements in generative models, particularly Variational Autoencoders (VAEs) [17] and Generative Adversarial Networks (GANs) [7], have made this task more achievable. VAEs are effective at learning smooth, probabilistic latent representations of data, but their reconstructions often suffer from blurriness and a lack of realism [17]. Conversely, GANs produce sharp and realistic outputs by leveraging a discriminator to refine results, but they are prone to mode collapse, limiting the diversity of generated data [7].

Both VAEs and GANs have demonstrated success in style transfer tasks across various domains, including image style transfer [6], timbre transfer in audio spectrograms [12], and symbolic music translation [1]. TimbreTron [12] utilizes spectrogram-based transformations to manipulate timbre across instruments, while MIDI-VAE [2] combines VAEs with symbolic music representations to model musical dynamics and instrumentation. Recent approaches, such as the diffusion-based model for multi-style conversions [13], further demonstrate advancements in music style transfer.

Although these models achieve state-of-the-art results, challenges such as preserving the essence of the original piece while adapting stylistic features still remains. This paper introduces a novel hybrid generative model, Variational Cyclic Generative Adversarial Network (VarCGAN), designed to translate songs from one genre to another. VarCGAN combines the probabilistic latent space modeling of VAEs with the adversarial optimization of GANs to overcome the limitations of each approach. By incorporating cyclic adversarial loss [25], the model ensures high-quality and realistic style-transferred outputs while preserving the core attributes of the original song. Additionally, the cyclic structure enables the generation of diverse variations of genre-transferred songs, capturing subtle stylistic nuances of the target genre.

The proposed VarCGAN model contributes to the growing body of work on generative models for music by addressing key challenges in genre translation and providing a framework that balances reconstruction quality and stylistic transformation.

2 Related Work

A significant body of research has been dedicated to neural style transfer, extending into the domain of music. Recent advancements in deep learning and audio-based models have facilitated the transfer of musical style from one genre to another. [25] introduced a cyclic consistency loss between two GANs to optimize the reconstruction of style-transferred images. Building upon this, [12] developed TimbreTron, focusing on musical timbre transfer by manipulating the timbre of a sound sample from one instrument to match another. They employed image domain style transfer on audio spectrograms and enhanced quality using the WaveNet synthesizer [21].

[4] highlighted the multi-modal characteristics of music, arguing that accounting for these aspects is crucial for improving the accuracy of music style transfer models. [18] concentrated on identifying which components of a song should be preserved versus modified for effective style transfer. [2] extracted symbolic music representations from MIDI files and input them into a variational autoencoder, utilizing separate recurrent encoder-decoder pairs with a shared latent space. Additionally, [1] applied CycleGAN to symbolic music genre transfer, demonstrating the feasibility of GAN-based models in this context.

Further contributions include [14], who introduced a model for musical composition style transfer via disentangled timbre representations, enabling rearrangement of music across genres. [13] proposed a diffusion model for music style transfer, achieving multi-to-multi style conversions with high-quality audio outputs. Recent methods, such as Gated-GAN [3], focus on improving adversarial architectures, while [24] extended GAN-based approaches for controllable style transfer in creative domains. These studies collectively advance the field of music style transfer, offering diverse methodologies and insights, and serve as a foundation for VarCGAN's novel hybrid approach.

3 Proposed Model

This paper proposes a novel Variational Cyclic Generative Adversarial Network (VarCGAN) for style transferring a song from genre A to genre B. The proposed model consists of two generators and two discriminators, as shown in Fig. 1. The generators are represented by G_A & G_B (G_A tries to style transfer from genre $B \rightarrow A$; G_B tries to style transfer from genre $A \rightarrow B$), and the discriminators are represented by D_A & D_B (D_A learns to distinguish between real and generated genre A songs, whereas D_B learns to distinguish between the real and generated genre B songs).

Both the generators are simple Variational Autoencoders (VAE) [16] that take the spectrogram representations of the songs as input and output the corresponding style transferred spectrogram representation. A spectrogram represents the spectral density of a signal over time at various frequencies [9]. G_A takes spectrogram X_B^r as input and outputs the genre style transferred spectrogram X_A^g which tries to trick the discriminator D_A, whereas G_B takes spectrogram X_A^r as input and outputs genre style transferred spectrogram X_B^g which tries to trick the discriminator D_B. Here the subscript A & B represents the genre of the song; superscript r represents that the songs are real; superscript g represents that the songs are generated.

As mentioned above, both the generators (G_A & G_B) are Variational Autoencoders (VAE) that are used for style transferring songs from one genre to another. Both the generators comprise an encoder ($q_\theta^{G_A}$ corresponding to G_A; $q_\theta^{G_B}$ corresponding to G_B) and a decoder ($p_\phi^{G_A}$ corresponding to G_A; $p_\phi^{G_B}$ corresponding to G_B). Both the encoders take a spectrogram representation as input (X_A^r

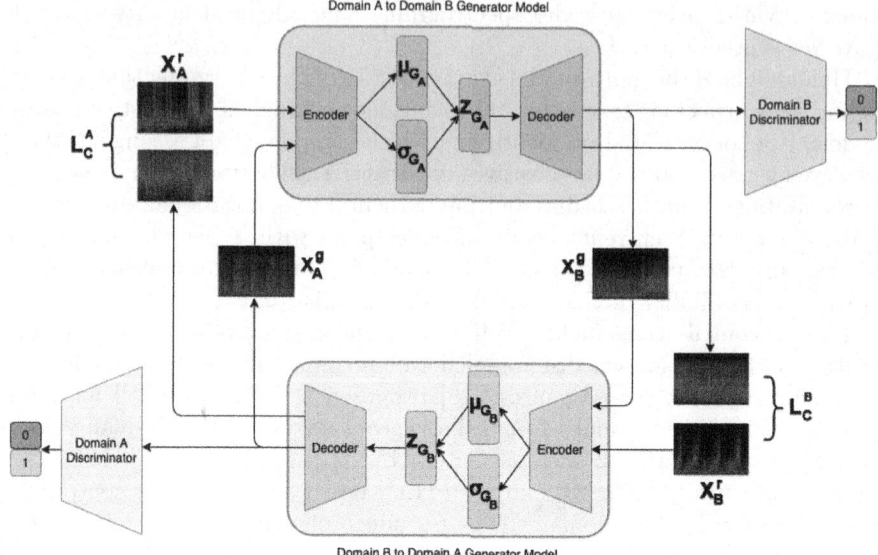

Fig. 1. Proposed Model Architecture

corresponding to G_A; X_B^r corresponding to G_B) and output a hidden latent representation (z_{G_A} corresponding to G_A; z_{G_B} corresponding to G_B). These lower-dimensional hidden latent spaces are stochastic in nature, and the decoders sample from these distributions and output the reconstructed style transferred spectrograms (X_B^g corresponding to G_A; X_A^g corresponding to G_B). The loss functions of the generators (G_A & G_B) are demonstrated by Eqs. (1) & (2).

$$L_{G_A} = \sum_{i=1}^{N} (\mathbf{E}_{x_{zG_A} \sim q_\theta^{G_A}(z_{G_A}|x_{A,i}^r)} [\log p_\phi^{G_a}(\\ x_{B,i}^g | z_{G_A})] + \mathbf{KL}(q_\theta^{G_A}(z_{G_A}|x_{A,i}^r) \| p(z_{g_A})) \quad (1)$$

$$L_{G_B} = \sum_{i=1}^{N} (\mathbf{E}_{x_{zG_B} \sim q_\theta^{G_B}(z_{G_B}|x_{B,i}^r)} [\log p_\phi^{G_B}(\\ x_{A,i}^g | z_{G_B})] + \mathbf{KL}(q_\theta^{G_B}(z_{G_B}|x_{B,i}^r) \| p(z_{g_B})) \quad (2)$$

In both of the above equations, the loss is calculated for N data points. In both the equations, the first term is the expectation of the encoders' distribution, and the second term is the Kullback-Leibler divergence between the encoders' distribution and $p(z_{g_A})$ or $p(z_{g_B})$ depending on the generator [16].

The proposed model also uses a cycle consistency loss apart from the above-mentioned generator loss to ensure that the generated style transferred spectrograms are meaningful and have sounds similar to those present in the original song. There are two cycles in the VarCGAN, one for the genre A songs and one for the genre B songs. In the case of genre A, spectrograms are given as input to G_B, which generates X_B^g, which again is given as input to G_A to generate a reconstruction of X_B^g in the genre A domain, whereas, in the case of genre B, spectrograms are given as input to G_A, which generates X_A^g, which again is given as input to G_B to generate a reconstruction of X_A^g in the genre B. These newly generated reconstructions in the original domain are used to compute the cycle consistency losses L_c^A & L_c^B for genre A & B respectively [26].

$$L_c^A = ||x_A^r - G_A(G_B(x_A^r))||_1 \tag{3}$$

$$L_c^B = ||x_B^r - G_B(G_A(x_B^r))||_1 \tag{4}$$

The total cyclic consistency loss is computed by summing L_c^A & L_c^B.

$$L_c = L_c^A + L_c^B \tag{5}$$

The final VarCGAN loss is computed by adding the total cyclic consistency loss with the adversarial loss of both the GANs present in the VarCGAN. The adversarial losses are represented by L_A & L_B.

$$L_{VarCGAN} = L_A + L_B + L_c \tag{6}$$

$$L_A = \max_{G_A} \min_{D_A} V(D_A, G_A)$$
$$= \mathbf{E}_{x_A \sim p_{data}(x_A)}[\log(D_A(x_A))] + L_{G_A} \tag{7}$$

$$L_B = \max_{G_B} \min_{D_B} V(D_B, G_B)$$
$$= \mathbf{E}_{x_B \sim p_{data}(x_B)}[\log(D_B(x_B))] + L_{G_B} \tag{8}$$

The adversarial losses are a summation of two terms, the first represents the discriminator classification loss, and the second represents the above-mentioned generator loss [8].

In short, the proposed VarCGAN comprises of two generative adversarial networks having variational autoencoders as its generator models. Each of the VAEs gets a spectrogram of the corresponding genre song as input [9]. After sampling from their respective latent distributions, these VAEs then try to reconstruct spectrograms of the corresponding parallel genres. These latent distributions learn the stylistic mapping between the genres while training. Then the discriminators take these translated spectrograms along with the real parallel spectrograms as inputs. The discriminators learn to classify whether the input was generated or real. This training process continues until both the VAEs perfectly learn to style transfer songs from one genre domain to another.

4 Experiments

The performance of the proposed model was tested based on the accuracy of the discriminators. It can be safely assumed that the generators are performing well in the genre style transfer task if the discriminator finds it challenging to distinguish between the generated and the real spectrograms. This evaluation metric aligns with standard practices in adversarial learning and style transfer tasks [8,26].

4.1 Dataset

Our main focus is to style transfer the music from one genre domain to another. For this purpose, we have used the *GTZAN Genre collection dataset* [23]. This dataset comprises of one thousand 30 s long audio tracks. In total, there are a hundred tracks for each of the ten genres. For the purpose of our experiment, we have focused on style transfer between classical, jazz, hip-hop, and rock. A total of 100 audio tracks were randomly chosen from all of these genres. All these audio tracks 22050 Hz monophonic 16-bit audio files that were converted to their corresponding spectrogram representations. Distribution of the dataset is shown in Table 1.

4.2 Training

The proposed model was trained on 90% of the data and the rest 10% was used for testing. 10% data out of the training set was used for validation purposes. All the song tracks were of the same duration so padding was not required. The encoders in the VAEs consisted of 8 convolutional layers having kernel size of 4, strides of 2 and number of filters equal to [64, 128, 256, 512, 512, 512, 512, 512] for each layer, whereas the decoders consisted of 8 deconvolutional layers having same kernel size & strides and the number of filters equal to [512, 512, 512, 512, 512, 256, 128, 64] for each layer. The discriminators were simple convolutional neural networks having the same kernel size & strides, and the number of filters equal to [512, 256, 128, 64, 1]. The output of each of all these convolutional and deconvolutional layers was passed through the *Leaky*ReLu [19] activation function having the value of α equal to 0.2. The model was trained for 10 epochs, with a batch size of 20. Learning rate was initially set to 0.001 and *Adam* optimizer was used for optimization.

Table 1. GTZAN Dataset Statistics

Genre	Number of Songs
Classical	100
Disco	100
Hip-Hop	100
Blues	100
Metal	100
Country	100
Rock	100
Reggae	100
Pop	100
Jazz	100

4.3 Experimental Results

The accuracy of the discriminators, when tested on the test data, was used as the evaluation metric. Lower accuracy indicates that the generators were able to produce genre-style-transferred spectrograms that successfully fooled the discriminators, a key goal in adversarial training [15]. The results in Table 2 report the accuracies achieved by both discriminators for four genre pairs, demonstrating the effectiveness of the proposed model.

Table 2. Results

Genre Pairs ($A - B$)	Acc (D_A)	Acc (D_B)
Classical - Hip Hop	88%	75%
Jazz - Rock	83%	82%
Hip Hop - Rock	76%	78%
Classical - Jazz	**63%**	**67%**

The above-demonstrated results show that the proposed model was only able to perform comparatively better on the less contrasting genres like *"Hip-Hop & Rock"* and *"Classical & Jazz"*. The proposed model was able to achieve the least discriminator accuracy of 63% in the case of *"Jazz → Classical"* genre style transfer.

In total, the proposed model was trained on 100 instances of each of the four genres (comparatively a small dataset). At the same, both VAEs & GANs are data-intensive. VAEs require more data to properly learn the probability distribution of the latent space. Despite these issues, the proposed model was able to generate sound style transferred reconstructions of the same song in a different genre.

In the proposed model, both the VAEs and the discriminators work together to create better continuous distribution of the latent space. The VAEs try to create a one-to-many mapping, as shown in Fig. 2, which is not the case with GANs. This approach enables the transformation of a single song from genre A into different variants of the same song in genre B, capturing the inherent stylistic mappings between the two genres.

5 Discussion

The results of the proposed VarCGAN model were evaluated using different types of assessment, the details of which are mentioned below.

- **Quantitative Assessment:** The Stability and performance of the proposed model were analyzed using Nash Equilibrium [20]. Section 5.1.1.

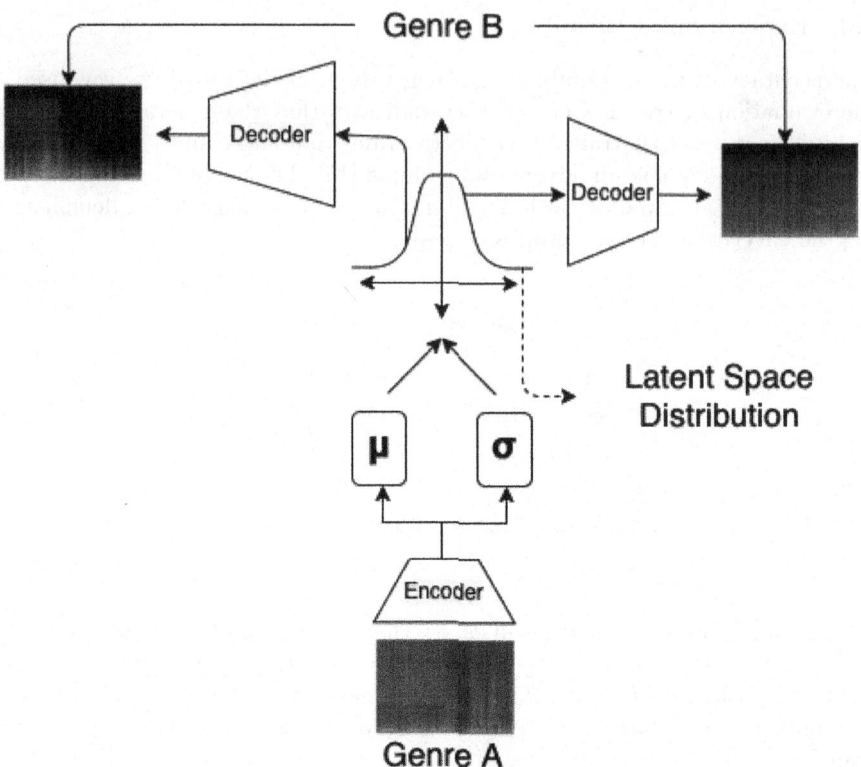

Fig. 2. The figure illustrates model's ability to generate different variants of genre style transferred songs

- **Qualitative Assessment:** Spectrogram visualizations [9] and human evaluations assessed fidelity, coherence, and originality of the proposed model. Section 5.2.2.
- **Genre Analysis:** The model's alignment was analyzed with known genre-specific features from the Music Genome Project [22]. Section 5.3.3.

These evaluations collectively demonstrate VarCGAN's effectiveness in achieving realistic and stylistically diverse music genre transfers.

This multi-faceted evaluation demonstrates VarCGAN's effectiveness in generating realistic, stylistically diverse, and musically coherent genre transfers.

Experimental results validate the model's ability to capture intricate musical nuances across different genre boundaries. The proposed approach represents a significant advancement in AI-driven music style transformation, bridging computational creativity with artistic expression.

5.1 Nash Equilibrium as an Evaluation Metric

We have redefined the evaluation score using Nash equilibrium [20] to capture the stability and quality of genre transfers. Each genre pair's payoff matrix is constructed, with payoffs representing the discriminators' accuracy (DA for Genre A's discriminator and DB for Genre B's discriminator). The Nash equilibrium highlights stable outcomes where neither the generator nor the discriminator can unilaterally improve performance [5]. This approach ensures that the model evaluation accounts for both the stability of adversarial training and the quality of genre-transferred outputs.

Payoff Table: The table quantifies the rewards or penalties associated with different strategic choices, allowing us to identify Nash equilibria.

Table 3. Payoff Table for Genre Pairs

Genre A\Genre B	Hip Hop	Rock	Jazz
Classical	(88, 75)	N/A	(63, 67)
Hip Hop	N/A	(76, 78)	N/A
Jazz	N/A	(83, 82)	N/A

Nash Equilibrium Analysis:

- **Classical vs. Hip Hop:** (88, 75) is a Nash equilibrium as neither genre can improve without the other worsening.
- **Jazz vs. Rock:** (83, 82) exhibits similar stability.
- **Hip Hop vs. Rock:** (76, 78) is stable for the same reasons (Table 3).
- **Classical vs. Jazz:** (63, 67) demonstrates stability despite the disparity in payoffs, showing the challenge of preserving stylistic fidelity.

These equilibria indicate the model's relative performance across genre pairs and the balance between discriminators and generators.

5.2 Qualitative Assessment

Spectrogram Visualizations:
The following figure shows the comparison of Rock and Hip Hop spectrograms (Fig. 3):

We have provided spectrograms [9] for each genre pair's original, transferred, and reconstructed songs. Below are the key observations:

- **Classical to Hip Hop:** The rhythmic density and high-frequency content increased, consistent with Hip Hop's style.

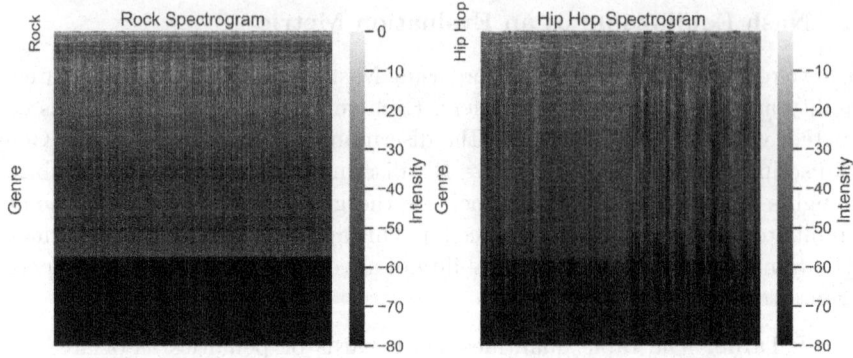

Fig. 3. Spectrogram Comparison between Rock and Hip Hop

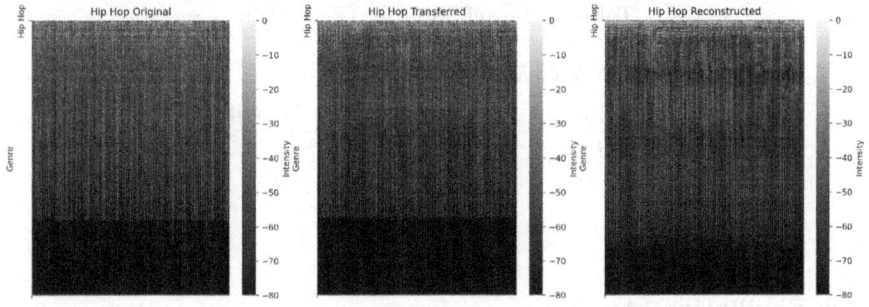

Fig. 4. Hip Hop Spectrogram Comparison

- **Jazz to Rock:** The harmonic structure preserved Jazz's complexity while introducing Rock's amplitude patterns (Fig. 4).

Human Evaluation:
We have conducted a survey with 20 listeners who were familiar with the genres. Participants rated the outputs on fidelity, coherence, and originality on a scale of 1 to 5. Table 4 tabulates the ratings given by the participants to each genre transfer.

Table 4. Human Evaluation Ratings

Metric	Classical-Hip Hop	Jazz-Rock	Hip Hop-Rock	Classical-Jazz
Fidelity	4.2	4.5	4.3	3.8
Coherence	4.0	4.4	4.1	3.7
Originality	4.1	4.6	4.2	3.9

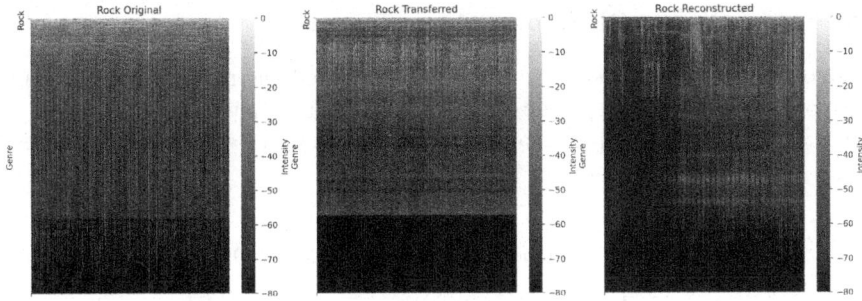

Fig. 5. Rock Spectrogram Comparison

Figure 6 shows a visual representation of the human evaluation ratings data from the table given above.

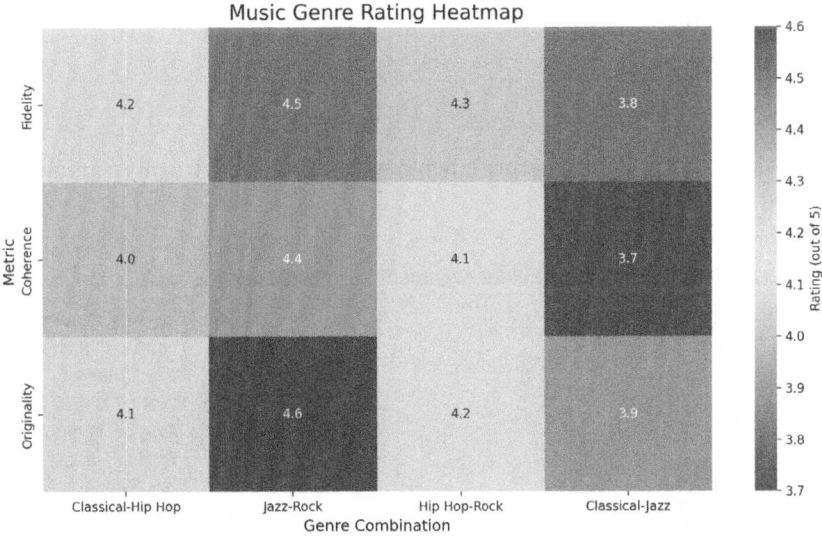

Fig. 6. Music Genre Rating Representation using Heat Map

Below are the key observations taken from the presented Table 4 and Fig. 6 of Human Evaluation Ratings:

- Genres with overlapping characteristics (e.g., Jazz-Rock) scored higher.
- Contrasting genres (e.g., Classical-Hip Hop) exhibited slight coherence issues but maintained originality (Fig. 5).

5.3 Comparison with Known Genre Features

Genre Feature Analysis: We have also evaluated the model's ability to preserve or introduce genre-specific characteristics using known attributes from the Music Genome Project [22]. Table 5 shows the actual vs obtained features of target genres with the obtained results.

Table 5. Feature Alignment of Target Genres and Obtained Values

Genre	Tempo (BPM)	Timbre Complexity	Chord Progression
Classical Actual	60–90	High (7–10)	Complex (8–10)
Classical Obtained	83	6	9
Hip Hop Actual	90–120	Medium (4–6)	Simple (2–4)
Hip Hop Obtained	108	3	4
Jazz Actual	70–100	High (7–10)	Complex (8–10)
Jazz Obtained	89	7	9
Rock Actual	100–140	Medium (4–6)	Moderate (5–7)
Rock Obtained	115	4	6

The Fig. 7 shows the Feature alignment of transferred genre samples with standard known attributes:

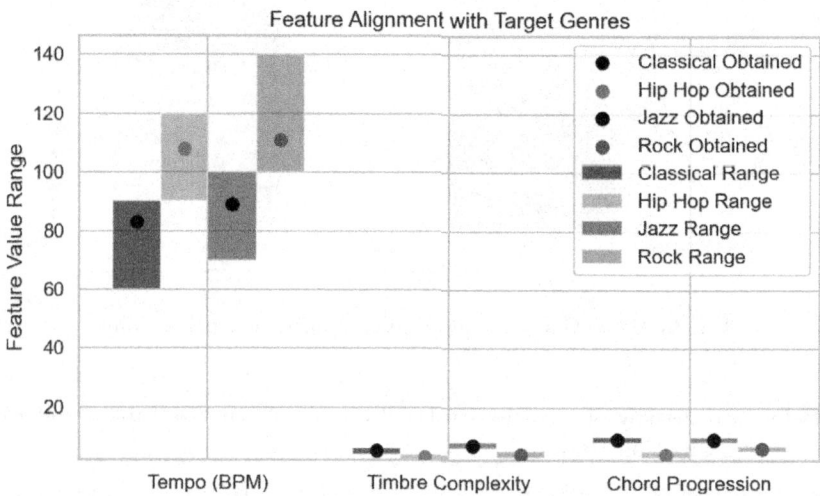

Fig. 7. Feature Alignment Chart

Key features analyzed include:

- **Tempo:** Hip Hop transfers exhibited increased tempo, aligning with its energetic style.
- **Instrumentation:** Classical to Jazz transfers preserved orchestral elements while introducing improvisational phrasing.
- **Rhythmic Complexity:** Jazz to Rock maintained syncopation while adapting to Rock's steady beats.

Insights gained from the figure above include the following:

- The model successfully aligned spectrogram features with target genres for Jazz-Rock and Hip Hop-Rock.
- Slight deviations occurred in Classical-Hip Hop transfers, indicating room for improvement.

6 Conclusion

This paper proposed VarCGAN, a novel hybrid generative model that combines the strengths of Variational Autoencoders (VAEs) and Generative Adversarial Networks (GANs) to achieve effective music genre style transfer. By integrating cyclic adversarial loss, VarCGAN addresses key challenges such as blurry reconstructions inherent in VAEs and mode collapse common in GANs. The model not only produces realistic genre-transferred outputs but also generates diverse variations of the same song, capturing subtle stylistic nuances of the target genre.

VarCGAN was evaluated on the GTZAN dataset across multiple genre pairs, including classical, jazz, hip-hop, and rock. Results demonstrated that the model performed particularly well on genres with overlapping stylistic characteristics, such as jazz-rock and hip-hop-rock, achieving discriminator accuracies of 82% and 78%, respectively. For more contrasting genres like classical-hip-hop, the model maintained stable performance, achieving a lowest discriminator accuracy of 63%, indicating meaningful and plausible style transfer even in challenging cases. Additionally, human evaluations rated the outputs highly on fidelity, coherence, and originality, with average scores above 4.0 across most genre pairs. The qualitative analysis of spectrograms further confirmed that VarCGAN successfully preserved key musical attributes of the original genre while introducing distinct stylistic elements of the target genre.

Despite these achievements, the model's performance was constrained by the relatively small size of the dataset, as both VAEs and GANs are data-intensive. Expanding the dataset and incorporating more diverse musical genres could further enhance the model's generalizability. Future work could also explore incorporating diffusion-based generative models [11] and disentangled latent spaces [10] to further refine the stylistic mappings and improve the quality of genre transfer.

Future Work: An important direction for future work is to explore the applicability of VarCGAN to sub-genres, such as jazz-rock or progressive metal, which inherently blend elements from multiple parent genres. These sub-genres present a more complex and nuanced stylistic landscape, requiring the model to disentangle and recombine fine-grained musical features. Evaluating VarCGAN in this context would provide insight into its ability to handle intricate genre boundaries and adapt to hybrid musical forms.

In summary, VarCGAN offers a promising approach to music genre style transfer, demonstrating the potential of hybrid generative models in achieving high-quality and diverse outputs. This framework has significant implications for music composition, production, and creative exploration, paving the way for future advancements in generative music systems.

Disclosure of Interests. The authors have no competing interests to declare that are relevant to the content of this article.

References

1. Brunner, G., Wang, Y., Wattenhofer, R.: Symbolic music genre transfer with cycle-GAN. In: Proceedings of the International Society for Music Information Retrieval Conference (ISMIR) (2018)
2. Brunner, G., Konrad, A., Wang, Y., Wattenhofer, R.: MIDI-VAE: modeling dynamics and instrumentation of music with applications to style transfer (2018)
3. Chen, X., Xu, C., Yang, X., Song, L., Tao, D.: Gated-GAN: adversarial gated networks for multi-collection style transfer. IEEE Trans. Image Process. **28**(2), 546–560 (2019)
4. Dai, S., Zhang, Z., Xia, G.: Music style transfer issues: a position paper. CoRR **abs/1803.06841** (2018). http://arxiv.org/abs/1803.06841
5. Daskalakis, C., Goldberg, P.W., Papadimitriou, C.H.: The complexity of computing a nash equilibrium. SIAM J. Comput. **39**(1), 195–259 (2009)
6. Gatys, L.A., Ecker, A.S., Bethge, M.: Image style transfer using convolutional neural networks. In: Proceedings of the IEEE Conference on Computer Vision and Pattern Recognition (CVPR) (2016)
7. Goodfellow, I., et al.: Generative adversarial nets. In: Ghahramani, Z., Welling, M., Cortes, C., Lawrence, N.D., Weinberger, K.Q. (eds.) Advances in Neural Information Processing Systems 27, pp. 2672–2680. Curran Associates, Inc. (2014). https://papers.nips.cc/paper/5423-generative-adversarial-nets.pdf
8. Goodfellow, I., et al.: Generative adversarial nets. In: Advances in Neural Information Processing Systems, pp. 2672–2680 (2014)
9. Griffin, D.W., Lim, J.S.: Signal estimation from modified short-time Fourier transform. In: ICASSP 1984. IEEE International Conference on Acoustics, Speech, and Signal Processing, pp. 236–239. IEEE (1984)
10. Higgins, I., et al.: beta-VAE: learning basic visual concepts with a constrained variational framework. In: International Conference on Learning Representations (2017)
11. Ho, J., Jain, A., Abbeel, P.: Denoising diffusion probabilistic models. arXiv preprint arXiv:2006.11239 (2020)
12. Huang, C.Z.A., Cooijmans, T., Roberts, A., Courville, A., Eck, D.: TimbreTron: a wavenet (cycleGAN (CQT (audio))) pipeline for musical timbre transfer. arXiv preprint arXiv:1811.09620 (2018)

13. Huang, W., Zhang, L., Yu, Z.: A diffusion model for music style transfer with multi-to-multi conversions. arXiv preprint arXiv:2301.01234 (2024)
14. Hung, H., Yang, Y.H.: Musical composition style transfer via disentangled timbre representations. In: Proceedings of the International Society for Music Information Retrieval Conference (ISMIR) (2019)
15. Isola, P., Zhu, J.Y., Zhou, T., Efros, A.A.: Image-to-image translation with conditional adversarial networks. In: Proceedings of the IEEE Conference on Computer Vision and Pattern Recognition, pp. 1125–1134 (2017)
16. Kingma, D.P., Welling, M.: Auto-encoding variational bayes. arXiv preprint arXiv:1312.6114 (2013)
17. Kingma, D.P., Welling, M.: Auto-encoding variational bayes. arXiv preprint arXiv:1312.6114 (2014)
18. Lu, W.T., Su, L., et al.: Transferring the style of homophonic music using recurrent neural networks and autoregressive model. In: ISMIR, pp. 740–746 (2018)
19. Maas, A.L., Hannun, A.Y., Ng, A.Y.: Rectifier nonlinearities improve neural network acoustic models. In: Proceedings of the ICML, vol. 30, p. 3 (2013)
20. Nash, J.: Equilibrium points in n-person games. Proc. Natl. Acad. Sci. **36**(1), 48–49 (1950)
21. van den Oord, A., et al.: WaveNet: a generative model for raw audio. arXiv preprint arXiv:1609.03499 (2016)
22. Prockup, M., Ehmann, A.F., Gouyon, F., Schmidt, E.M., Òscar Celma, Kim, Y.E.: Modeling genre with the music genome project: comparing human-labeled attributes and audio features, pp. 31–37 (2015)
23. Tzanetakis, G., Cook, P.: Musical genre classification of audio signals. IEEE Trans. Speech Audio Process. **10**(5), 293–302 (2002)
24. Yang, S., Wang, Z., Wang, Z., Xu, N., Liu, J., Guo, Z.: Controllable artistic text style transfer via shape-matching GAN. In: Proceedings of the IEEE International Conference on Computer Vision, pp. 4442–4451 (2019)
25. Zhu, J.Y., Park, T., Isola, P., Efros, A.A.: Unpaired image-to-image translation using cycle-consistent adversarial networks. In: 2017 IEEE International Conference on Computer Vision (ICCV) (2017)
26. Zhu, J.Y., Park, T., Isola, P., Efros, A.A.: Unpaired image-to-image translation using cycle-consistent adversarial networks. In: Proceedings of the IEEE International Conference on Computer Vision, pp. 2223–2232 (2017)

Innovative Framework for Early Estimation of Mental Disorder Scores to Enable Timely Interventions

Himanshi Singh[1]($\boxtimes$), Sadhana Tiwari[1], Ritesh Chandra[1], Sonali Agarwal[1], Sanjay Kumar Sonbhadra[2], and Vrijendra Singh[1]

[1] IIIT Allahabad, Prayagraj, Uttar Pradesh, India
{prf.himanshi,rsi2018507,rsi2022001,sonali}@iiita.ac.in
[2] Siksha 'O' Anusandhan University, Bhubaneswar, Odisha, India
vrij@iiita.ac.in

Abstract. Individuals' general well-being is greatly impacted by mental health conditions, including depression and Post-Traumatic Stress Disorder (PTSD), underscoring the importance of early detection and precise diagnosis to facilitate prompt clinical intervention. An advanced multimodal deep learning system for the automated classification of PTSD and depression are presented in this paper. Utilizing textual and audio data from clinical interview datasets, the method combines features taken from both modalities by combining the architectures of Long Short-Term Memory (LSTM) and Bidirectional Long Short-Term Memory (BiLSTM). Although text features focus on speech's semantic and grammatical components, audio features capture vocal traits including rhythm, tone, and pitch. This combination of modalities enhances the model's capacity to identify minute patterns connected to mental health conditions. Using test datasets, the proposed method achieves classification accuracies of 92% for depression and 93% for PTSD, outperforming traditional unimodal approaches and demonstrating its accuracy and robustness.

Keywords: Mental Health Diagnosis · PTSD · BiLSTM · LSTM

1 Introduction

PTSD and depression are prevalent mental health issues that have a substantial effect on people and put a burden on healthcare systems, particularly in underprivileged areas. Because of their limited accessibility and subjective character, traditional diagnostic instruments frequently fall short. New developments in deep learning (DL) and machine learning (ML) technology have shown great promise in addressing these issues. By evaluating many data modalities, including text, audio, and even visual clues, these automated systems offer scalable, and effective methods of recognizing mental health disorders. This study proposes a new deep learning method for the early diagnosis of PTSD and depression using the DAIC-WOZ dataset, which contains text, audio and expert-labeled data.

The structure of the paper is as follows: Sect. 2 reviews related work, Sect. 3 explains the proposed method, Sect. 4 presents the results, and Sect. 5 discusses future directions.

2 Literature Review

Recent developments in affective computing have advanced multimodal approaches for depression detection. Jo et al. [1] improved accuracy by combining text and audio features using CNN and Bi-LSTM. Jung et al. [2] introduced HiQuE, reflecting clinical interview structure, while Ding et al. [3] proposed IntervoxNet using AMST and BERT-CNN for text-audio fusion. Wang et al. [4] enhanced speech emotion recognition with 3D CNN and BiGRU. Models like AVTF-TBN [5] and RoBERTa-BiLSTM [7] further boosted performance using multimodal data. Mai et al. [6] presented AIGC-Brain, generating brain-guided multimodal inputs. These studies highlight the effectiveness of deep learning in mental health assessments.

To adequately capture the intricacies of mental health, existing techniques frequently rely on single data modalities. They struggle to generalize across different populations because of imbalanced datasets and poor temporal emotion tracking. Physiological cues remain poorly understood, and data privacy receives insufficient priority. Furthermore, the lack of real-time systems and standardized tools restricts clinical integration.

3 Methodology

This section describes how to deploy deep learning models, extract significant features, and integrate multimodal data to accurately classify mental health disorders.

3.1 Process Workflow

Figure 1 shows how the depression detection algorithm collects text and audio input.

After processing, the model extracts key properties and analyzes them to detect depressive symptoms. The model may identify emotional patterns and reveal mental health by using an organized technique.

3.2 Model for Detecting Depression

The model for diagnosing PTSD and depression uses a dual-branch configuration to handle audio and text inputs independently. Audio features (1,193) are processed by an LSTM with dropout and dense layers, while text features (1,768),

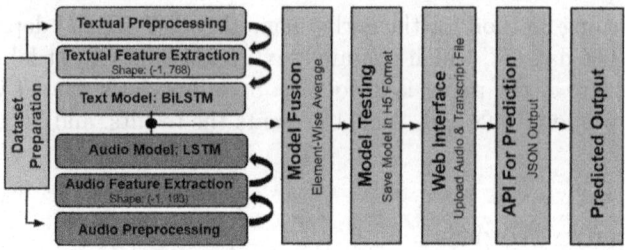

Fig. 1. Multimodal PTSD and Depression Detection Pipeline

are processed by a bidirectional LSTM, dense layers, and dropout. To improve prediction accuracy, the combined outputs are averaged and then run through a sigmoid-activated layer for binary classification, which successfully combines linguistic and acoustic information.

Long Short-Term Memory and Bidirectional LSTM Networks. By employing gated cells to regulate information flow, LSTMs are made to handle sequential input and get around problems like vanishing gradients that regular RNNs have. Applications like time-series analysis and speech recognition have shown them to be effective. Their performance is improved by BiLSTMs, which process sequences both forward and backward, allowing for the simultaneous collection of past and future context. They are therefore ideal for natural language tasks that call for thorough sequence comprehension.

PTSD Detection Model Architecture and Training Procedure. Using multimodal fusion of text and audio inputs processed by 64-unit LSTM and BiLSTM layers, the PTSD detection model is designed similarly to the depression model. Following input normalization and scaling, a sigmoid output layer is used to classify the features and merge them by element-wise averaging. TensorFlow and Keras are used to implement the Adam optimizer, a batch size of eight, binary cross-entropy loss, a learning rate of 0.001, ten training epochs, and an 80/20 split with a dropout rate of 0.3 to minimize overfitting. The features used from text and audio data to predict PTSD and depression are shown in Table 1, along with pertinent classification-related attributes.

To support mental health research, the DAIC-WOZ dataset [8] provides synchronized multimodal data from human-controlled virtual interviews. It includes high-quality 16 kHz audio recordings and time-aligned text transcripts, enabling in-depth examination of speech and language characteristics. This structured dataset is perfect for creating machine learning models that are centered on psychological assessment.

The Algorithm 1 provides the sequential steps of the novel framework for estimating the scores of early mental disorders and the architecture of the system, outlines the multimodal technique of combining text and auditory information to predict PTSD and depression, is depicted in Fig. 2.

Table 1. Extracted Features from Audio and Text Data

Feature	Shape	Description
Audio Features (193)		
MFCCs	(13,)	Captures main spectral features
Delta MFCCs	(13,)	Tracks MFCC variations
Delta2 MFCCs	(13,)	Measures MFCC acceleration
Chroma	(12,)	Energy across 12 pitch classes
Mel Spectrogram	(128,)	Frequency intensity on a perceptual scale
Contrast	(7,)	Highlights spectral peaks/valleys
Tonnetz	(6,)	Encodes tonal and harmonic info
Pitch	(1,)	Dominant frequency measure
Text Features (768)		
BERT Embedding	(768,)	Contextual word representations

Algorithm 1. Early Estimation of Mental Disorder Scores

1: **Load Data:** Import text/audio features and binary labels.
2: **Preprocess:** Reshape features for LSTM input; split into train/test sets (80/20).
3: **Model Architecture:**
4:   **Text Branch:** TextInput → BiLSTM → Dropout → Dense
5:   **Audio Branch:** AudioInput → LSTM → Dropout → Dense
6:   **Fusion:** Element-wise average of both outputs.
7: **Compile:** Use Adam optimizer (lr=0.001) and binary cross-entropy loss.
8: **Train:** Batch size = 8, epochs = 8, validation split = 0.2.
9: **Evaluate:** Assess model on test set (accuracy, loss).
10: **Predict:** Generate scores for new inputs.
11: **if** score > threshold **then**
12:   Trigger Early Intervention
13: **else**
14:   Schedule Regular Monitoring
15: **end if**

Feature Extraction. Gathering audio and related text data is the first step. Then features will be extracted from spectral descriptors and MFCCs (mel-frequency cepstral coefficients) after segmenting and removing silence from the audio. BERT and other models create embeddings, while lemmatization, tokenization, and stopword removal clean the text. Models for identifying depression and PTSD are then trained using these data from both modalities.

The feature extraction pipeline ensures scalability by leveraging Spark for parallel processing and HDFS for distributed storage. It consists of two main components: audio and text processing. Librosa extracts audio features such as pitch, chroma, and MFCCs, while BERT generates text embeddings after tokenization and lemmatization. Interim results are managed in local storage before final data is stored in HDFS for further analysis or modeling.

The integration approach combines semantic text embeddings with audio features such as MFCCs, chroma, pitch, Mel spectrogram, and Tonnetz to form structured feature sets for each audio segment. This fusion captures both acoustic and semantic cues, enhancing the detection of mental health indicators. The

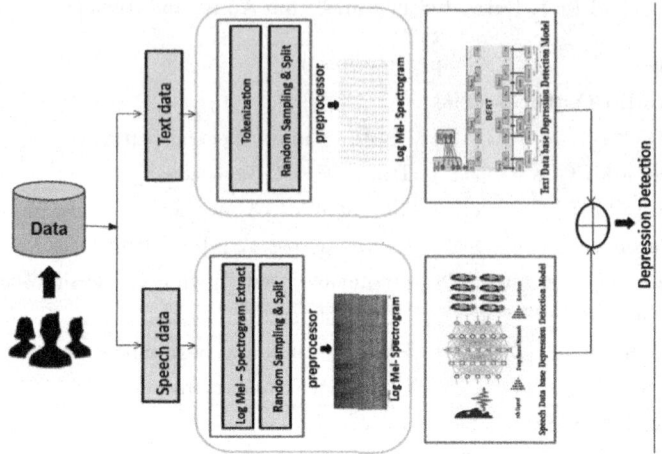

Fig. 2. Multimodal Architecture for Mental Disorder

system operates via two scripts: `process-dataset.scala` handles file downloads, feature extraction, and data uploads; `extractfeatures.py` uses BERT for text embeddings, Librosa for audio analysis, and Pandas for data handling. The output is a set of CSV files containing features for downstream classification.

4 Results and Discussion

Performance variations between classes are highlighted in Table 2, which summarizes class-wise precision, recall, F1-score, and support for PTSD and depression diagnoses. While the overall evaluation measures, such as accuracy, and the weighted and macro averages of precision, recall, and F1-score for both mental health problems are shown in Table 3.

Table 2. Classification Metrics for PTSD and Depression Detection

Class	Condition	Precision	Recall	F1-Score	Support
0	PTSD	0.93	0.96	0.95	344
	Depression	1.00	0.84	0.91	371
1	PTSD	0.94	0.90	0.92	239
	Depression	0.78	1.00	0.88	212

The performance of different models is summarized in Table 4, where missing metrics are indicated as (-).

A web application built with Flask is created to predict PTSD and depression by utilizing multimodal inputs, such as text and audio files. The system combines text and audio features, using pre-trained models (depressed.h5, ptsd.h5)

Table 3. Overall Classification Metrics for PTSD and Depression Detection

Condition	Precision	Recall	F1-Score	Support	Test Accuracy
PTSD Accuracy	–	–	0.93	583	0.93
PTSD Macro Avg	0.94	0.93	0.93	583	–
PTSD Weighted Avg	0.93	0.93	0.93	583	–
Depression Accuracy	–	–	0.90	583	0.92
Depression Macro Avg	0.89	0.92	0.90	583	–
Depression Weighted Avg	0.92	0.90	0.90	583	–

Table 4. Comparison of Models for Classification

S.No.	Model	Acc.	Prec.	Rec.	F1
1	SVM & BiLSTM [7]	–	0.74	0.66	0.69
2	CNN, SVM, KNN, RF, LR [9]	0.77	0.78	0.98	0.87
3	3D-CBHGA, RF [4]	0.77	0.63	–	0.64
4	AMST + BERT CNN [3]	0.86	0.88	0.92	0.90
5	LSTM, BiLSTM (Proposed)	–	–	–	0.92

for reliable predictions. Librosa processes the audio, while a BERT tokenizer processes the text. JSON answers are used to communicate the results.

5 Conclusion and Future Work

This study leverages BiLSTM and LSTM architectures to integrate temporal features from text and audio, demonstrating the effectiveness of multimodal deep learning in detecting mental health conditions such as depression and PTSD. The model achieves high classification accuracy by combining semantic and prosodic cues. Future work will incorporate additional modalities–gestures, facial expressions, and physiological signals–to enhance predictive performance and diagnostic depth. Expanding cross-lingual support and integration with healthcare systems will improve real-world usability. Addressing challenges such as data imbalance, diagnostic complexity, and privacy concerns is essential for ensuring fairness, reliability, and user trust.

Acknowledgment. This research was supported by the Council of Science and Technology, Uttar Pradesh (CSTUP), under Sanction No. CST/D-71, Project ID 3965. The authors thank CSTUP for the funding and IIIT Allahabad, Prayagraj, for providing infrastructure and institutional support.

References

1. Jo, A.-H., Kwak, K.-C.: Diagnosis of depression based on four- stream model of bi-LSTM and CNN from audio and text information. IEEE Access **10**, 134 113–134 135 (2022). https://doi.org/10.1109/ACCESS.2022.3231884
2. Jung, J., Kang, C., Yoon, J., Kim, S., Han, J.: HiQuE: hierarchical question embedding network for multimodal depression detection. In: Proceedings of the 33rd ACM International Conference on Information and Knowledge Management, pp. 1049–1059 (2024). https://doi.org/10.1145/3627673.3679797
3. Ding, H., et al.: IntervoxNet: a novel dual-modal audio-text fusion network for automatic and efficient depression detection from interviews. Front. Phys. **12**, 1430035 (2024). https://doi.org/10.3389/fphy.2024.1430035
4. Wang, H., Liu, Y., Zhen, X., Tu, X.: Depression speech recognition with a three-dimensional convolutional network. Front. Hum. Neurosci. **15**, 713823 (2021). https://doi.org/10.3389/fnhum.2021.713823
5. Zhang, Z., et al.: Multimodal sensing for depression risk detection: integrating audio, video, and text data. Sensors **24**(12), 3714 (2024). https://doi.org/10.3390/s24123714
6. Mai, W., Zhang, J., Fang, P., Zhang, Z.: Brain-conditional multimodal synthesis: a survey and taxonomy. IEEE Trans. Artif. Intelli. (2024). https://doi.org/10.1109/TAI.2024.3516698
7. Zhang, Y., He, Y., Rong, L., Ding, Y.: A hybrid model for depression detection with transformer and bi-directional long short-term memory. In: 2022 IEEE International Conference on Bioinformatics and Biomedicine (BIBM), pp. 2727–2734. IEEE (2022). https://doi.org/10.1109/BIBM55620.2022.9995184
8. E-daic: EEG dataset for AI classification (2024). Accessed 10 Jan 2025. https://dcapswoz.ict.usc.edu/wwwedaic/
9. Kanoujia, S., Karuppanan, P.: Depression detection in speech using ML and DL algorithm. In: 2024 IEEE International Conference on Interdisciplinary Approaches in Technology and Management for Social Innovation (IATMSI), vol. 2, pp. 1–5. IEEE (2024). https://doi.org/10.1109/IATMSI60426.2024.10503510

A Hybrid Approach to Estimating AI Carbon Emissions

Salvatore Borraccia, Elio Masciari, and Enea Vincenzo Napolitano[✉]

DIETI, University of Naples Federico II, Naples, Italy
{elio.masciari,eneavincenzo.napolitano}@unina.it

Abstract. Measuring the environmental impact of computational processes is a crucial step towards developing more sustainable digital technologies. While artificial intelligence (AI) is playing a transformative role in various fields, its energy-intensive nature gives way to concerns about carbon emissions. In this paper, we present a hybrid approach that combines a weighted average framework and a dynamic model matching methodology to improve the accuracy and adaptability of carbon footprint estimates. The weighted average method integrates multiple carbon tracking tools and assigns weights based on reliability, accuracy and applicability to provide a unified emissions estimate. Meanwhile, the dynamic method categorizes AI models based on their computational characteristics and assigns them to the most appropriate emissions estimation framework. Future research will focus on refining weighting strategies, integrating real-time energy consumption data, and embedding sustainability considerations into AI development workflows. Our proposed methodology contributes to a more standardized and comprehensive assessment of the environmental impact of AI, encouraging responsible AI innovation.

Keywords: AI Carbon Footprint Estimation · Green AI · Carbon Emission Tracking

1 Introduction

Climate change is one of the most pressing global challenges, with greenhouse gas (GHG) emissions contributing significantly to rising global temperatures. Among the many sources of emissions, Information and Communication Technology (ICT) plays a crucial role. As digital infrastructure expands, the energy consumption of data centers, cloud computing, and artificial intelligence (AI) systems has become a growing concern [2]. The increasing computational demands of AI models, particularly in deep learning and large-scale data processing, lead to significant carbon footprints. Understanding and mitigating the environmental impact of these technologies is essential for achieving sustainability goals [13]. AI models require substantial computational resources, often utilizing high-performance GPUs and TPUs in data centers powered by energy sources with varying carbon intensities. The training and deployment of AI systems contribute to CO_2 emissions, making it necessary to assess their environmental impact systematically. Carbon footprint estimation tools have been developed to quantify emissions from AI workloads, enabling researchers and practitioners to make informed decisions about optimizing model efficiency while minimizing energy consumption and emissions. However, current methods face several limitations that hinder accurate and comprehensive impact assessment.

While various tools, such as CodeCarbon, MLCO2, Experiment-Impact-Tracker (EIT), CarbonTracker, and Green Algorithms, have been proposed to estimate emissions from AI workloads, they present several challenges. Many tools are designed for specific AI tasks, such as deep learning or cloud-based training, making them less adaptable to a broader range of applications. Some frameworks rely on fixed conversion factors and energy intensity values that do not account for dynamic grid carbon intensity variations. Additionally, estimations often depend on vendor-specific APIs, limiting applicability across different hardware architectures. There is also no standardized methodology for comparing emissions across different AI models, leading to inconsistencies in reported values.

To address these limitations, this study introduces two novel methodologies for estimating AI-related carbon emissions. The first is a weighted average framework that integrates multiple carbon tracking tools, assigning weights based on their reliability, precision, and applicability to various AI models. The second is a dynamic matching approach that categorizes AI algorithms based on computational requirements and matches them with the most appropriate carbon estimation frameworks. These approaches aim to improve the accuracy, adaptability, and comprehensiveness of AI carbon footprint estimations, providing a more holistic perspective on the environmental impact of AI technologies.

2 Related Work and Background

2.1 Overview of Carbon Emission Estimation Methods

The expansion of AI applications has led to significant environmental sustainability challenges [15]. As AI models continue to grow in complexity and computational demands, the need for accurate and accessible carbon footprint estimation tools has become more pressing. Various methodologies have been proposed to assess and mitigate the environmental impact of AI workloads, leveraging both theoretical models and real-time energy monitoring.

Existing estimation approaches can be broadly categorized into three main types:

- **Empirical Measurement-Based Methods:** These methods rely on direct hardware monitoring tools, such as power meters, Running Average Power Limit (RAPL) interfaces [5], and GPU telemetry, to track real-time energy consumption [9]. While highly accurate, their applicability is often limited to controlled environments with access to specialized hardware.

- **Model-Based Estimations:** This category includes analytical models that estimate energy usage based on system specifications, execution time, and historical data on power consumption.
- **Cloud and Grid-Aware Approaches:** With the increasing reliance on cloud computing for AI training, several tools incorporate geolocation-based carbon intensity metrics to estimate the carbon footprint based on the energy sources powering cloud data centers [12]. These methods emphasize regional differences in energy generation and their impact on emissions.

Several established tools have been widely recognized in the literature, including *CarbonTracker*, *MLCO2*, *Experiment-Impact-Tracker (EIT)*, *Cumulator*, *ECO2AI*, *CodeCarbon*, and *Green Algorithms*. Additionally, newer solutions such as *OpenCarbonEval* and *LLMCarbon* have emerged to address specific computational workloads and provide advanced tracking capabilities.

2.2 Existing Tools for Carbon Emission Estimation

CodeCarbon. *CodeCarbon* [10] is a widely adopted Python-based library designed to track the energy consumption of ML training phases. It estimates carbon emissions using the Running Average Power Limit (RAPL) interface for Intel processors and integrates geolocation-based carbon intensity metrics. The tool relies on public databases such as eGRID [11] to determine energy sources' environmental impact and offers detailed energy consumption breakdowns. CodeCarbon is particularly useful for cloud-based ML workloads, as it supports integration with major cloud providers to estimate data center-specific emissions.

MLCO2. *MLCO2* [7] provides estimates of carbon footprint for ML models using metadata from the cloud infrastructure and energy consumption rates. It enables researchers to quantify emissions based on hardware configurations, runtime, and geographical location. This tool is instrumental in highlighting the role of sustainable cloud computing practices by allowing users to compare emissions between different service providers and execution environments.

Experiment-Impact-Tracker (EIT). *EIT* [6] is a flexible monitoring tool that records AI model energy usage and emissions throughout the ML training process. It integrates with various power measurement interfaces, including Intel RAPL and NVIDIA-SMI, to capture CPU and GPU consumption. Additionally, EIT provides interpretability features, making results accessible to non-expert audiences. Its modularity allows for extension to various ML frameworks, enabling broad applicability.

CarbonTracker. *CarbonTracker* [1] is specifically designed for deep learning models, estimating energy usage and carbon emissions during model training. The framework accounts for hardware characteristics and employs a proactive forecasting approach to minimize computational inefficiencies. By estimating future energy demands, researchers can make informed decisions about computational budget allocation.

Cumulator. *Cumulator* [14] is an open-source API for measuring the carbon footprint of ML models. It integrates seamlessly with research environments and supports multiple computational settings. The tool is widely adopted in AI applications within scientific and medical research domains, where precise carbon accounting is necessary for sustainability assessments.

Eco2AI. *Eco2AI* [3] is another Python-based tool for tracking AI-related energy consumption. It continuously logs hardware resource usage and calculates emissions for GPUs, CPUs, and RAM. The framework supports real-time monitoring and regional energy intensity adjustments, providing accurate carbon footprint estimates. Unlike other tools, Eco2AI also incorporates historical data for trend analysis, helping organizations track the effectiveness of sustainability initiatives over time.

Green Algorithms. *Green Algorithms* [8] is a general-purpose carbon emission estimator applicable across various computational workloads. It incorporates real-time electricity grid data and allows comparisons between different execution environments. The framework includes a pragmatic scaling factor (PSF) to normalize emissions across diverse AI tasks, making it particularly useful for cross-disciplinary AI applications.

OpenCarbonEval and LLMCarbon. *OpenCarbonEval* [16] and *LLMCarbon* [4] are recent tools designed for large-scale AI models, particularly in natural language processing (NLP) and large language models (LLMs). These frameworks offer specialized support for cloud-based computations and extensive dataset processing, making them highly relevant for tracking emissions from foundation models. Given the increasing size and computational cost of LLMs, these tools address the urgent need for transparent and standardized emissions reporting in large-scale AI applications.

The discussed tools highlight various strategies for measuring AI-related emissions. Several factors influence the accuracy of these estimates, including: Hardware Configuration, Computational Infrastructure, Model Complexity, Geographical Location, Energy Efficiency Strategies.

This section provides a foundational understanding of existing carbon tracking frameworks, serving as the basis for subsequent evaluations and methodology development in this study.

3 Weighted Average of Existing Tools

The first novel method introduced in this work involves determining the weighted average of the ten methodologies previously discussed. The assignment of weights to each framework depends on multiple factors, which are summarized as follows:

1. **Authority:** The scientific and academic recognition of the proposed tracker, which determines its relevance and weight compared to others.
2. **Precision:** The accuracy of the results provided, ensuring reliable estimations.
3. **Completeness:** The ability of the methodology to estimate emissions across various AI applications rather than focusing on specific cases.
4. **Real-time Data Availability:** The presence of real-time information, which enables continuous updates and adaptability to changing conditions.
5. **Ease of Implementation:** The existence of a dedicated open-source Python library that facilitates implementation and accessibility.
6. **Result Accessibility:** Transparency and public availability of results on relevant platforms or websites.

3.1 Weighted Average Computation

The first step of the framework in constructing the new measure involves normalizing all resulting values, as they may use different units of measurement. However, all tools considered already report emissions in gCO_2e, so no mathematical conversion is necessary. The calculation is carried out as follows:

1. Assign weights (ω_i) to each framework based on the six criteria defined above.
2. Compute the weighted average of carbon emissions using the following formula:

$$E_{weighted} = \frac{\sum_{i=1}^{n} \omega_i E_i}{\sum_{i=1}^{n} \omega_i} \qquad (1)$$

where E_i is the emission value (in gCO_2e or $kgCO_2e$) of the i-th framework.

The assignment of weights (ω_i) to each tool was conducted through a semi-structured expert evaluation process. Each criterion was scored on a scale from 0 to 5 for each framework. The final weight was obtained by normalizing the total score of each framework across all six criteria. This process ensured that tools recognized in the academic community, offering high precision and ease of integration, received greater influence in the weighted average. The assigned weights are presented in Table 1.

4 Dynamic Methodology Based on AI Model Type

The goal of this methodology is to construct a systematic process for recognizing specific algorithms or models within AI domains and assigning them to the

Table 1. Assigned Weights to Frameworks

Tracker	Authors	Weight (ω)
CodeCarbon	Lottick et al. (2019)	0.142
MLCO2	Lacoste et al. (2019)	0.128
Experiment Impact Tracker	Henderson et al. (2020)	0.096
CarbonTracker	Anthony et al. (2020)	0.131
Cumulator	Trebaol et al. (2020)	0.090
Eco2AI	Budennyy et al. (2022)	0.099
Green Algorithms	Lannelongue et al. (2020)	0.148
LLMCarbon	Faiz et al. (2023)	0.044
OpenCarbonEval	Yu et al. (2024)	0.041
Cloud Instances	Dodge et al. (2022)	0.081

corresponding framework among the ten considered in this study. The selected framework will be the most appropriate for measuring the environmental impact, particularly carbon emissions, associated with a given algorithm's computational workload. To ensure adaptability and dynamism, it is essential to define examples of algorithms that operate within the primary AI fields:

1. Machine Learning (ML)
2. Deep Learning (DL)
3. Natural Language Processing (NLP)
4. Large Language Models (LLMs)

All LLMs fall under the broader NLP category, meaning a framework optimized for large-scale language models is also suitable for standard NLP tasks. Similarly, Deep Learning is a subset of Machine Learning, and frameworks tailored for DL often support ML tasks as well.

The methodology consists of two primary phases:

1. **Phase 1:** Define matching principles that relate AI model characteristics to the appropriate measurement frameworks.
2. **Phase 2:** Apply the principles through practical examples to validate the methodology.

4.1 Matching Rules Between Frameworks and AI Models

To achieve effective matching, we developed a set of association rules using statistical matching techniques. The two primary groups considered are:

1. The ten carbon measurement frameworks: CodeCarbon, MLCO2, Experiment Impact Tracker (EIT), CarbonTracker, Cumulator, Eco2AI, Green Algorithms, LLMCarbon, OpenCarbonEval, and Cloud Instances.

2. AI models classified into ML, DL, NLP, and LLMs, considering infrastructure usage (e.g., cloud-based computation vs. local GPU/CPU usage).

The matching process involves evaluating the following factors:

- AI model type (ML, DL, NLP, LLM).
- Computational infrastructure: single/multiple GPUs, cloud-based execution, or local hardware deployment.
- CPU and RAM involvement in computations.
- Dataset type and size for training.
- Framework support for multi-GPU environments.
- Framework compatibility with different AI model architectures.

Table 2 summarizes the compatibility between AI domains and measurement frameworks.

Table 2. AI Model-Framework Matching

Framework	ML	DL	NLP	LLM	Cloud Support
CodeCarbon	Specific	Yes	Adaptable	Adaptable	Yes
MLCO2	Specific	Yes	No	No	Yes
EIT	Specific	Yes	Adaptable	Adaptable	No
CarbonTracker	Yes	Specific	Adaptable	Adaptable	No
Cumulator	Specific	Yes	No	No	No
Eco2AI	Yes	Specific	Yes	Specific	No
Green Algorithms	Yes	Yes	Yes	No	Yes
LLMCarbon	No	No	Yes	Specific	Yes
OpenCarbonEval	No	No	Adaptable	Specific	Yes
Cloud Instances	Specific	Yes	Specific	No	Specific

5 Conclusion and Future Work

This paper presents a hybrid approach to estimating AI-related carbon emissions, combining a weighted average framework with a dynamic model matching methodology. By integrating multiple carbon tracking tools and assigning weights based on their reliability, accuracy and applicability, we achieved a more comprehensive and standardized assessment of emissions. In particular, the dynamic methodology provides an adaptable framework for matching AI models with the most appropriate emissions estimation tools based on computational characteristics and infrastructure.

The weighted average approach enables a view of emissions by mitigating discrepancies between different tools, while the dynamic model matching strategy enhances usability by matching AI workload characteristics with appropriate estimation frameworks.

Future work should focus on refining the weighting process by incorporating machine learning techniques to dynamically adjust weights based on real-world usage patterns and empirical accuracy. In addition, expanding the dataset to include a wider range of estimation frameworks and real-time energy consumption data will improve the generalizability of our approach.

Another promising direction is to integrate our methods into existing AI development workflows to facilitate automated and real-time carbon tracking. This could include embedding the framework into popular machine learning libraries, cloud platforms, or CI/CD pipelines to provide continuous emissions monitoring and optimization recommendations.

In addition, exploring energy-efficient AI model design strategies such as model pruning, quantization, and knowledge distillation in conjunction with carbon tracking can help minimize emissions while maintaining computational performance. Assessing the environmental impact of emerging AI paradigms, such as federated learning and edge AI, will also be critical to promoting sustainable AI development.

Acknowledgement. This work has been supported by the project "SMIMI – Security in Modern Information Management Infrastructure - PE00000014 SERICS - Spoke 10".

References

1. Anthony, L.F.W., Kanding, B., Selvan, R.: Carbontracker: tracking and predicting the carbon footprint of training deep learning models. arXiv preprint arXiv:2007.03051 (2020)
2. Bolón-Canedo, V., Morán-Fernández, L., Cancela, B., Alonso-Betanzos, A.: A review of green artificial intelligence: towards a more sustainable future. Neurocomputing, 128096 (2024)
3. Budennyy, S.A., et al.: eco2AI: carbon emissions tracking of machine learning models as the first step towards sustainable AI. Doklady Math. **106**, S118–S128 (2022)
4. Faiz, A., et al.: LLMCarbon: modeling the end-to-end carbon footprint of large language models. arXiv preprint arXiv:2309.14393 (2023)
5. Garcia, J.A.: Exploration of energy consumption using the intel running average power limit interface. In: 2019 IEEE Space Computing Conference (SCC), pp. 1–10. IEEE (2019)
6. Henderson, P., Hu, J., Romoff, J., Brunskill, E., Jurafsky, D., Pineau, J.: Towards the systematic reporting of the energy and carbon footprints of machine learning. J. Mach. Learn. Res. **21**(248), 1–43 (2020)
7. Lacoste, A., Luccioni, A., Schmidt, V., Dandres, T.: Quantifying the carbon emissions of machine learning. arXiv preprint arXiv:1910.09700 (2019)
8. Lannelongue, L., Grealey, J., Inouye, M.: Green algorithms: quantifying the carbon footprint of computation. Adv. Sci. **8**(12), 2100707 (2021)

9. Latif, I., et al.: Empirical measurements of AI training power demand on a GPU-accelerated node. arXiv preprint arXiv:2412.08602 (2024)
10. Lottick, K., Susai, S., Friedler, S.A., Wilson, J.P.: Energy usage reports: environmental awareness as part of algorithmic accountability. arXiv preprint arXiv:1911.08354 (2019)
11. Meydani, A., Shahinzadeh, H., Ramezani, A., Moazzami, M., Nafisi, H., Askarian-Abyaneh, H.: Comprehensive review of artificial intelligence applications in smart grid operations. In: 2024 9th International Conference on Technology and Energy Management (ICTEM), pp. 1–13. IEEE (2024)
12. Ozawa, Y., et al.: Grid-aware energy management by data center workload control across multiple data centers. In: Proceedings of the 25th International Middleware Conference: Demos, Posters and Doctoral Symposium, pp. 13–14 (2024)
13. Regona, M., Yigitcanlar, T., Hon, C., Teo, M.: Artificial intelligence and sustainable development goals: systematic literature review of the construction industry. Sustain. Cities Soc., 105499 (2024)
14. Trébaol, T.: CUMULATOR–a tool to quantify and report the carbon footprint of machine learning computations and communication in academia and healthcare (2020)
15. Verdecchia, R., Sallou, J., Cruz, L.: A systematic review of green AI. Wiley Interdisc. Rev. Data Mining Knowl. Discovery **13**(4), e1507 (2023)
16. Yu, Z., Wu, Y., Deng, Z., Tang, Y., Zhang, X.P.: OpenCarbonEval: a unified carbon emission estimation framework in large-scale AI models. arXiv preprint arXiv:2405.12843 (2024)

A Data Product Classification by Technical and Machine Learning Aspects

Laura Schuiki[1](✉)[iD], Ulf Schreier[2], Holger Schwarz[1][iD], and Bernhard Mitschang[1][iD]

[1] IPVS/AS, University of Stuttgart, Stuttgart, Germany
{laura-sophie.schuiki,holger.schwarz,
bernhard.mitschang}@ipvs.uni-stuttgart.com
[2] Furtwangen University, Furtwangen im Schwarzwald, Germany
ulf.schreier@hs-furtwangen.de

Abstract. Similar to the software services methodology, data products (DP) can be seen as a kind of data services that specify all important issues for data provisioning and data consumption. DPs come in many different varieties stretching from simple data pipelines to complex machine learning models and model inferences and, above all, typically result in complex data networks. It is time to come up with a useful categorization and structuring of the DP topic in order to conquer complexity. In this paper, we present and assess a basic classification approach that focuses on DP characteristics and thus provides the basis for blueprinting and architectural discussions.

Keywords: Data Product · Classification Hierarchy · ML

1 Introduction

Large enterprises are on one hand confronted with distributed and heterogeneous IT systems and on the other hand with the need for federation and integration of individual systems and platforms, for sharing of relevant data. As data sharing has become increasingly important, new data architectures such as data mesh have been proposed to facilitate data sharing among multiple participants [9]. It was first coined by Dehghani [2,3] and has gained business and industrial relevance as well as popularity, inspiring further scientific research and other publications that address the associated challenges and that provide initial guidance on how to best utilize this promising concept [8].

As part of the data mesh approach the concept of data products (DPs) defines the cooperation link and all relevant data sharing issues between data consumers and data producers. The three main internal components of a DP are data, code and metadata: Data is collected from multiple source systems, e.g., applications that provide data as well as existing DPs. It is typically aligned, transformed, or analyzed, and the result as well as metadata is made available to other DPs as well as applications that further process and visualize it [6].

Practical experience and literature clearly indicate that DPs come in many different varieties and nuances that stretch different dimensions and characteristics.

In order to prevent an uncontrolled proliferation of DPs accompanied by ever-increasing complexity, it is necessary to identify the central characteristics of DPs, including machine learning (ML) models, and to derive typical DP classes. These classes serve as prototypes for DP design, are the basis for DP implementation guidelines and support the selection of and access to DPs by data consumers.

In this paper, we analyze the central characteristics of DPs, derive a classification of DPs and give characteristic examples for each class. The DP classification helps DP providers as well as DP consumers to identify the DP class that best suits their use case.

The remainder of this paper is structured as follows: Sect. 2 introduces a crisp classification for important types of DPs as well as characteristic DP examples. A comparison to related work and an assessment of our approach is given in Sect. 3. Section 4 concludes the paper.

2 Classification

Data products come in different varieties and have different characteristics. These characteristics influence how a DP is implemented and how it is made available to consumers. Hence, it is important for DP providers as well as for DP consumers to know different classes and to use them to provide or select a DP class that is appropriate for their use case.

As a first step towards a classification, we collected all characteristics of DPs that were mentioned in the considered literature and extended this list by considering further aspects. Finally, we evaluated the importance of all characteristics. To do so, we investigated if they describe behavior that consumers expect from a DP and how they influence the implementation of a DP. We also considered dependencies between them. This evaluation helped us to identify the minimum set of characteristics for a comprehensive classification of DPs.

Since we are especially interested in technical aspects, we kept characteristics that have an impact on the implementation of DPs. The type of transformations is relevant as the software stack needed to derive ML models is quite different from the one required for typical ETL-like transformations. In a similar way, the interface to a DP that only delivers data on request, is different from the interface to a DP that continuously delivers data by means of data streaming technology. We further focused on characteristics that allow consumers to discern whether or not a DP fits their use case. To this end we on one hand kept distinctions that are visible to and have an impact on consumers, for example, whether a DP automatically delivers updates or not. On the other hand, we removed factors that do not impact the consumers at all, for example, which artifacts have to be stored in order to realize a DP.

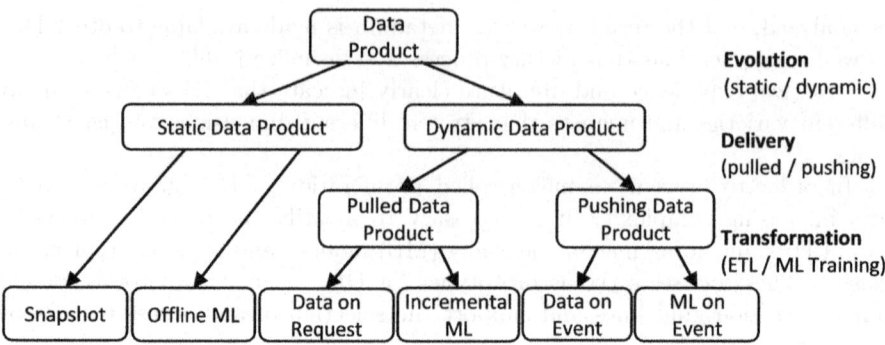

Fig. 1. Decision tree classifying six different types of DPs based on: Whether they change, what delivery strategy they use, and what data transformations are applied.

We also excluded characteristics that do not allow to distinguish between classes of DPs, for example, whether the DP is managed by persons or not and the storage technology used for storing data within the DP. Lastly, some distinctions that are mentioned in literature can be considered as sub-aspects of the characteristics in Fig. 1 that would reduce clarity and are therefore not further considered. For example, if ETL-like transformations applied to data are simple or more advanced. In the following, we present the characteristics of DPs that we identified as most relevant as a classification hierarchy.

2.1 Classification Hierarchy for Data Product Classes

The decision tree shown in Fig. 1 helps DP providers decide which class or classes of DPs fit their use case. In our evaluation we decided on three basic characteristics. Each corresponds to a layer in the tree, resulting in six different classes of DPs.

The root node represents generic DPs. The first layer of inner nodes represents the first characteristic, which revolves around the *evolution* of the DP. A DP can be static or dynamic. If a DP is *static*, its output will never change once it is deployed. This is typically the case when the input data will not change in the future, or when the output of the DP is not intended to change in the future. If a DP is *dynamic*, its output can change as the input data changes.

The second layer describes how the DP handles the *delivery* of data to the consumers of the DP. This second characteristic focuses on the two strategies that can be used to update the output of a DP and therefore can only be applied to dynamic DPs. For a *pulled* DP, consumers of the DP must actively request the updated output, i.e., as long as the consumers do not ask for updates, they will not receive current information. In contrast, *pushing* DPs actively deliver an update to their consumers after a defined event occurs. Updates require recalculations, which may for example occur due to changes in the input data or after a specified time interval.

The third characteristic focuses on the types of data *transformations* used in a DP. *ETL-like transformations* include DPs that cleanse, merge, unify and aggregate data, by applying SQL and statistical functions. In this case, one output port can provide all stored data and another port can support an SQL interface enabling views on stored data. Such ETL-like transformations are used in the three classes of DPs that we call *Snapshot*, *Data on Request*, and *Data on Event*. The second type of DPs resulting from the third characteristic uses input data for training an *ML model*, which can also be considered a transformation. In this case, one output port can deliver predictions made by the model while another may provide the whole model file. An ML model is used in *Offline ML*, *Incremental ML*, and *ML on Event* DPs. In the following, the six resulting classes are described in more detail.

2.2 DP Classes

In this subsection, we describe the six DP classes resulting from the decision tree including their characteristics as well as an example for each of the classes.

The first class is called *Snapshot*. It consists of static DPs that use classic ETL-like data transformations to provide data that will not change in the future. An example is the average energy consumption of a manufacturing plant in the last ten years before 2025 per day.

DPs from the second class *Offline ML*, comprise a trained AI model that will never be retrained in the future. Consider as an example an AI model that predicts the energy consumption of a manufacturing plant for the next day. If there is no need to change the model, a static DP is appropriate here. Note that model prediction takes as input the current dynamic situation and can be applied as often as requested. The preprocessed training data for this model is provided by another DP, that needs no change and therefore belongs to the *Snapshot* class.

The third class is called *Data on Request*. It consists of dynamic DPs that use ETL-like transformations and provide an updated output upon consumer request. Calculating the average sales values on an ever-growing dataset of a large company is an example here. Recalculations may occur every few seconds, e.g., always when a new sales activity is recorded. The ETL transformation procedure itself remains the same. While the output could be updated after each recalculation, this may result in too much network traffic. Instead, consumers need to request the updates when they need them.

The fourth class of DPs consists of dynamic DPs that contain an ML model and provide updates when consumers request them. We call them *Incremental ML* DPs, because they cover situations that are a use case for incremental ML algorithms [4]. Consider an ML model that predicts the customer group of new customers as an example. It is regularly retrained as new purchasing behavior is added to the input training data. Consumers of this DP request access to the updated model when they need it, rather than receiving frequent updates.

Dynamic DPs that use ETL-like data transformations and provide updates to consumers whenever the output is recalculated belong to the fifth class called

Data on Event. Recalculations are processed after defined events occur, e.g., when a certain condition is met or the input data is updated. As an example, consider a dashboard that visualizes the current production of each manufacturing plant. The input data changes once a day because all plants report their production data at the end of each day, which is the triggering event. Therefore, the DP is also updated daily and pushes the updated output to its consumers.

The sixth class is called *ML on event*. It consists of dynamic DPs that contain an ML model and provide model updates whenever the output is recalculated based on emerging events. As before, this can happen when a certain condition is met or whenever the input data is updated. One example is predictive maintenance based on a model that uses training data provided by machines. New data is used to improve the model once a day, e.g. with incremental ML, making the predictions more accurate over time. To prevent consumers from using old versions, updated predicted maintenance times for all observed machines are pushed to them immediately.

3 Related Work

In this section, we briefly assess and compare our classification approach to related work. Dehghani [3] classifies DPs by their alignment between data sources and data consumers focusing on organizational considerations. Thereby *Source-aligned* DPs, *Aggregate* DPs and *Consumer-aligned* DPs, are defined on an abstract business or organizational level. Please note that in contrast to Dehghanis classes our classes differentiate more technical details and therefore help much better towards architectural patterns and blueprints.

The classification by Loukides [7] is data oriented but only classifies based on whether data is part of the output of a DP (*Overt* DP) or not (*Covert* DP). Thereby, the dividing line between the two classes is not clearly defined, which makes classifying a particular DP difficult. Hence, DPs could be both overt and covert. Therefore, all our classes match with both *Overt* and *Covert*.

Hasan and Legner [5] classify DPs based on the complexity of the transformations applied to data within the DP. This results in *Basic*, *Analytical* and *Advanced-analytical* DPs. Our approach matches well with these classes, however as multiple of our classes match with each of their three classes, our classes are more fine granular: *Basic* DPs match our *Snapshot* DPs as well as *Data on Request* and *Data on Event*. All three of our ML related classes can be matched with the *Advanced-analytical* DPs, but express a more fine granular description.

The six classes of Acceldata [1] classify DPs based on their intended functionality. For clarity we divided them into two groups of similar classes. The first group contains *Visualization Dashboard* and *Search Engine* DPs, which both use ETL-like transformations. This group matches well with our three ETL classes *Snapshot*, *Data on Request*, and *Data on Event*. The second group includes *Predictive Model*, *Recommendation Systems*, *Anomaly Detectors* and *Conversationally Intelligence Systems* which are all using ML to transform data. Therefore, they match our three ML classes *Offline ML*, *Incremental ML* and *ML on Event*.

As can been seen, our classes are significantly more fine granular than those from related work. Using three characteristics (Evolution, Delivery and Transformation), enables a more detailed and comprehensive characterisation of DPs, thereby improving upon related work. Furthermore, to the best of our knowledge, we are the first to use the characteristics *Evolution* and *Delivery* in the context of DP classification. In contrast to related work, we do not consider the complexity of transformations or the alignment of the DPs as characteristics. Instead, we assessed these characteristics as not fitting to technically classify DPs, as they are not precisely definable and do not hint towards implementation issues. Additionally, Dehghanis point of view is mainly from a business and organizational perspective. She also considers the embedding of DPs in the data mesh environment via the alignment classification. However, this is not our goal as we focus on the technical aspects in order to move towards implementation considerations.

4 Conclusion

In this paper, we present and assess a basic classification approach that uses more fine grained categorization parameters than related work. Our classification hierarchy defines three basic categorization parameters that indicate six characteristic classes sufficient to cover the full spectrum of DPs mentioned in literature. This is a good starting point for further work on architectural and implementation issues and it further helps in conquering the complexity of networks of DPs as known from data mesh applications.

Acknowledgments. We thank Corinna Giebler and Eva Hoos for their support in the context of the Industrial Data Lab collaboration with Robert Bosch GmbH.

Disclosure of Interests. The authors have no competing interests to declare that are relevant to the content of this article.

References

1. Acceldata Product Team: What are the different types of data products? (2023). https://www.acceldata.io/blog/what-are-the-different-types-of-data-products
2. Dehghani, Z.: How to move beyond a monolithic data lake to a distributed data mesh, May 2019. https://martinfowler.com/articles/data-monolith-to-mesh.html
3. Dehghani, Z.: Data Mesh: Delivering Data-Driven Value at Scale. O'Reilly Media, Sebastopol, CA, USA (2022)
4. Gepperth, A., Hammer, B.: Incremental learning algorithms and applications. In: European Symposium on Artificial Neural Networks, pp. 357–368 (2016)
5. Hasan, M.R., Legner, C.: Understanding data products: motivations, definition, and categories. In: ECIS 2023 Proceedings (2023)
6. Hasan, M.R., Legner, C.: Improving consumer-provider interaction with data products: insights from traditional industries. In: ECIS 2024 Proceedings (2024)

7. Loukides, M.: The evolution of data products (2011). https://www.oreilly.com/radar/evolution-of-data-products/
8. Majchrzak, J., Balnojan, S., Siwiak, M.: Data Mesh in Action. Manning Publications, Shelter Island, NY, USA (2023)
9. Priebe, T., Neumaier, S., Markus, S.: Finding your way through the jungle of big data architectures. In: 2021 IEEE International Conference on Big Data (Big Data), pp. 5994–5996. IEEE (2021)

Classification Techniques

Discovering Voting Power for Ensemble Methods

Pratik Karmakar[1,2], Angelo Saadeh[2], Pierre Senellart[2,3,4,5(✉)], and Stéphane Bressan[1,4]

[1] National University of Singapore, Singapore, Singapore
pratik.karmakar@u.nus.edu
[2] CNRS@CREATE LTD, Singapore, Singapore
angelo.saadeh@cnrsatcreate.sg, pierre@senellart.com
[3] DI ENS, ENS, PSL University, CNRS, Inria, Paris, France
[4] IPAL, CNRS, Singapore, Singapore
[5] Institut Universitaire de France, Paris, France

Abstract. Ensemble methods aggregate the predictions of multiple models by some form of weighted voting. In this work, we consider the impact of the choice of the assignment of voting power to every individual model on the performance of ensemble methods. We empirically and comparatively evaluate the accuracy and running time of the different power voting ensemble methods using standard classifiers and mainstream classification benchmarks. The results show that power ensemble voting outperforms the equal-power baseline, and that unsupervised learning of the voting power can be competitive with respect to supervised learning; within supervised approaches, learning voting power through Shapley values and regression outperforms simply using accuracy.

Keywords: Ensemble Methods · Voting Power · Shapley Values · Inverse Entropy · Truth Discovery · Classification

1 Introduction

Ensemble methods aggregate the predictions of multiple models, to sometimes obtain better performance than any single model [29]. The four main classes of ensemble methods are *boosting* [11], *bagging* [2], *stacking* [32] and *voting*. In this work, we focus on *voting* because it is the setting in which models are trained independently. In addition, we also focus on classification tasks.

Most existing ensemble voting methods assign equal voting power to each model [25]. But treating all models equally may yield less accurate results since expert and non-expert models are treated equally. For this reason, some works have considered assigning a different voting power to each model based on some

---

Stéphane Bressan tragically passed away recently, after contributing to this work.

© The Author(s), under exclusive license to Springer Nature Switzerland AG 2026
R. Wrembel et al. (Eds.): DEXA 2025, LNCS 16046, pp. 347–362, 2026.
https://doi.org/10.1007/978-3-032-02049-9_28

criterion, allowing models with a higher voting power to influence the final prediction more. The voting power of a model is a weight that is used for weighted voting. This usually outperforms equal voting power, as shown in [17,20].

Most works use the models' accuracy or a regression on their predictions to assign voting powers. These two techniques require an auxiliary labelled dataset. Therefore, we refer to them as *supervised*. Our work presents other ways to assign voting powers to the models. The first is to use the game-theoretic solution known as *Shapley values* [28]. A Shapley value is a number computed using a set function; in our example, it would indicate the influence of a model on the accuracy of the aggregation of predictions. In general, Shapley values take exponential time to compute. A more computationally efficient but similar algorithm is *Leave-One-Out* [14], which has polynomial-time complexity. Whether regression, accuracy, or Shapley values are used to compute voting powers, these supervised methods require additional information other than the predictions: a ground truth to compare to the models' predictions. Therefore, the next technique we use to compute the voting powers is *truth discovery* (or truth finding) [12,33], which does not require any other information except the models' predictions. An example state-of-the-art truth discovery algorithm is CRH [18], which assigns weights to the models by minimizing some objective function, these weights serving as voting powers. Finally, we introduce using the *inverse of the entropy* of each prediction vector, based on a vector of weights assigned to each class by each model and used to compute how confident the models are about their predictions. Note that Shapley values and inverse entropy for ensemble voting are original contributions.

In ensemble voting for classification tasks, the set of candidates are target classification classes, the models' predictions serve as votes, and the set of voters is the set of models. For example, the set of candidates in a logistic regression in the MNIST data set for the optical character recognition of digits [7] is $\mathcal{C} = \{0, 1, \ldots, 9\}$. In voting theory, voting amounts to ordering the set of candidates according to the voter's preferences. Votes are aggregated using a voting mechanism such as plurality voting or Borda count [1]. More precisely, in a plurality voting mechanism, the candidate who gets the most first-choice votes is the winner. However, in Borda count, each candidate is given some points based on rank of preference. The Borda winner is the candidate who gets the most points once the points across all the voters are summed up. In our work, in addition to plurality voting, we use a customized version of the Borda voting because the output prediction of a model naturally assigns a weight to each class (candidate). Although other voting mechanisms could be used in ensemble voting, these two are the most commonly used for their simplicity and accuracy [31].

The paper is as follows: After presenting related work in Sect. 2, we define and explain the different methods we use to compute and assign voting powers in Sect. 3. In Sect. 4 we empirically evaluate the accuracy and running time of the methods using standard classifiers and mainstream classification benchmarks in the areas of computer vision and Web classification. The code required to repeat our experiments is available at https://github.com/pratik2358/voting_power_ensemble_anon.

2 Related Work

Ensemble methods aim at better predictive performance than the one obtained from one model [9]. In some ensemble techniques, like boosting, models are trained sequentially. Each model's training depends on the performance of the previous model. An example of a Boosting algorithm is AdaBoost [11], which has many variants [27]. Bagging [2], unlike boosting, does not require interactions between models. However, bagging involves training multiple models on different subsets of the training data using the same training algorithm. Then predictions from these models are aggregated using averaging for regressions and majority voting for classification tasks. Bagging is different from voting in that voting does not necessarily rely on the same type of model. Another primary ensemble technique is stacking [32], where predictions from multiple models are aggregated using weights learned by regressing the predictions to a target prediction. Our work focuses on ensemble *voting* where the models need not be trained using the same algorithm or on the same data distribution.

Different voting mechanisms can be used in ensemble voting for classification tasks. The most common one is plurality voting where the winner is the candidate that gets the most vote. It is also loosely referred to as majority voting (which *stricto sensu* is when the candidate gets more than 50% of the votes). Plurality and majority voting are used for most ensemble voting [25]. Examples of other voting algorithms are Condorcet [3], instant run-off method, and Borda count [1]. In Borda count, each candidate is assigned some points, then the candidate that gets the most points is the winner. Borda count is also used in ensemble voting [22]. The authors of [31] compare voting mechanisms in ensemble voting, and they conclude that plurality voting and Borda count have the lowest complexity-to-performance ratio.

Previously discussed ensemble voting techniques assign equal voting power to the models. However, there are ways to assign different voting powers to models. The first approach is to use the accuracy of each model as its voting power. For instance, [13] uses the accuracy of a model as its voting power. Another strategy is to include only models that meet a minimum threshold accuracy and then use it as their voting power [23]. In other approaches also based on accuracy [8,26], the models that correctly make predictions incorrectly classified by most other classifiers are given more voting power. In other words, the focus is more on true positives than overall accuracy. The second approach consists of assigning voting powers as the weights learned from a regression of the predictions to fit a target prediction. The authors of [19] do so using a linear regression model. Other works [10] use a custom loss function that assigns voting powers based on the reliability of each model for specific output classes. These two approaches require an auxiliary labelled dataset, which we call supervised. Truth discovery is also used to assign voting powers [16] in an unsupervised manner. In summary, existing works mainly use regression or the accuracy of models to assign a voting power for each model. Our work defines other techniques to discover and assign voting powers. Moreover, we compare them by evaluating their accuracy and running time.

Table 1. Classification of power finding methods.

Prediction format	Unsupervised	Supervised
Vector format	Inverse entropy	Regression
Label format	Truth discovery (CRH)	Accuracy
		Shapley values
		Leave-one-out (LOO)

3 Methodology

In our work, we assign voting powers to models and use plurality and Borda voting for ensemble voting for an ℓ-class classification problem (for simplicity, we write the classes as $0, \ldots, \ell - 1$). To compute and assign the voting powers, we use supervised power learning mechanisms like accuracy, regression, and Shapley values, as well as unsupervised mechanisms like truth discovery and inverse entropy. These mechanisms can be further categorized with respect to the required format of the predictions, which can either be *soft* classifications, where the output of the classifier is a vector of ℓ scores assigned to all classes, or *hard* predictions of the most common label. For example, for a "cat" (0) or "dog" (1) binary classification task, a vector output would be $(0.8, 0.2)$, and a label output would be $0 = \arg\max_{0,1}(0.8, 0.2)$. The different power-assigning techniques we consider are summarized in Table 1. We write $M(q) \in \{0, \ldots, \ell - 1\}$ for the hard prediction of model M for class q and $\boldsymbol{M}(q) \in \mathbb{R}_+^\ell$ for the soft vector of scores.

Supervised methods require a labelled set of queries $\mathcal{F}$ to query the models and have their predictions compared with a ground truth. Unsupervised methods do not require such an auxiliary labelled dataset – but the computation of voting power for unsupervised method uses an auxiliary set of queries, with no access to ground truth, also called $\mathcal{F}$ here.

We note that in the case where a Borda voting mechanism is used to aggregate the predictions after the voting power computation, then choosing a method from Table 1 that requires a vector as a voting format would not require more information than the methods that require only the label as a prediction; this is because the Borda voting mechanism requires a vector format anyway.

As previously mentioned, for all methods, we assume access to a certain number of queries, which are used as auxiliary information to compute the model voting powers, and we call this set of m queries $\mathcal{F}$ (for supervised methods, we need access to the ground-truth label of each query in $\mathcal{F}$). If we have n different models $M_1, M_2, \ldots, M_n$, the goal is to obtain the voting powers $\boldsymbol{w} = (w_1^*, w_2^*, \ldots, w_n^*)$ which represent the voting power of each model. The voting powers need not sum up to 1; the equal-power case is where all w_i^* are identical.

3.1 Voting Mechanisms

Plurality voting. In plurality voting (with voting power), we assume each classifier M_i outputs its prediction label $M_i(q)$ for query q. The winning prediction for query q is the one which maximizes the amounts of votes, weighted by their voting power: $\arg\max_{0 \leq k < \ell} \sum_{i=1}^{n} w_i^* \delta_{k, M_i(q)}$ with ties randomly resolved and $\delta_{ij} := 1$ if $i = j$, and 0 otherwise.

Borda Voting. In Borda voting (with voting power), each classifier M_i outputs a vector $\boldsymbol{M}_i(q)$ of scores for each label, for query q. The winning prediction for query q is the one which maximizes the total score, weighted by the voting power of each model: $\arg\max_{0 \leq k < \ell} \sum_{i=1}^{n} w_i^* (\boldsymbol{M}_i(q))_k$.

3.2 Supervised Methods

For supervised methods, we assume the ground truth $\gamma(q)$ of every query $q \in \mathcal{F}$.

Accuracy. Accuracy of the models is a naïve choice for the model voting powers. We compute the accuracy of each model M_i on $\mathcal{F}$ as: $w_i^* := \frac{1}{|\mathcal{F}|} \sum_{q \in \mathcal{F}} \delta_{M_i(q), \gamma(q)}$.

Regression. In both Plurality and Borda settings, it is not desired to make the mechanisms perform optimally only for one query. We are rather interested in optimizing the performance of all $\mathcal{F}$. For any q, let $\mathbf{M}(q)$ be the $n \times \ell$ matrix: $\mathbf{M}(q) := \left(\boldsymbol{M}_1(q)^\mathrm{T} \cdots \boldsymbol{M}_n(q)^\mathrm{T}\right)$. We aim to minimize the empirical risk by choosing as voting power: $\boldsymbol{w}^* := \arg\min_w \mathbb{E}_{q \sim \mathcal{F}} \left[\|(w \times \mathbf{M}(q)) - \gamma(q)\|_2^2\right]$, where $\gamma(q)$ is represented as a one-hot vector. This, now, is a convex optimization (regression) problem that we solve using gradient descent. In our case, having access to the set of queries $\mathcal{F}$, we minimize the empirical risk of $\mathcal{F}$, assuming i.i.d.

Shapley Values. Shapley values [28] are used to assess the contribution of a party (or model in our case) to a specific task. Computing the Shapley value requires us to define the task through a set function called *value function* given by $v : 2^{\{1,\dots,n\}} \to \mathbb{R}$. We take the value function v to be the accuracy of equal power plurality voting on $\mathcal{F}$: $v := S \mapsto \frac{1}{|\mathcal{F}|} \sum_{q \in \mathcal{F}} \delta_{\gamma(q), \arg\max_k \sum_{i \in S} \delta_{k, M_i(q)}}$. The Shapley value of model M_i is: $w_i^* := \frac{1}{n!} \sum_{S \subseteq \{1,\dots,n\} \setminus \{i\}} |S|!(n - |S| - 1)!\,(v(S \cup i) - v(S))$ which, for each combination of models, measures the impact of model M_i. Shapley values are chosen as a power-assigning technique, for it satisfies four significant axioms: Efficiency, Symmetry, Dummy, and Additivity [21], which makes it an attribution method for fair reward distribution. However, their exponential computational complexity limits the use of Shapley values.

Leave-One-Out (LOO). The size of the powerset used to compute Shapley values increases exponentially with the number of models, making the Shapley values impossible to compute. A more computationally efficient similar algorithm is leave-one-out [4,14], which has polynomial time complexity. The leave-one-out value for a model M_i is given by: $\tilde{w}_i^* := v(\{1,\dots,n\} \cup i) - v(\{1,\dots,n\})$ with v

defined as for Shapley values. While not as well-grounded as Shapley values for power assignment, this method is much less resource-intensive, making it a more practically feasible attribution method for some instances.

3.3 Unsupervised Methods

We now assume a set $\mathcal{F}$ of queries, but no knowledge of the ground truth.

Inverse Entropy. The entropy of a vector defining a discrete probability distribution $P = (p_0, p_1, \ldots, p_{\ell-1})$ is defined as $\mathrm{H}(P) = -\sum_{k=0}^{\ell-1} p_k \log(p_k)$. Then, the inverse entropy for model M_i is defined as $w_i^* := \frac{|\mathcal{F}|}{\sum_{q \in \mathcal{F}} \mathrm{H}(M_i(q))}$. The rationale behind using the inverse mean entropy on $\mathcal{F}$ as the model's weight is quite straightforward: entropy is a measure of uncertainty of a model's decision. The higher the model's uncertainty, the higher the entropy value. Thus, inverse entropy assigns a higher weight to the models, which are more certain about their decisions on $\mathcal{F}$. Note that if a model is fully certain, i.e., if $\mathrm{H}(\boldsymbol{M}(q))$ is 0 for all q, then the inverse entropy is $+\infty$ for this model, meaning this model's predictions will be selected as the ensemble one; this extreme case does not occur in practice with the classifiers we considered in our experiments.

CRH Weights. We use the CRH [18] truth discovery algorithm to assign voting power, which attempts to optimize voting power w_i^* and predictions (*truth values*) y_q for each query $q \in \mathcal{F}$. This is done iteratively: first, y_q are set to the equal-power plurality vote; then, voting powers are first estimated by minimizing the following loss function: $\boldsymbol{w}^* := \arg\min_w \sum_{i=1}^n w_i \sum_{q \in \mathcal{F}} \delta(y_q, M_i(q))$, with constraint $\sum_i \exp(-w_i) = 1$. Then, the y_q votes are re-estimated using plurality vote with voting power $\boldsymbol{w}^*$; new values for $\boldsymbol{w}^*$ are then iteratively computed. This is repeated until some convergence criterion (see [18]) is met. The optimization is solved using Lagrange multipliers [18].

4 Performance Evaluation

Experimental Setup. We performed our experiments using four different standard datasets, with a variety of classifiers trained over these: MNIST [7] handwritten digits, CINIC-10 [5] image classification, DMOZ URL Classification[1], and Webpage Phising Detection [15].

We need models of varying accuracy to evaluate different power-assigning methods in our setting. We assess model quality by evaluating accuracy on a test set distinct from the training set. To simulate models with varying levels of quality, we employ two strategies. In *Label Flipping*, the training dataset is partitioned into subsets where labels are flipped with varying probabilities; more precisely, for a specific model M_i, a probability $p_i \in [0, 1]$ is drawn and M_i is

---

[1] https://www.kaggle.com/datasets/shawon10/url-classification-dataset-dmoz.

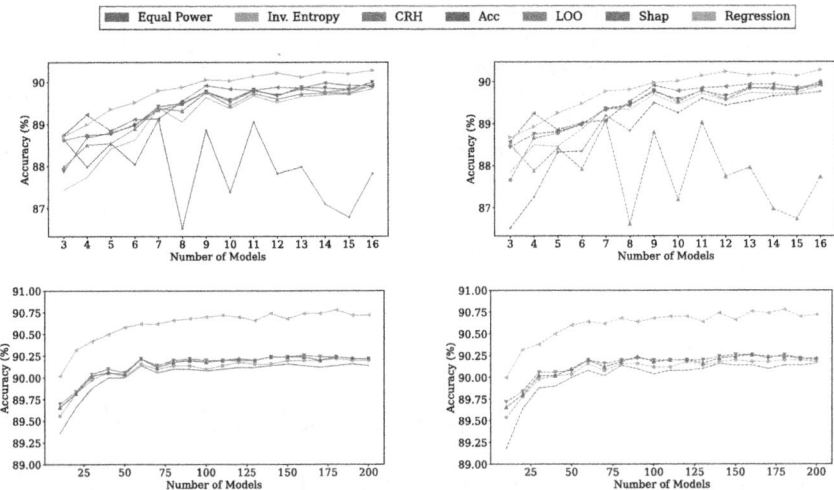

Fig. 1. Performance of the power-assigning methods in Borda (left) and Plurality (right) settings for up to 16 models (top) and up to 200 models (bottom) (MNIST data).

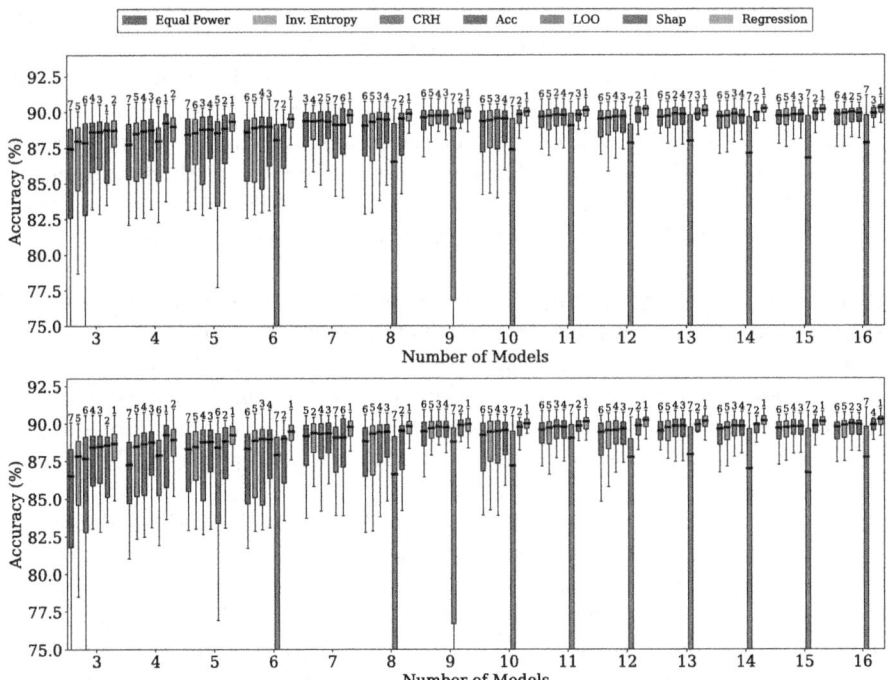

Fig. 2. Borda (top) and Plurality (bottom) voting with different voting power assigning methods (MNIST data).

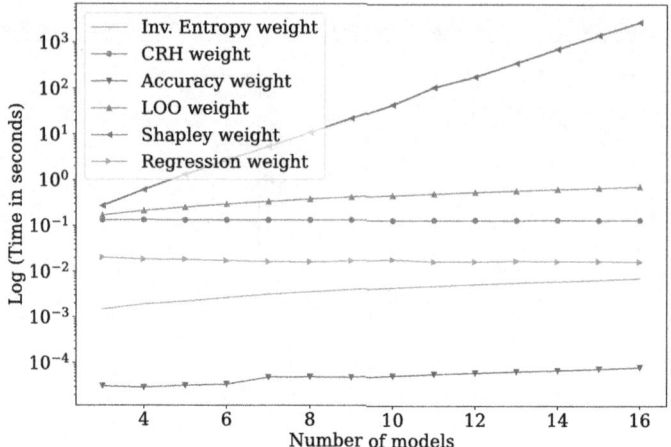

Fig. 3. Time comparison of the power-assigning methods (MNIST data). The y-axis uses a logarithmic time scale.

then trained on a dataset where a fraction p_i of the labels has been flipped. In *Class Imbalance*, the training dataset is partitioned into subsets with varying degrees of class imbalance; models trained on these subsets demonstrate differing accuracy based on the extent of class imbalance.

Every dataset is split into a training set (used to train the classifiers, possibly on different subsets), a validation set (playing the role of the set $\mathcal{F}$ used for computing the voting power for every power-assigning technique), and a test set (used to evaluate the overall performance of the ensemble methods). To ensure statistically significant results, the experiments on MNIST data are conducted 75 times, and experiments on other datasets are conducted 20 times.

MNIST Dataset. Our experiments use a logistic regression classifier, trained using PyTorch [24]. We randomly pick 5000 samples from the MNIST dataset to train each model. These random subsets are not necessarily disjoint. We do this to keep the number of training data points constant throughout all the experiments' models. We simulate varying model quality using label-flipping noise in the training subsets. As expected, higher label flipping results in lower model accuracy.

We vary the number of models. To better evaluate the voting power computation techniques, we repeat the experiments multiple times with a different training dataset for each model, for each number of models to be aggregated. More precisely, we use fifteen different label-flipping probability assignments and each label flipping is used in five different experiments. For each number of models to be aggregated, we thus repeat the experiment seventy-five times on different training data.

In Figs. 1 and 2, we present the performance of different methods used for Borda and Plurality voting, respectively, with varying numbers (3–16) of models to aggregate. The ranks of the methods for every set of model numbers are indicated above the boxes in Fig. 2. In Fig. 1, we also present the experimental results for up to 200 models (bottom). In the case of 200 models, we do not show the results of using Shapley values as voting powers due to the exponential computational cost, leading to resource constraints. Figure 2 compares the variance of the voting power assigning methods and the effect of the number of models on the variance of aggregate decisions. In Fig. 3, we compare the time complexities (in log-scale) of finding the voting powers using different methods. The equal-power method is not presented here as we do not compute any voting power in this case, and thus, it is the least costly method.

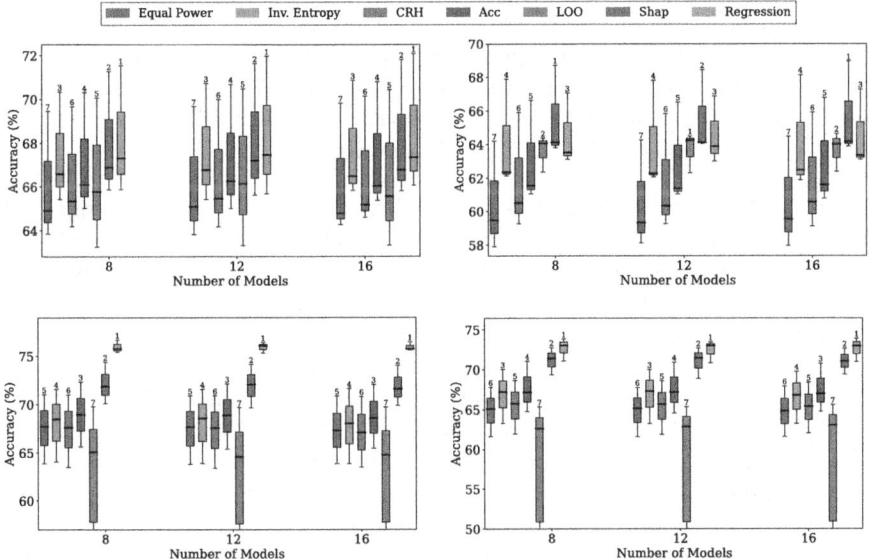

Fig. 4. Performance of the power-assigning methods in Borda (left) and Plurality (right) settings for label-flipping noise (top) and class-imbalance noise (bottom) (CINIC-10 data)

Our empirical results on this dataset indicate that the voting powers found using regression perform better than the others among the chosen methods of voting power assignment. Shapley's value as voting power outperforms the rest of the methods in most cases. Among the unsupervised methods, CRH performs better than the inverse entropy method, showing a higher median and lower variance in performance in most cases, but it is of higher time complexity. The performance of LOO is poor and exhibits a high variance.

CINIC-10 Dataset. The CINIC-10 dataset consists of 60,000 images from CIFAR-10 and 210,000 images from ImageNet [6], all categorized using the same classes as CIFAR-10. We have used models of two different architectures here:

a Convolutional Neural Network (CNN) with three convolutional layers and VGG16 [30]. We train $2n$ CNN models and $2n$ VGG16 models. Specifically, n models are trained on n subsets of the CIFAR-10 training subset of CINIC-10, and the other n models are trained on n subsets of the ImageNet training subset of CINIC-10. These $4n$ models, exhibiting different qualities, are then used for our experiments. In our experiments, $n \in \{2, 3, 4\}$. The validation and test splits consist of images from CIFAR-10 and ImageNet, ensuring a comprehensive model performance evaluation. We apply label flipping and class imbalance separately to create distinct models for separate experiments. We use this strategy to split the dataset so that our models of different architectures learn from two different distributions and are then tested on a subset containing data from both.

We present the empirical results in Fig. 4. In most cases, the ensembles using voting powers outperform the equal-power ensemble. Powers found using regression and Shapley values perform the best throughout. While in the case of label-flipping noise, the equal-power ensemble performs the worst, in class imbalance, voting powers found using LOO perform the worst.

DMOZ URL Classification Dataset. The URL classification dataset from DMOZ used in our study contains URLs and corresponding classes in the DMOZ catalogue for 1,562,978 web pages spanning 15 categories. We split the data into disjoint partitions: 70% for training, 15% for testing, and 15% for validation. We use Multinomial Naive Bayes (MNB) models. To train each model, we randomly sample 90% of the training data and introduce varying levels of label-flipping noise. This process ensures that resulting models exhibit different levels of accuracy. Our experiments are conducted with ensembles of 3, 5, 7, and 10 models.

As shown in Fig. 5, voting powers derived from Shapley values outperform all other methods. Conversely, powers determined using LOO perform the worst. Notably, all other power voting methods surpass the equal-power voting approach. We also notice that the inverse entropy method does not perform significantly better than equal-power voting.

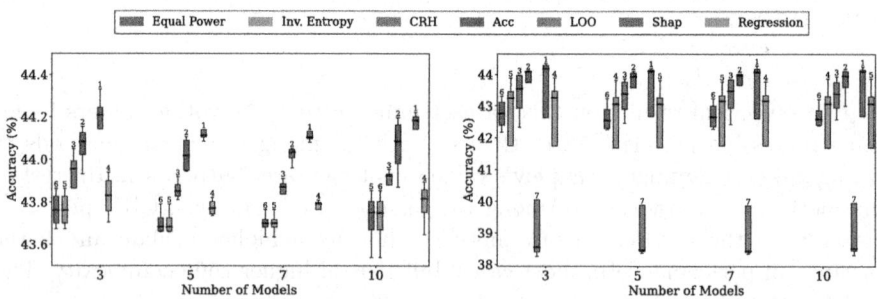

Fig. 5. Performance of the power-assigning methods in Borda (left) and Plurality (right) settings for label-flipping noise (DMOZ data)

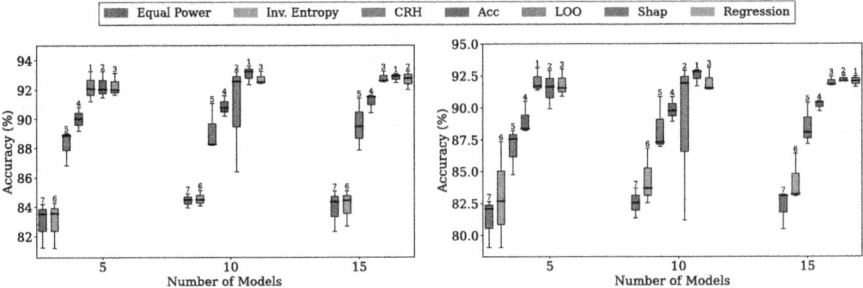

Fig. 6. Performance of the power-assigning methods in Borda (left) and Plurality (right) settings for label-flipping noise (Phishing data)

Webpage Phising Detection Dataset. This dataset comprises data from 11,430 web pages, each characterized by 88 features, including the URL, URL length, hostname length, domain age, Google index, and page rank, among others. The data is divided into disjoint partitions for training (60%), validation (20%), and testing (20%). We use a 3-layer neural network for this experiment. We divide the training partition into n disjoint subsets and then add varying amounts of label-flipping noise to the subsets to train models of different quality. In our experiments $n \in \{5, 10, 15\}$.

The empirical results are shown in Fig. 6. All voting power methods outperform the equal-power approach except the inverse entropy method, particularly in the Borda voting scenario. Shapley values as voting powers consistently outperform all other methods across Borda and Plurality voting strategies.

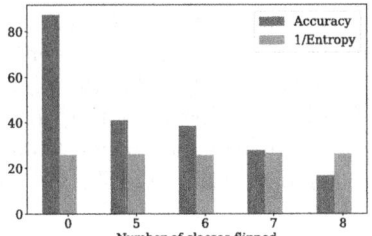 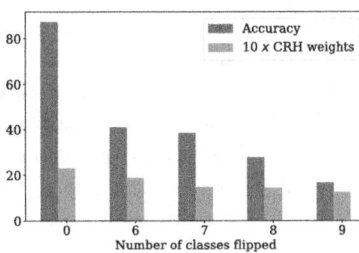

Fig. 7. Comparison of the effect of label-interchange in training data on inverse entropy weights (left) and CRH weights (right) (MNIST data)

Other Voting Mechanisms. Six techniques were employed to compute and assign voting powers to sources, which were then utilized in power voting. In addition to Plurality and Borda, we evaluated other voting mechanisms, too.

First is Condorcet voting, in which a voter ranks their vote from highest preference to lowest preference, and a candidate is iteratively eliminated until a

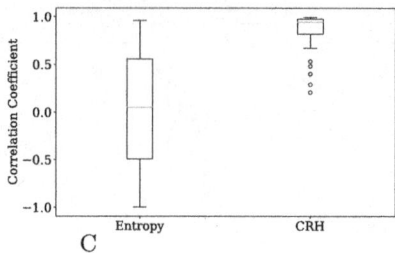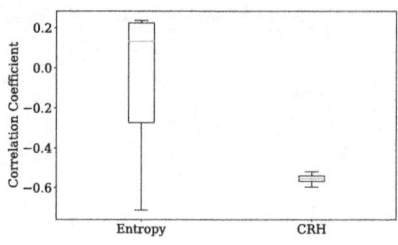

Fig. 8. Correlation between model accuracy and model voting powers computed using inverse entropy and CRH when the complete label interchange pattern is random (left) or deterministic (right) across different training datasets (MNIST data)

majority winner prevails. This means that a candidate is the most preferred candidate of half of the voters. We tested this voting mechanism, which performed badly but have not reported its results. Indeed, there is a loss of granularity when a vector of votes is transformed from probability of preference (like in Borda) to ranked candidates with no probabilities or values assigned to the candidates. The lost information is crucial for the voters because they might have a first-choice candidate with 100% confidence in Borda and all other candidates with a value of zero. Second, we tried taking into account the vote of the voter with the highest voting power and ignoring the other voters, which we do not report here. The tests also showed a low performance for the method, which shows that a collective decision would be better than an individual one. More particularly, taking into account the vote of the model with the best accuracy is not better than taking the average prediction of all the models. On one hand, this could partially be explained by the fact that more models considered means more data was used in the training process, yielding a better generalization. On the other hand, there is a loss of diversity in opinion, i.e., some voters may train their models differently, making their insight important for the aggregation result. In summary, the choice between Borda and plurality voting does not significantly affect the aggregation result, as seen in the experiments. However, choosing aggregate votes using Condorcet voting or choosing the "best" voter's predictions drastically reduces accuracy.

Voting Power Computation and Assignment. It was shown that most voting power computation techniques give similar results. The difference lies in the information required to compute these powers. Indeed, using truth discovery (CRH) or the entropy techniques yields almost the same results as using Shapley or the local accuracy of the models. However, computing the latter two powers requires access to labelled data to assess the models.

The name truth discovery is misleading because "truth" is defined as what the majority thinks on average. Therefore, if most models are wrong on average, then truth discovery will not yield good results. Usually, this is an improbable scenario, and even in this case, the other techniques will not be better. Furthermore, while truth discovery only uses prediction from the sources, entropy requires having the prediction as a probability vector, which is more information than a simple label for truth discovery. Nevertheless, if the aggregator would like to use Borda voting, then the aggregator would need the full vector of probabilities anyway. Therefore, entropy would not require more information than truth discovery.

The inverse entropy method has a lower computational cost than the CRH method, as seen in Fig. 3, but we may have situations of trade-off between the two. We explain one instance partly explaining the performance gap between the two methods. Using inverse entropy can go severely wrong as it depends on the models' confidence. To illustrate this phenomenon, we train five logistic regression models on five subsets of the MNIST handwritten digits dataset. The first subset has all its true labels intact. From the second subset onward, we randomly interchange class labels of 5, 6, 7, and 8 classes. In Fig. 7, we see the decrease in accuracy when models trained on these datasets are tested on a subset disjoint from these training subsets. Inverse entropy values remain the same for all models. Even though models trained on interchanged class labels have learned the wrong labels, the learning degree is not affected by risk minimization on the training sets.

The similar quality disorder in training data does not affect the CRH method as much as it affects the inverse entropy method. In Fig. 7, we present one instance of the setup mentioned above to compare the effect of the data quality flaw mentioned above in assigning voting power using the inverse entropy method (left) and CRH method (right). In Fig. 8 (left), we show the correlation coefficient between the accuracy of the models trained on data with 0, 5, 6, 7, and 8 randomly interchanged labels and the voting powers were computed using inverse entropy and CRH (20 runs). We notice a significantly higher correlation between the voting powers and the accuracy of the models in the case of CRH than in the case of inverse entropy. The robustness of CRH lies in the fact that to find the power of a model, this method has to consider votes from other models in the ensemble, too. Thus, if all the models do not deviate from the truth, there's still a probability of the truth found using CRH converging to the ground truth. Thus, for CRH to fail deterministically, the label interchange must be deterministic, not random. However, this is a very unlikely phenomenon as there is rarely a chance that all the models are being trained on different datasets where the interchange of labels has taken place similarly across all training datasets. We demonstrate this in Fig. 8 (right).

Results show that using regression voting powers yields the best accuracy. Among the other informed methods, Shapley's voting powers outperform other methods. However, this technique is very resource-intensive, and its computation complexity increases with the number of voters. Even though there are better approximations than the one we use, LOO, they are also computationally expensive. LOO performs very poorly, especially when the number of models is high because it assesses the added value of a model on all the other models. In plurality and Borda voting, the probability that one model changes the aggregation result becomes very low as the number of models grows.

5 Conclusion and Future Work

This work extends ensemble voting techniques by exploring multiple ways of assigning voting powers to individual classifiers (including original approaches such as inverse entropy and Shapley values) paired with voting mechanisms like plurality and Borda. The results confirm that assigning voting powers to classifiers outperforms the equal-power baseline. Within supervised power-assigning methods, the classical accuracy-based voting powers are less effective than regression and Shapley values. Unsupervised power-assigning methods often perform competitively with supervised ones without a labelled ground truth.

Moving forward, we are applying this research to knowledge transfer frameworks that aggregate predictions via plurality voting from multiple trained models to label an unlabelled dataset that is then used to train a new privacy-preserving model. The obtained model would not reveal information on the data used because calibrated noise is added during the aggregation process. Since our techniques could potentially be applied at the aggregation step of this scenario, our future work consists of doing a privacy analysis on our aggregation techniques by computing their sensitivity, adding noise accordingly, and evaluating how each method affects the privacy and performance of the new model.

Acknowledgments. This research is part of the program DesCartes and is supported by the National Research Foundation, Prime Minister's Office, Singapore under its Campus for Research Excellence and Technological Enterprise CREATE) program. This work is made in memory of our co-author, colleague, and friend Stéphane Bressan, who unexpectedly passed away in December 2024.

References

1. de Borda, J.C.: Mémoire sur les élections au scrutin. Histoire de l'Académie Royale des Sciences (1781)
2. Breiman, L.: Bagging predictors. Mach. Learn. **24**(2), 123–140 (1996)
3. de Condorcet, N.: Essai sur l'application de l'analyse à la probabilité des d écisions rendues à la pluralité des voix. Cambridge University Press (2014)
4. Cook, R.D.: Detection of influential observation in linear regression. Technometrics **42**(1), 65–68 (2000)
5. Darlow, L.N., Crowley, E.J., Antoniou, A., Storkey, A.J.: CINIC-10 is not ImageNet or CIFAR-10. arXiv preprint arXiv:1810.03505 (2018)
6. Deng, J., Dong, W., Socher, R., Li, L.J., Li, K., Fei-Fei, L.: ImageNet: a large-scale hierarchical image database. In: 2009 IEEE Conference on Computer Vision and Pattern Recognition, pp. 248–255. IEEE (2009)
7. Deng, L.: The MNIST database of handwritten digit images for machine learning research [best of the web]. IEEE Sig. Process. Mag. **29**(6), 141–142 (2012)

8. Dogan, A., Birant, D.: A weighted majority voting ensemble approach for classification. In: 2019 4th International Conference on Computer Science and Engineering (UBMK), pp. 1–6 (2019)
9. Dong, X., Yu, Z., Cao, W., Shi, Y., Ma, Q.: A survey on ensemble learning. Front. Comput. Sci. **14**(2), 241–258 (2020)
10. Ekbal, A., Saha, S.: Weighted vote-based classifier ensemble for named entity recognition: a genetic algorithm-based approach. ACM Trans. Asian Lang. Inf. Process. **10**(2), 9:1–9:37 (2011)
11. Freund, Y., Schapire, R.E.: Experiments with a new boosting algorithm. In: Saitta, L. (ed.) Machine Learning, Proceedings of the Thirteenth International Conference (ICML 1996), Bari, Italy, 3–6 July 1996, pp. 148–156. Morgan Kaufmann (1996)
12. Galland, A., Abiteboul, S., Marian, A., Senellart, P.: Corroborating information from disagreeing views. In: Proceedings of the Third ACM International Conference on Web Search and Data Mining, pp. 131–140 (2010)
13. Georgiou, H.V., Mavroforakis, M.E.: A game-theoretic framework for classifier ensembles using weighted majority voting with local accuracy estimates. CoRR abs/1302.0540 (2013)
14. Ghorbani, A., Zou, J.: Data Shapley: equitable valuation of data for machine learning. In: International Conference on Machine Learning, pp. 2242–2251. PMLR (2019)
15. Hannousse, A., Yahiouche, S.: Web page phishing detection. Mendeley Data V3 (2021)
16. Jin, Y., Yang, Z., He, Y., Bao, X., Wu, G.: Ensemble classification method based on truth discovery. In: 2019 IEEE International Conference on Big Knowledge (ICBK), pp. 122–128. IEEE (2019)
17. Kuncheva, L.I., Rodríguez, J.J.: A weighted voting framework for classifiers ensembles. Knowl. Inf. Syst. **38**(2), 259–275 (2014)
18. Li, Y., et al.: Conflicts to harmony: a framework for resolving conflicts in heterogeneous data by truth discovery. IEEE Trans. Knowl. Data Eng. **28**(8), 1986–1999 (2016)
19. Li, Y., Luo, Y.: Performance-weighted-voting model: an ensemble machine learning method for cancer type classification using whole-exome sequencing mutation. Quant. Biol. **8**(4), 347–358 (2020). https://doi.org/10.1007/s40484-020-0226-1
20. Mahendran, N., et al.: Sensor-assisted weighted average ensemble model for detecting major depressive disorder. Sensors **19**(22), 4822 (2019)
21. Molnar, C., Casalicchio, G., Bischl, B.: Interpretable machine learning – a brief history, state-of-the-art and challenges. In: Koprinska, I., et al. (eds.) ECML PKDD 2020. CCIS, vol. 1323, pp. 417–431. Springer, Cham (2020). https://doi.org/10.1007/978-3-030-65965-3_28
22. Niyas, K.P.M., Paramasivan, T.: Improving Alzheimer's classification using a modified Borda count voting method on dynamic ensemble classifiers. Knowl. Inf. Syst. **66**(8), 4755–4787 (2024)
23. Okuboyejo, D.A., Olugbara, O.O.: Classification of skin lesions using weighted majority voting ensemble deep learning. Algorithms **15**(12), 443 (2022)
24. Paszke, A., et al.: Automatic differentiation in Pytorch. In: NIPS 2017 Autodiff Workshop (2017)
25. Raza, K.: Improving the prediction accuracy of heart disease with ensemble learning and majority voting rule. In: U-Healthcare Monitoring Systems, pp. 179–196. Elsevier (2019)

26. Rojarath, A., Songpan, W.: Cost-sensitive probability for weighted voting in an ensemble model for multi-class classification problems. Appl. Intell. **51**(7), 4908–4932 (2021). https://doi.org/10.1007/s10489-020-02106-3
27. Shahraki, A., Abbasi, M., Haugen, Ø.: Boosting algorithms for network intrusion detection: a comparative evaluation of real AdaBoost, gentle AdaBoost and modest AdaBoost. Eng. Appl. Artif. Intell. **94**, 103770 (2020)
28. Shapley, L.S.: A value for n-person games. In: Kuhn, H.W., Tucker, A.W. (eds.) Contributions to the Theory of Games II, pp. 307–317. Princeton University Press, Princeton (1953)
29. Shwartz-Ziv, R., Armon, A.: Tabular data: deep learning is not all you need. Inf. Fusion **81**, 84–90 (2022)
30. Simonyan, K., Zisserman, A.: Very deep convolutional networks for large-scale image recognition. arXiv preprint arXiv:1409.1556 (2014)
31. Torres-Sospedra, J., Hernandez-Espinosa, C., Fernandez-Redondo, M.: A comparison of combination methods for ensembles of RBF networks. In: Proceedings. 2005 IEEE International Joint Conference on Neural Networks, vol. 2, pp. 1137–1141. IEEE (2005)
32. Wolpert, D.H.: Stacked generalization. Neural Netw. **5**(2), 241–259 (1992)
33. Yin, X., Han, J., Yu, P.S.: Truth discovery with multiple conflicting information providers on the web. In: Proceedings of the 13th ACM SIGKDD International Conference on Knowledge Discovery and Data Mining, pp. 1048–1052 (2007)

Classifying Public and Private Documents Using Context-Based Predictions

Abrar Hasin Kamal[1] and Anne V.D.M. Kayem[2](✉)

[1] University of Potsdam, Potsdam, Germany
abrar.hasin.kamal@uni-potsdam.de
[2] University of Exeter, Exeter, UK
a.v.kayem@exeter.ac.uk
https://www.uni-potsdam.de/en/university-of-potsdam ,
https://www.exeter.ac.uk

Abstract. Unstructured data, such as textual data, is prevalent on digital platforms such as social media, with users posting content that sometimes includes sensitive information. For instance, in March of 2025, a press story by journalist Jeffrey Goldberg indicated that secret Trump administration plans for an air strike on Yemen had inadvertently been leaked to him (U.S. national-security leaders included me in a group chat about upcoming military strikes in Yemen...). Despite regulations, and the existence of access control policies data leaks such as this are persistent. One of the reasons for this is that contextually analysing unstructured data to determine whether or not sensitive information exists therein, is a challenging problem. Existing approaches identify individual elements of sensitive information such as Personally Identifiable Information (PII), but few works look at the issue of document sensitivity based on the occurrence of sensitive information. In this paper we propose a novel approach to classifying documents as either public or private, based on the occurrence of, and contextual interpretation of sensitive information within the document. We employ an ensemble model composed of transformer-encoder models and standard machine learning models to support the classification process. Our empirical study, conducted on a curated dataset composed of ENRON emails and Tweets of U.S. Congress members, indicates that the Random Forest algorithm when combined with the BERT model classifies all public documents correctly (i.e. 100% accuracy), and achieves a 0.98 recall score for private documents.

Keywords: Classification · PII Detection · Named Entity Recognition · Transformer Models · BERT · Contextual Data

1 Introduction

According to global digital reports, the volume of multimodal data generated daily is growing exponentially, with social networking sites and emails contributing approximately 63% of the global data share which by estimates in 2024, stood at 120 Zettabytes [1,2]. Yet, this availability of data raises challenges in protecting sensitive data from exposure [3–6,10]. With 95% of the Internet's multimodal data existing in unstructured formats (e.g. email, customer reviews,...) and the growing number of data breaches, it is crucial to protect documents that contain personal information from being released on public channels [7,8,27]. As per statistics from 2024, the average total cost of a data breach is roughly 4.88 million U.S. Dollars, which is a 10% increase from what it was in 2023 [9]. Cybercriminals who have successfully gained access to Personally Identifiable Information (PII) have used the sensitive information to foster criminal activity such as forgery and blackmail [5].

Studies have found that anonymised personal information (PII) in free text can be reverse-engineered to link to an individual through inferences drawn from contextual interpretations of the text [11,13–18,31]. One way of addressing this issue is to identify documents containing PII and determine whether or not the levels and/or types of PII in the document qualifies it as "sensitive". However, classifying documents based on sensitivity (e.g. as either `public` or `private` depending on the presence of PII), using standard machine learning classification algorithms alone, raises a high number of false positives. This is mainly due to the failure of these algorithms to account for contextual information. Failed classifications require manual verifications which is a time-consuming process especially for very large datasets. Therefore, having a high accuracy document classification model, with a low false positive rate, can make an important contribution to reducing sensitive information exposure.

This paper presents a novel two-step hybrid scheme to classify public and private documents based on contextualised predictions. In the first step, we do the following: (1) We pre-process a text corpus to support our empirical study in Sect. 4. The corpus consists of both ENRON emails [12] and Tweets of U.S. Congress data. (2) We manually label PIIs in the dataset with BILOU tagging to support the Named Entity Recognition tasks. (3) With the support of an ensemble model composed of transformer-encoder models such as BERT, DeBERTa, and RoBERTa, we use the PII labels to support PII detection. Finally (4) we employ a context-based classification metric to decide whether or not a PII occurrence is classifiable as `public` or `private`. In the second step, we employ an ensemble classification model, which accepts input data from the transformer-encoder ensemble model, to classify a document as either public or private based on the rate of occurrence of private PII within the document. Our experiments show a 99.3% accuracy during training and a 97.29% validation accuracy for the BERT model, given the variation of data in our dataset. For the classification task, the Random Forest Algorithm achieves the lowest number of false negatives and an accuracy of 97%.

The rest of the paper is structured as follows: in Sect. 2, we discuss the definition of sensitive information, personally identifiable information (PII) and the different types of PII as well as related work on PII detection. In Sect. 3, we present our methodology and our implementation pipeline. In Sect. 4, we present results from our empirical analysis as well as a discussion of the implications of our findings. We offer concluding statements and offer directions for future work in Sect. 5.

2 Related Work

Work in the field of PII detection in unstructured and semi-structured documents, using machine learning algorithms was pioneered by Korba, et al. [19] with an approach that combined Named Entity Recognition (NER) with supervised machine learning algorithms. The rule-based system checks for the presence of an exact match between each word in the unstructured document and the dictionary, or for a similar pattern that can be matched with a regular expression. While this approach promises a fail-safe approach towards detecting various names and patterns it raises the risk of not identifying the entities that it has not encountered. Assumptions about the structure of the dataset entries also pose a further drawback for this approach. For instance, the regular expressions assume that phone numbers start with the prefix PHONE: at the beginning followed by the country code, which implies that phone numbers that do not obey this syntax are ignored. Since large unstructured text documents have a high word count having to cross-check each false-positive/negative case is manually infeasible.

To alleviate this caveat, McDonald, G. et al. present two studies using classifiers [23] and part-of-speech n-grams [24] to enable detection of sensitive and private data. Both solutions focused on government records that are made public and the importance of a system that is able to identify and prioritize records that contain personal data or information that could potentially disrupt international relations. In [23], a Support Vector Machine (SVM) classifier is used to classify the documents and achieves a maximum accuracy of 73%. In their second study [24], the authors use a dataset of 143 government documents and use part-of-speech N-grams augment their work to measure the 'sensitivity load' of a document to flag any documents that have a higher load in comparison to other documents. However, in both studies, the authors focus on a limited subset of PII, namely personal names, which are direct identifiers[1], and limits the scope of the study.

Geetha et al.'s [22] work on analysing sensitive private information disclosure on Twitter indicates that for approaches such as [23,24] to be extensible to large unstructured datasets, a large amount of labeled data is required to train traditional machine learning algorithms. Using machine learning techniques such as Naive Bayes, Support Vector Machines (SVM) and Gradient Boosting Methods (GBM), Geetha et al. classify twitter posts into personal, professional and health domains from which they further categorised as public or private posts. The Naive Bayes model performs comparatively well, with a maximum accuracy

---

[1] Elements of personal information that support one-to-one mappings for re-identifications.

of 84%, in detecting `personal`, `professional` and `health` related tweets compared to the SVM and GBM models. For the other topics, all proposed methods perform rather poorly and have low precision and recall values.

In a similar vein, textual data analysis to determine subjective information (Sentiment Analysis) has been deeply researched and explored through the past few decades. For instance, Aisopos et al. [26] conducted a comparative analysis between content and context in classifying unstructured social media data. They showed that while n-gram models can be used to detect textual patterns when classification is only dependent on content, context is equally important in enriching the information extracted from the classification process.

PrivacyBot addresses the issue of private sensitive information (PSI) in social media posts, namely in Twitter and the risks associated with leakage of the private data into public channels [25]. The PSI classification system aims to classify private entities in unstructured text into granular PSI categories. Similar to previous work, the authors leverage the Twitter Streaming API to download social media posts over the span of three months, capturing 319120 posts. The results across various classifiers (SVM, Decision Trees, Logistic Regression, Naive Bayes, and Random Forest) using Precision, Recall and F-1 scores for each class shows that Random Forests (Average F-1 score of 88%) outperform the other classifiers in the multi-label classification task. However, for sparse texts in the data that do not have enough context, the Term Frequency - Inverse Document Frequency (TF-IDF) vector does not provide enough information for the classifier to learn from, leading to less precise results. Furthermore, relationships between entities in unstructured text tend to be non-linear and may not be well represented with only TF-IDF vectors and their corresponding labels.

Bioglio and Pensa [20] presented a novel method to analyze and classify privacy-sensitive content in social media posts by comparing different deep learning models for the classification task. A dataset of 10,000 posts collected from the myPersonality project [21] were labeled as sensitive or non-sensitive by human experts for the study. The work shows that transformer-based models (BERT) offer the best performance compared to other sequential neural networks, such as Convolutional Neural Networks (CNN), Gated Recurrent Units (GRU), and Long Short-Term Memory (LSTM) networks. However, the authors reported the F-1 measure. While the F-1 measure provides a holistic measure of the model's performance by calculating a harmonic mean of the precision and recall of the model, it does not reveal how the model performs for each individual label. The dataset used also suffers from a class imbalance issue in that there are more non-sensitive labels compared to sensitive labels. The BERT model presented in the paper reports an F-1 score of 0.89 but it is unclear whether or not classifying sensitive content is negatively impacted by the class imbalance in the training dataset.

3 Hybrid Two-Step Document Classification

Our hybrid document classification scheme works with unstructured textual data. As mentioned before, the reason for this choice is the growing volume of sensitive data (e.g. personal messages, official documents, announcements through media platforms) that is available on public digital platforms and the potential for exploiting such data to provoke online harms. Our unstructured dataset is composed of a mixture of ENRON data (>1.5 million emails) [28] and the 'Tweets of Congress' dataset [29].

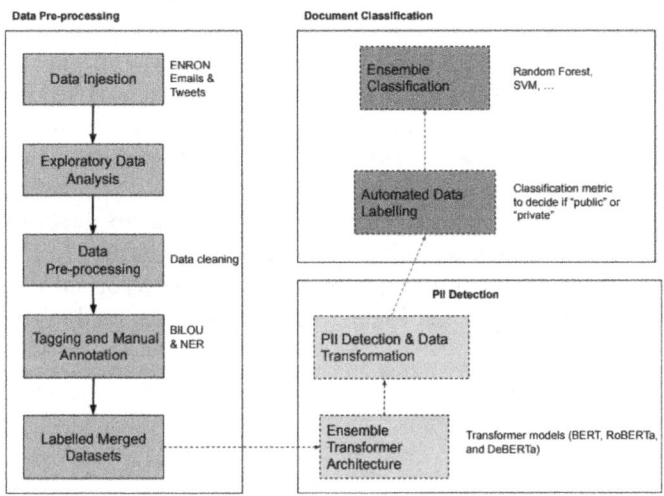

Fig. 1. Hybrid Classification Scheme - Pipeline

Figure 1 depicts the high level overview of the work presented in this paper. The architecture of the pipeline for our proposed approach follows a modular structure as shown. This allows us to define each section independently and utilize the learning from existing work to use the appropriate tools and techniques for them. We implement a script using Python that queries the repositories, extracts the data and stores it as CSV files for the downstream tasks. Prior to data preprocessing, we invest time and research to get an understanding of the data. This step helps us to learn about various aspects of the dataset such as the length of the documents in the dataset, the temporal dimensions of the documents, and the actual content of the emails and tweets in the dataset. Information about the dataset obtained in the exploration step helps us in defining the transformations we apply to prepare it for labeling and the rest of the pipeline. We then use an open source tool [30] for the manual data labelling process followed by the finetuning of Transformer models to equip them with the knowledge required for PII detection. Using the results, we apply data transformations that allow us to use an automated script which bins the data into public or private classes. Finally, our pipeline concludes with the training and validation of an ensemble of classification algorithms that predict if a document is public or private using the contextualised predictions. We preprocessed the data to identify and eliminate duplicate entries by assigning keys to entries based on an analysis of the content of the entry. Entries that occur more than once are processed to eliminate the extra ones. The Enron dataset is already unique and requires no duplicates handling. The duplicates are discarded from the Tweets dataset, reducing the data to approximately 4 million tweets and to approximately 5 million records in our total pool of data. The removal of duplicates eliminates the chance of any email or tweet being sampled more than once which would then introduce bias and the risk of overfitting. Furthermore, tweets that have an irregular identifier ($n = 13$) (IDs that are null values or have pieces of text as identifiers) as the key in the raw data set are also removed. In line with existing work, we discard any body of text that is less than 50 characters. We found 144 emails from the Enron dataset, and 67 tweets to be in violation of this threshold and removed them to minimize noise in our data. Furthermore, we sample about 4000 records from the Enron ($n = 2000$) and the Tweets of Congress ($n = 2000$) datasets for manual annotation since it is not feasible to annotate 5 million records without a large team of expert annotators. In terms of content extraction, the Congress for Tweets API returns data in a JSON format, the content of the tweet is accessible with ease. Figure 2 shows the segmented email separating metadata and body sections.

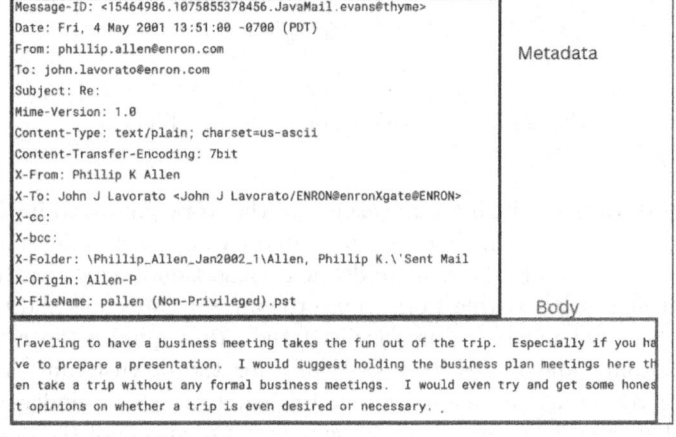

Fig. 2. A raw email segmented into metadata and body parts

Following this, we cleaned the data to eliminate special characters, whitespace, and repeat characters. After which we manually annotated the data to identify sensitive information elements. We used manual data labeling because of its correctness and the capability of transformers to fine-tune themselves on a smaller subset of the data. A human annotator along with the lead author of the paper participated in the data labeling procedure using an open source tool known as Label Studio. To label the data, we focus on sensitive data that can be used, directly or in combination with other sensitive information, to reveal a person's identity. Taking inspiration from existing work, and privacy regulation guidelines, we label 7 types of Named Entities using the BILOU tagging scheme (Table 1).

Table 1. PII Data and their Interpretations

Entity Tag	Description
PER	Person (Individual names)
ORG	Organization (Companies, institutions, etc.)
LOC	Location (Physical locations, such as landmarks, buildings, etc.)
GPE	Geo-political Entity (Countries, cities, states, etc.)
EMAIL	Email Address
PHONE	Phone Number
USER	Username or Online Handle

Each annotator labeled roughly 1200 records randomly sampled from our preprocessed dataset of 4000 records. After reviewing the annotated sets, we were left with 2,502 records that served as the training dataset. In order to create a labeled dataset of public and private entries, to support the second step of the classification process, we define a threshold of occurrence of sensitive information in an entry (document) to decide if it is belongs in the *public* or *private* class.

Within the scope of this paper, each document with any piece of sensitive information along with a direct identifier is initially classified as a `public` document in the scope of this paper. When there are direct identifiers but no additional sensitive information to infer an individuals identity, we label it as `public`. For instance, since many people can share similar names no sensitive information is disclosed simply by revealing a name. Lastly, if there are no direct identifiers but only entities like addresses, phone numbers, organization names, then we also label those documents as *public* since they do not directly identify an individual and therefore are not strictly private information.

Finally, the last step of the data labeling module in our pipeline handles tokens that have not been assigned a tag during the manual labeling process. These are tokens that are assigned the O (Outside) tag and do not belong to any entity. The importance of assigning O-tags to tokens lies in enabling models to distinguish between entity tokens and non-entity tokens. The assignment of O tags also enables models to learn the boundaries of groups of named entities as **B** and **L** tags are expected to have no O-tag in between.

The labeled data obtained after pre-processing is tokenized using the model-specific tokenizer, that breaks down each sequence to the token level and builds the token embeddings and vocabulary for the finetuning step. The transformer models, BERT, RoBERTa, and DeBERTa, are then finetuned using the tokenization process to generate context-based predictions of the sensitivity levels of the data which can then be used to support the classification process (classifying the data into `public` and `private` class labels). Figure 3 provides an example of the prediction in its raw format from a DeBERTa model that we fine-tuned for our task.

Each document in our dataset therefore receives an array-like structure that contains dictionaries with each detected PII value, the token label (entity_group) and the probability of the word belonging to the label.

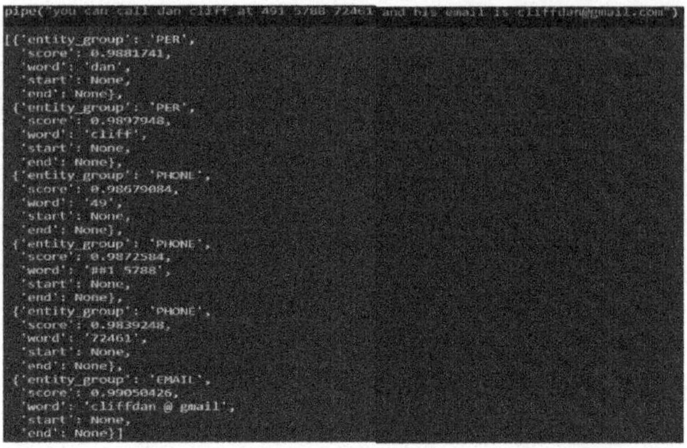

Fig. 3. Example of a prediction returned by the DeBERTa model

The first step in generating the prediction dataset is assigning identifiers to the documents in the original dataset with their true labels so that we are able to map the prediction from the Transformers back to the original document. A simple row count function returns an integer value as the row identifier that is added to each row of the dataset. The tokenisers for each transformer model are downloaded and then the finetuned models are loaded to define a dictionary that maps the tokeniser and its corresponding transformer model. Using the defined dictionary of tokenisers and their models, we apply functions on a columnar level to perform two operations:

- Add a new column to the dataset for the transformer model being used.
- Generate a prediction for each row in the sentences column that contains the unstructured textual data.

The transformation step at this stage of our pipeline is mainly responsible for creating the datasets that enable the classification layer to assign documents to the public or private class depending on the contextualized predictions generated by the transformers in the preceding layer.

The labeled and tokenized dataset is reloaded and only the document identifier, sentence and word label columns are used in this step. As our transformer predictions aggregate the positional indicators to provide a simplistic classification of each token, we apply a string replace function, where the positional prefixes (B-, L-, I-, U-) are replaced with empty spaces. Initially, the BILOU tagging scheme is used to encode the positional information of a token along with the entity that the token belongs to. The BILOU tagging scheme consists of Beginning (Beginning of a token), Inside (Token inside a multi token entity), Last (The last token of the entity), Outside (All tokens that do not belong to any entity), and Unit (Single representations of an entity) tag representations. Tokens that belong together (e.g. full names) that occur in sentences together can be tagged using the appropriate positional tag (B/I/L/U) followed by the type of the named entity (PER). This applies to all named entities and enables the recognition of named entities to be grouped into spans that represent a single named entity. BILOU gives more detailed structure, which can help models learn entity boundaries more accurately. This enables us to create a 1:1 mapping of the tokens we extract from the predictions and our originally labeled tokens. Since O-tags are non entity type tokens, we remove the O tags from the label sequence and apply a counter on the remaining named entity tokens in the sequence. This results in a map containing the PII token and the frequency count of each token in the corresponding token. A pivot is then applied to each map to set the count value of each token on a columnar basis, creating a dataset that consists of the identifier, the document, and the named entities counts for each row.

As discussed earlier, we use a set of rules to determine if a document can be classified as *public* or *private*. The assigned label is induced from the count of the type of PII tokens detected by the transformer model. Algorithm 1 provides

---

Algorithm 1. Assign Public or Private Label

Require: A row with attributes PER, USER, LOC, GPE, ORG, PHONE, and EMAIL.
Ensure: A label: public, private, or unknown.
1: **if** (PER = 0 ∧ USER = 0) **and** (LOC > 0 ∨ GPE > 0 ∨ ORG > 0 ∨ PHONE > 0 ∨ EMAIL > 0) **then**
2:     return public
3: **else if** (PER > 0 ∨ USER > 0) **and** (LOC > 0 ∨ GPE > 0 ∨ ORG > 0 ∨ PHONE > 0 ∨ EMAIL > 0) **then**
4:     return private
5: **else if** (PER > 0 ∨ USER > 0) **and** (LOC = 0 ∧ GPE = 0 ∧ ORG = 0 ∧ PHONE = 0 ∧ EMAIL = 0) **then**
6:     return public
7: **else if** PER = 0 ∧ USER = 0 ∧ LOC = 0 ∧ GPE = 0 ∧ ORG = 0 ∧ PHONE = 0 ∧ EMAIL = 0 **then**
8:     return public
9: **else**
10:    return unknown
11: **end if**

---

the pseudocode followed by the function that accepts each row from our data set as input and assigns a label based on the defined conditions of the class to which the document belongs. The algorithm assigns a document with the public label in the following scenarios:

- When there are no Names and User identifiers but quasi-identifiers such as locations, phone numbers, email addresses, or organization names exist. (line 1 of Algorithm 1)
- When only names or user identifiers exist in a document without any quasi-identifier that can be used to single out an individual. (line 5 of Algorithm 1)
- When no direct or indirect identifiers exist in the document (line 7 of Algorithm 1)

A document is labeled as private when there exists at least one direct identifier along with at least one of the quasi identifiers. We perform a count of the labels assigned to address the topic of class imbalance that could result in a classifier with a bias towards one particular class over the other. We find that our dataset is almost equally balanced with 1299 public documents and 1199 private documents. We accept the slight imbalance, as it does not create a strong bias towards the public class over the private class. Furthermore, as shown in pseudo-code, the presence of direct and quasi-identifiers strictly labels documents as private.

Our classification layer is responsible for attributing a `public` or `private` label to documents based on the contextualised predictions produced by the transformers. The data transformations applied on the predictions from the transformers, allows us to obtain a dataset with structured data in rows and columns. Datasets generated from our transformation step in the preceding layer are loaded before performing an 80:20 train-test split. We make use of k-fold cross validation while performing a grid search over a set of parameters to obtain the best model. All models are defined inside a map data structure where the keys are the names of the model, and the value for each model is a map containing parameter names as keys and a set of possible values that are randomly generated.

4 Empirical Analysis

We now report on our experimental results to validate our proposed two-step hybrid classification scheme. The implementation was done using the Python programming language in Google Collaboratory Notebooks. To prevent *Out of memory* issues during computationally heavy tasks such as row-wise data operations and finetuning our transformer models. We experimented with three transformer based models anemly, BERT, RoBERTa, and DeBERTa, for the Named Entity Recognition task that provides the following classification layer with the data it needs to classify documents. The BERT model was finetuned based on an evaluation of the training and validation loss every 100 steps to monitor the models convergence. The training loss for the BERT model is reported at 0.6% with a training accuracy of 99.4%. The finetuned BERT model performed relatively well on the validation set as well, in which the model had a validation accuracy of 97.3% and a loss of 3.2%. We also report a multi class classification report to report the evaluation metrices for each Named Entity in our study.

Table 2. BERT Classification Report

Class	Precision	Recall	F1-Score
EMAIL	0.95	0.96	0.95
GPE	0.70	0.78	0.74
LOC	0.25	0.22	0.24
ORG	0.63	0.61	0.62
PER	0.88	0.90	0.89
PHONE	0.95	0.95	0.95
USER	0.98	0.98	0.98
Micro Avg	**0.80**	**0.81**	**0.81**
Macro Avg	**0.76**	**0.77**	**0.77**
Weighted Avg	**0.80**	**0.81**	**0.80**

Table 3. RoBERTa Classification Report

	Precision	Recall	F1-Score
EMAIL	0.15	0.51	0.24
GPE	0.65	0.76	0.70
LOC	0.33	0.20	0.25
ORG	0.61	0.68	0.65
PER	0.89	0.91	0.90
PHONE	0.71	0.95	0.81
USER	0.98	0.97	0.97
Micro Avg	**0.71**	**0.81**	**0.76**
Macro Avg	**0.62**	**0.71**	**0.65**
Weighted Avg	**0.76**	**0.81**	**0.78**

Table 2 encapsulates the Precision, Recall, and F1-scores for each Named Entity in the documents.

The RoBERTa model incurs a training loss value of 0.8% and a training accuracy of 99%. On the validation set, the RoBERTa model reports a validation loss of 3.72% and a validation accuracy of 96.34%. The classification report for the RoBERTa model in the validation set has been reported in Table 3.

The DeBERTa model (deberta-base) reports a training accuracy of 99.3% after 10 epochs and the training loss at 0.7%. DeBERTa also reaches a validation accuracy of 97.2% with a validation loss of 3.29% on the test set. We report the precision, recall and F-1 score values for the DeBERTa model in Table 4.

In the classification step, we experimented with 7 classification algorithms with the objective of obtaining a classifier that can perfectly distinguish between a *public* and a *private* document based on the detection of PII tokens collected

Table 4. DeBERTa Classification Report

Entity	Precision	Recall	F1-Score
EMAIL	0.94	0.94	0.94
GPE	0.70	0.75	0.72
LOC	0.26	0.20	0.23
ORG	0.64	0.66	0.65
PER	0.88	0.91	0.89
PHONE	0.86	0.93	0.90
USER	0.96	0.95	0.96
Micro avg	**0.80**	**0.82**	**0.81**
Macro avg	**0.75**	**0.76**	**0.76**
Weighted avg	**0.79**	**0.82**	**0.81**

by transformer models (BERT, RoBERTa, and DeBERTa). Table 6 reports the evaluation results of the classification algorithms along with their accuracy in the test set.

Table 5. Accuracy of Classification Models on Transformer generated Dataset

Model	BERT	RoBERTa	DeBERTa
Naive Bayes	.70	.68	.69
Decision Tree	.96	.93	.91
Random Forest	.97	.94	.92
kNN	.96	.93	.91
Logistic Regression	.81	.80	.79
SVM	.97	.94	.92
Multilayer Perceptron	.97	.94	.92

Table 5 reports the accuracy of our classification algorithms on each of the PII datasets generated by the BERT, DeBERTa and RoBERTa model. Naive Bayes and Logistic Regression perform poorly mainly due to assumptions about the data, such as feature independence which is unlikely in context based datasets, and linearly separable data which does not apply since there are underlying relationships between the different PII entities that occur in the data. We perform a 3-fold cross validation for each model with the Grid Search method enabled and extract the best parameters for each classification algorithm. Using the best parameters we then evaluate the classification algorithm first on the test set by evaluating the model on the golden dataset we created using our classification label assignment guidelines earlier. Table 6 is used to report the evaluation results of the classification algorithms along with their accuracy in the test set. To refine it further, we also inspect the classification report of each of the algorithms. The confusion matrix has been interpreted as follows:

- Private documents are interpreted as positives.
- Public documents are interpreted as negatives.

This translates to interpreting False positives as public documents wrongly classified as private, and False negatives referring to the inverse. In our scope, the classification algorithm classifying a document as public when it should be private is critical and therefore we would want for the classification models to minimize the number of False negatives, whilst also keeping the number of False positives in check to ensure minimum human interference in the process. The Decision Tree, Random Forest, and SVM algorithms classify 1 private document as public from the test set while in kNN the number of false positives were higher (n=11). The Multilayer perceptron classified all instances correctly and the confusion matrix reported no False positives or False negatives. The other classification algorithms had a higher number of false positives, as induced by their classification reports. With the models evaluated on the test set using the best parameters obtained, we apply the classification on the predicted datasets from BERT, RoBERTa and DeBERTa and compare the predicted label with the true label to generate corresponding classification reports and confusion matrices.

Table 6. Classification Report for Classification Algorithms on Golden dataset

	Public			Private			Accuracy
	Precision	Recall	F-1 score	Precision	Recall	F-1 score	
Naive Bayes	.65	.84	.73	.78	.55	.64	.70
Decision Tree	1.00	.99	.99	.99	1.00	.99	.99
Random Forest	1.00	1.00	1.00	1.00	1.00	1.00	1.00
kNN	.96	.98	.97	.98	.96	.97	.97
Logistic Regression	.74	.89	.81	.86	.69	.77	.79
SVM	1.00	1.00	1.00	1.00	1.00	1.00	1.00
Multilayer Perceptron	1.00	1.00	1.00	1.00	1.00	1.00	1.00

The classification reports of the transformers reported in Table 2, and Table 4 indicate that the DeBERTa model edges slightly past the BERT and RoBERTa models. However, this is heavily influenced by the class labels for the different datasets where we observe BERT exhibiting a much more stable performance across different types of PII tokens in comparison to the other models. This leads to BERT generating more simpler and linearly separable embeddings that aide in the classification phase for more traditional Machine Learning models. While RoBERTa and DeBERTa makes use of more advanced techniques and are also trained on larger datasets, they also produce more complex higher dimensional embeddings that possibly introduces overlaps between similar PII tokens such as LOC, and GPE. This makes it difficult for the classification algorithms to draw a distinction between them, hence resulting in reduced accuracy (see Table 5).

Table 7. Classification Report for on BERT dataset

	Public			Private		
	Precision	Recall	F-1 score	Precision	Recall	F-1 score
Naive Bayes	0.66	0.84	0.74	0.75	0.54	0.63
Decision Tree	0.97	0.97	0.97	0.97	0.96	0.97
Random Forest	0.97	0.98	0.97	0.97	0.96	0.97
kNN	0.95	0.97	0.96	0.97	0.94	0.95
Logistic Regression	0.78	0.89	0.78	0.85	0.72	0.83
SVM	0.97	0.98	0.97	0.97	0.96	0.97
Multilayer Perceptron	0.97	0.98	0.97	0.97	0.96	0.97

We first evaluate the classification models on our BERT generated dataset. The Naive Bayes classifier reports a considerable number of False positives (549) and False negatives (211). We observe slightly lesser numbers of False positives (333) and False negatives (147) for the Logistic Regression model. The Decision Tree algorithm and the Random Forest classifier to be quite similar in it's results with 44 and 43 False positives; and 37 false negatives in both respectively. Support Vector Machines also exhibit similar performance to that of the Random Forest classifier. GridSearchCV reports 3 as the optimal number of neighbors for the kNN algorithm that reports 73 False positives and 35 False negatives in the confusion matrix. Lastly, we find the 3 layer Multilayer Perceptron to report slightly the least number of False positives (42) and False negatives (32). The accuracy of each classification model on the validation set is reported in Table 5. We also report the Precision, Recall and F-1 scores for our classification models in Table 7.

On the RoBERTa dataset, the accuracy of the classification algorithms followed a similar pattern, as reported in Table 5. Random Forests and the Multi Layer Perceptron models shared identical confusion matrices with 72 False Positives and 83 False Negatives. Following close, the SVM classifier had only one extra False Positive but also had one False negative less. The Decision Tree algorithm also reports similar results in its confusion matrix with 74 False positives and 97 False negatives. kNN reports 100 false positives, while Logistic regression and Naive Bayes report 312 and 527 false positives respectively. Table 8 records precision, recall and F-1 scores for the classifiers on the RoBERTa dataset.

On the DeBERTa generated dataset, we find the classification algorithms to report a higher number of False positives on the DeBERTa dataset. The Naive Bayes classifier reports 547 False positives and 223 False negatives. Following the

Table 8. Classification Report for on RoBERTa dataset

	Public			Private		
	Precision	Recall	F-1 score	Precision	Recall	F-1 score
Naive Bayes	0.66	0.80	0.72	0.72	0.56	0.63
Decision Tree	0.94	0.93	0.93	0.92	0.94	0.93
Random Forest	0.94	0.94	0.94	0.93	0.94	0.94
kNN	0.92	0.94	0.93	0.93	0.92	0.92
Logistic Regression	0.78	0.86	0.82	0.83	0.74	0.78
SVM	0.94	0.94	0.94	0.94	0.94	0.94
Multilayer Perceptron	0.94	0.94	0.94	0.94	0.94	0.94

Table 9. Classification Report for on DeBERTa dataset

	Public			Private		
	Precision	Recall	F-1 score	Precision	Recall	F-1 score
Naive Bayes	0.66	0.83	0.74	0.75	0.54	0.63
Decision Tree	0.91	0.94	0.92	0.93	0.89	0.91
Random Forest	0.91	0.94	0.92	0.93	0.90	0.92
kNN	0.89	0.94	0.91	0.93	0.87	0.90
Logistic Regression	0.76	0.87	0.81	0.83	0.71	0.77
SVM	0.91	0.94	0.92	0.93	0.90	0.91
Multilayer Perceptron	0.91	0.94	0.92	0.93	0.90	0.91

pattern observed earlier, the Logistic regression model reports 346 False positives and 173 False negatives. In the kNN algorithm, we find 158 False positives and 173 False negatives. Decision Trees, Multi layer perceptron and the SVM models report similar number of False positives (126, 123, 124) and False negatives (82, 79, 78) respectively while the Random Forest classifier reports the lowest number of False positives (121) and False negatives (77) on the validation dataset. The accuracy results for the models are reported in Table 5. Precision, Recall and F-1 scores of the algorithms are reported in 9.

BERT and DeBERTa are most effective in detecting PII tokens in the unstructured data, as indicated by the overall macro and weighted average scores in Tables 2 and 4. While BERT demonstrates overall stable performance with an average recall score of 0.81 across all classes in both weighted and unweighted settings, DeBERTa demonstrates slightly better performance with an average Recall score of 0.82. RoBERTa demonstrates similar recall scores but shows slightly poor performance in terms of Precision and F-1 scores. In terms of the overall accuracy, we find the BERT model to show slightly better accuracy than the DeBERTa model.

The 97.3% accuracy on the validation set indicates that our transformer model is accurate in terms of picking up PII tokens, as well as detecting non-PII tokens and classifying them as such. Combined with the Random Forest classifier, we find the pipeline to report a Recall score of 0.98, which indicates the effectiveness of our proposed solution in classifying private documents.

5 Conclusion

In this paper, we present a novel approach to classifying documents as `public` or `private` depending on contextualized predictions that is automatically generated by a finetuned BERT model. A manual data labeling process across two different data sources results in 2502 records originating from the Enron and Tweets for Congress datasets. We experiment with three different foundational models (BERT, RoBERTa and DeBERTa) for the task of PII detection where we find BERT and DeBERTa models to be quite effective. Using predictions generated by the transformer models in the first step, we use the results to train an ensemble model of classification algorithms that assign a class label (`public` or `private`) to the documents. While we find the BERT + Random Forest combination (97.3% accuracy in detecting PII and non-PII tokens and 0.98 recall score) to be the effective solution for its computational simplicity, we also suggest the use of a Multilayer Perceptron or SVMs as they provide similar results.

The unavailability of public labeled datasets was the reason we created our own labeled dataset for the work, which is a limitation in terms of benchmarking our results against other solutions on a standard dataset. However, the creation of our own dataset aides us in defining the guidelines by inspecting and taking into considerations the most recent privacy regulation guidelines. We believe that a larger training corpus, labeled by more human experts with domain knowledge can be beneficial for future work.

References

1. We Are Social & Meltwater: Digital 2024 Global Overview Report (2024). https://datareportal.com/reports/digital-2024-global-overview-report. Accessed 07 Oct 2024
2. Rydning, D.R.J.G.J., Reinsel, J., Gantz, J.: The digitization of the world from edge to core. Framingham: International Data Corporation, 16, 1–28 (2018)
3. Gandomi, A., Haider, M.: Beyond the hype: Big data concepts, methods, and analytics. Int. J. Inf. Manage. **35**(2), 137–144 (2015)
4. Tsesis, A.: The right to erasure: privacy, data brokers, and the indefinite retention of data. Wake Forest L. Rev. **49**, 433 (2014)
5. Poyraz, O.I., Bouazzaoui, S., Keskin, O., McShane, M., Pinto, C.A.: Cyber-assets at risk (CAR): the cost of personally identifiable information data breaches. In: In ICCWS 2020 15th International Conference on Cyber Warfare and Security, vol. 402, March 2020
6. Sei, Y., Okumura, H., Takenouchi, T., Ohsuga, A.: Anonymization of sensitive quasi-identifiers for l-diversity and t-closeness. IEEE Trans. Dependable Secure Comput. **16**(4), 580–593 (2017)

7. Edwards, B., Hofmeyr, S., Forrest, S.: Hype and heavy tails: a closer look at data breaches. J. Cybersecur. **2**(1), 3–14 (2016)
8. Zou, Y., Mhaidli, A. H., McCall, A., Schaub, F.: " I've Got Nothing to Lose": consumers' risk perceptions and protective actions after the Equifax data breach. In: Fourteenth Symposium on Usable Privacy and Security, SOUPS 2018, pp. 197–216 (2018)
9. Ponemon Institute: Cost of a Data Breach 2024 (IBM) (2024)
10. Sarjito, A.: Data security and privacy in the digital era: challenges for modern government. JIAN-Jurnal Ilmiah Administrasi Negara **8**(3), 01–13 (2024)
11. Teo, T.W., Choy, B.H.: STEM education in Singapore. In: Tan, O.S., Low, E.L., Tay, E.G., Yan, Y.K. (eds.) Singapore Math and Science Education Innovation. Empowering Teaching and Learning through Policies and Practice: Singapore and International Perspectives, vol. 1. Springer, Singapore (2021). https://doi.org/10.1007/978-981-16-1357-9_3
12. Enron Corp & Cohen, William W.: Enron Email Dataset. United States Federal Energy Regulatory Commissioniler, comp [Philadelphia, PA: William W. Cohen, MLD, CMU] [Software, E-Resource] Retrieved from the Library of Congress (2015). https://www.loc.gov/item/2018487913/
13. Schwartz, P.M., Solove, D.J.: Pii 2.0: privacy and a new approach to personal information. Privacy and Security Law Report (2012)
14. Ren, J., Rao, A., Lindorfer, M., Legout, A., Choffnes, D.: ReCon: revealing and controlling PII leaks in mobile network traffic. In: Proceedings of the 14th Annual International Conference on Mobile Systems, Applications, and Services, pp. 361–374, June 2016
15. Rysavy, M.D., Michalak, R.: Data privacy and academic libraries: non-PII, PII, and librarians' reflections (Part 1). J. Libr. Adm. **59**(5), 532–547 (2019)
16. Voigt, P., Von dem Bussche, A.: The EU General Data Protection Regulation (GDPR). A Practical Guide, 1st edn. Springer, Cham (2017). 10(3152676), 10-5555
17. Office for Civil Rights: Summary of the HIPAA privacy rule (2003)
18. Jaar, D., Zeller, P.E.: Canadian privacy law: the personal information protection and electronic documents act (PIPEDA). Int. In-House Counsel J. **2**, 1135 (2008)
19. Korba, L.: Private data discovery for privacy compliance in collaborative environments. In: Luo, Y. (ed.) CDVE 2008. LNCS, vol. 5220, pp. 142–150. Springer, Heidelberg (2008). https://doi.org/10.1007/978-3-540-88011-0_18
20. Bioglio, L., Pensa, R.G.: Analysis and classification of privacy-sensitive content in social media posts. EPJ Data Sci. **11**(1), 1–24 (2022). https://doi.org/10.1140/epjds/s13688-022-00324-y
21. Stillwell, D.J., Kosinski, M.: myPersonality Project website (2015)
22. Geetha, R., Karthika, S., Kumaraguru, P.: Tweet-scan-post: a system for analysis of sensitive private data disclosure in online social media. Knowl. Inf. Syst. **63**(9), 2365–2404 (2021). https://doi.org/10.1007/s10115-021-01592-2
23. McDonald, G., Macdonald, C., Ounis, I., Gollins, T.: Towards a classifier for digital sensitivity review. In: de Rijke, M., et al. (eds.) ECIR 2014. LNCS, vol. 8416, pp. 500–506. Springer, Cham (2014). https://doi.org/10.1007/978-3-319-06028-6_48
24. McDonald, G., Macdonald, C., Ounis, I.: Using part-of-speech n-grams for sensitive-text classification. In: Proceedings of the 2015 International Conference on the Theory of Information Retrieval, pp. 381–384, September 2015

25. Tesfay, W. B., Serna, J., Rannenberg, K.: PrivacyBot: detecting privacy sensitive information in unstructured texts. In: 2019 Sixth International Conference on Social Networks Analysis, Management and Security (SNAMS), pp. 53–60. IEEE, October 2019
26. Aisopos, F., Papadakis, G., Tserpes, K., Varvarigou, T.: Content vs. context for sentiment analysis: a comparative analysis over microblogs. In: Proceedings of the 23rd ACM Conference on Hypertext and Social Media, pp. 187–196, June 2012
27. Cyber Security Intelligence: British railway passengers attacked. Cyber Security Intelligence, 27 September 2024. https://www.cybersecurityintelligence.com/blog/british-railway-passengers-attacked-7979.html
28. Shetty, J., Adibi, J.: The Enron email dataset database schema and brief statistical report. Inf. Sci. Inst. Tech. Rep. Univ. Southern California **4**(1), 120–128 (2004)
29. Litel, A.: Tweets of Congress. GitHub (2020). https://github.com/alexlitel/congresstweets. Accessed 3 Nov 2024
30. Tkachenko, M., Malyuk, M., Holmanyuk, A., Liubimov, N.: Label studio: data labeling software (2020). https://github.com/heartexlabs/label-studio
31. Wasserman, L.: All of Statistics: A Concise Course in Statistical Inference. Springer, New York (2010). ISBN: 9781441923226 1441923225

Author Index

A

Agarwal, Sonali I-98, I-322
Akter, Mst Shapna I-122
Al Zubaer, Abdullah II-197
Alam, Syeda Sadia I-122
Ali, Dildar I-276, II-147
Ali, Ushtar I-68
Amagasa, Toshiyuki I-68, I-169, II-115
Angelopoulos, Sotirios II-163
Auer, Dagmar II-212
Ayoub, Naeem II-305

B

Banerjee, Suman I-276, II-147
Baron, Mickaël I-83
Bell, Morris II-34
Bellatreche, Ladjel II-360
Bitsakos, Constantinos I-260
Borraccia, Salvatore I-329
Borse, Harsh I-130
Bortolaso, Christophe II-367
Bou, Savong I-169
Boukharouba, Ikram II-367
Bressan, Stéphane I-347
Breuillard, Hugo II-287
Burero, Saifullah I-23

C

Cao, Yixuan II-80
Cavalcante, Claudio I-199
Chandra, Ritesh I-98, I-322
Chang, Qiong II-136
Chatziantoniou, Damianos II-163
Chaudhari, Payal II-313
Cuzzocrea, Alfredo I-122

D

D'orazio, Laurent II-178
de Almeida, Eduardo C. II-376

Dechambenoit, Guillaume II-287
Dignös, Anton I-23, II-66
Ding, Cheng I-139
Djoulde, Barry Amadou II-254
do Nascimento, Thiago Germano II-293
Docini, Ruanitto II-376
Duan, Yijun II-3
Dzierwa, Piotr M. II-299

E

El Outa, Faten II-287
Elmi, Sayda II-34

F

Feng, Jiaji I-139

G

Gajoch, Sandra II-19
Gamper, Johann I-23, II-66
Gauly, Matthias II-66
Gharote, Mangesh II-338
Goda, Kazuo II-329
Granitzer, Michael II-197
Groppe, Jinghua I-53
Groppe, Sven I-53
Gruenwald, Le II-178
Guermouche, Nawal II-254
Gumienny, Grzegorz II-19

H

Hariri, Ali I-83
Huang, Hsiao-Ting I-245

I

Inoue, Masaharu I-154
Ito, Yuki I-230
Iwaihara, Mizuho I-113

J

Jahan, Raya I-184
Jaśkowiec, Krzysztof II-19
Jean, Stéphane I-83

K

Kalambe, Dhruv II-313
Kamal, Abrar Hasin I-363
Kantere, Verena II-163
Karmakar, Pratik I-347
Kasahara, Hidekazu I-154
Kaspar, Janette Christin II-383
Kawakami, Shun I-169
Kayem, Anne V. D. M. I-363
Khandelwal, Ankita I-307
Kitahata, Tetsuya I-291
Kobiela, Jaroslaw I-207
Kobiela, Jarosław II-299
Koh, Jia-Ling I-245
Konstantinou, Ioannis I-260
Kostrubiec, Viviane II-254
Koziris, Nectarios I-260
Küng, Josef II-212
Kurumatani, Akifumi II-329

L

Lanz, Oswald II-66
Leung, Carson K. I-35
Lien, Yin-Ju I-245
Lifschitz, Sergio I-199
Liu, Chang I-215
Liu, Yifan II-80
Lodha, Sachin II-338
Loizou, Andreas II-237
Luo, Ping II-80
Lynden, Steven I-68

M

Ma, Qiang I-154, I-230, II-3, II-51
Małysza, Marcin II-19
Manglani, Priyanshi II-313
Marcjan, Łukasz II-19
Marquet, Andreas I-53
Martins, Pedro Henrique Malheiros Costa II-293
Masciari, Elio I-329

Masuda, Tadashi II-115
Matono, Akiyoshi I-68
Matsumoto, Marin II-99
Matt, Dominik II-66
Mishra, Dhruv I-307
Mitra, Bivas I-130
Mitschang, Bernhard I-338
Miyazaki, Jun II-136
Mohamed, Aya II-212
Mondal, Mainack I-130
Mondal, Sutapa II-338
Monizza, Gabriele Pasetti II-66
Moraiti, Ioanna II-163
Mouakher, Amira II-305
Mozaffari, Maryam II-66
Muradi, Bahara II-212

N

Nabeoka, Takuto II-3
Nadamoto, Akiyo I-291
Napolitano, Enea Vincenzo I-329
Navuluru, Sai Karthik II-34
Ndao, Mouhamadou Lamine II-270
Niang, Ndèye II-270
Noleto, Jaqueline Donin Noleto II-293
Nonn, Simon Alexander II-197

O

Oguchi, Masato II-99
Ogura, Yudai II-115
Ohtani, Ryusei II-228
Oliveira, Luiz S. II-376
Ordonez, Carlos II-360
Oyama, Satoshi II-228

P

Pan, Su I-139
Paubel, Pierre Vincent II-254
Peng, Yiqiang I-215
Phan, Thuong-Cang II-178
Prasad, Yamuna I-276, II-147

Q

Quadir, Hashirul II-360

Author Index

R
Rajan, M. A. II-338
Roy, Shubhro II-338
Ruberg, Nicolaas I-199

S
Saadeh, Angelo I-347
Sahu, Pankaj II-338
Sakai, Yuto II-51
Sakurai, Yuko II-228
Sangma, Jerry W. I-23
Saporta, Gilbert II-270
Sasaki, Yuta II-130
Sato, Shota II-130
Satpathy, Utkalika I-130
Schmid, Daniel II-212
Schreier, Ulf I-338
Schuiki, Laura I-338
Schwarz, Holger I-338
Seabra, Antony I-199
Sèdes, Florence II-367
Seki, Kazuhiro I-291
Senellart, Pierre I-347
Shah, Nish II-313
Shiraishi, Ryuta II-228
Shiraishi, Yuhki II-130
Sienkiewicz, Mariusz I-3
Singh, Himanshi I-322
Singh, Pooja I-307
Singh, Vrijendra I-322
Sonbhadra, Sanjay Kumar I-322

T
Takagi, Shun II-99
Takahashi, Tsubasa II-99
Tekfa, Bechir Ben II-305
Tiwari, Sadhana I-98, I-322
Tran, Phan-An-Truong II-178
Tsoumakos, Dimitrios I-260, II-237

V
Varghese, Robin II-360
Verhaeghe, Benoit II-367

W
Walz, Annabel I-53
Wendlinger, Lorenz II-197
Wilk-Kołodziejczyk, Dorota II-19

Y
Yamashita, Fumihiro II-136
Yoshimoto, Hiromasa II-329
Youness, Genane II-270
Yu, Feng I-184

Z
Zakrzewicz, Maciej II-347
Zhang, Jianwei II-130
Zhang, Ying I-113
Zhang, Yongpan I-139
Zhu, Yan I-215
Zouni, Konstantina II-163

Made in the USA
Monee, IL
03 May 2026